Higher BIOLOGY

SECOND EDITION

Team Co-ordinator: James Torrance

Writing Team:

James Torrance

James Fullarton

Clare Marsh

James Simms

Caroline Stevenson

Diagrams by James Torrance

HODDER
GIBSON

AN HACHETTE UK COMPANY

The Publishers would like to thank the following for permission to reproduce copyright material:

Photo credits
Pages 1–88 (running graphic) © Andrei Tchernov/istockphoto.com; Pages 89–242 © Gregory Spencer/istockphoto.com; Pages 243–348 © Kate Leigh/istockphoto.com; Figure 1.2 © gamscholtte/www.bioplek.org; Figure 1.4 © Dr David Patterson/Science Photo Library; Figure 2.2 © Biophoto Associates/Science Photo Library; Figure 2.18 © Kevin Schafer/Alamy; Figure 4.3 © ISM/Science Photo Library; Figure 4.7 © TM-Photo/zefa/Corbis; Figure 5.11 © Visuals Unlimited/Corbis; Figure 7.1 © Andrew Syred/Science Photo Library; Figure 8.4 © Steve Gschmeissner/Science Photo Library; Figure 8.8 © Visuals Unlimited/Corbis; Figure 8.9 © Science Photo Library; Figure 9.2 © Prof. K.Seddon & Dr. T.Evans, Queen's University Belfast/Science Photo Library; Figure 9.4 © Phototake Inc./Alamy; Figure 10.1a © Visuals Unlimited/Corbis; Figure 10.1b © Eye of Science/Science Photo Library; Figure 10.1c © Omikron/Science Photo Library; Figure 10.1d © BSIP, Cavallini James/Science Photo Library; Figure 10.2 © Mike Baldwin/Cornered/www.cartoonstock.com; Figure 10.4 © Medical-on-Line/Alamy; Figure 11.2 © Phototake Inc./Photolibrary; Figure 11.8 © Matt Meadows/Peter Arnold Inc./Science Photo Library; Figure 11.9 © Elmer Parolini/www.cartoonstock.com; Figure 12.3 © Gene Cox/Science Photo Library; Figure 13.4 © Wayne Hutchinson/Alamy; Figure 15.4 © Steve Allen/Brand X/Corbis; Figure 17.10 © Biomedical Imaging Unit, Southampton General Hospital/Science Photo Library; Figure 19.1 © Celia Mannings/Alamy, © David Hosking/Alamy, © Miguel Castro/Science Photo Library; Figure 20.2 © Reproduced with permission of Punch Ltd., www.punch.co.uk; Figure 20.6 Charlotte Thege/Das Fotoarchiv; Figure 21.13 © Scimat/Science Photo Library; Figure 21.15 © Stan Eales/www.cartoonstock.com; Figure 21.16 © Dr Jeremy Burgess/Science Photo Library; Figure 22.5 © Galen Rowell/Corbis; Figure 24.3 © Scott Camazine/Alamy; Figure 24.5 © Juniors Bildarchiv/Photolibrary; Figure 24.9 © M.H. Sharp/Science Photo Library; Figure 24.10 © The Photolibrary Wales/Alamy; Figure 24.14 © George McCarthy/Corbis; Figure 24.17 © Ron Sanford/Corbis; Figure 26.9 © Joe McDonald/Corbis; Figure 26.11 © dani/jeske/BIOS/Still Pictures; Figure 27.2 © Phototake Inc/Photolibrary; Figure 27.4 © M.I. Walker/Science Photo Library; Figure 29.9 © Mike Baldwin/Cornered/www.cartoonstock.com; Figure 30.23 © Robert Thompson/www.cartoonstock.com; Figure 31.8 © Biophoto Associates/Science Photo Library; Figure 34.7 © Alan Carey/Science Photo Library; Figure 35.2 © tbkmedia.de/Alamy; Figure 35.6 © Daniel Heuclin/NHPA; Figure 35.11 © Lucky Look/Alamy; Figure Ap 1.2 © S. Harris /www.cartoonstock.com.
All other photos by the author.

Acknowledgements
Every effort has been made to trace all copyright holders, but if any have been inadvertently overlooked the Publishers will be pleased to make the necessary arrangements at the first opportunity.

Although every effort has been made to ensure that website addresses are correct at time of going to press, Hodder Gibson cannot be held responsible for the content of any website mentioned in this book. It is sometimes possible to find a relocated web page by typing in the address of the home page for a website in the URL window of your browser.

Hachette's policy is to use papers that are natural, renewable and recyclable products and made from wood grown in sustainable forests. The logging and manufacturing processes are expected to conform to the environmental regulations of the country of origin.

Orders: please contact Bookpoint Ltd, 130 Milton Park, Abingdon, Oxon OX14 4SB. Telephone: (44) 01235 827720. Fax: (44) 01235 400454. Lines are open 9.00–5.00, Monday to Saturday, with a 24-hour message answering service. Visit our website at www.hoddereducation.co.uk. Hodder Gibson can be contacted direct on: Tel: 0141 848 1609; Fax: 0141 889 6315; email: hoddergibson@hodder.co.uk

Cover photo © Steve Bloom Images/Alamy
Illustrations by James Torrance
Typeset in 11pt Minion by Fakenham Photosetting Ltd, Fakenham, Norfolk
Printed in Italy

A catalogue record for this title is available from the British Library

Without Answers edition:

Impression number   5 4 3 2 1
Year   2012 2011 2010 2009 2008

ISBN-13: 978 0340 95913 8

With Answers edition:

Impression number   5 4 3
Year   2012 2011 2010 2009

ISBN-13: 978 0340 95914 5

Contents

Preface

This book has been written to articulate closely with Standard Grade Biology and Intermediate 2 Biology. It is intended to act as a valuable resource for pupils studying Higher Grade Biology.

The book provides a concise set of notes which adheres closely to the SQA Higher Still syllabus for Higher Grade Biology. Each section of the book matches a unit of the syllabus; each chapter corresponds to a content area. The book contains a variety of special features:

Practical activities and reports

Assignments designed to match the required performance criteria and provide students with opportunities to develop *Practical Abilities* and to write *Scientific Reports* which include description of procedure, recording of results, analysis and presentation of results, drawing of conclusions and evaluation of procedure. These reports form a mandatory part of SQA assessment.

Testing your knowledge

Key questions incorporated into the text of every chapter and designed to continuously assess *Knowledge and Understanding*. These are especially useful as homework and as instruments of diagnostic assessment to check that full understanding of course content has been achieved.

What you should know

Summaries of key facts and concepts as 'cloze' tests accompanied by appropriate word banks. These feature at regular intervals throughout the book and provide an excellent source of material for consolidation and revision prior to the SQA examination.

Applying your knowledge

A variety of questions at the end of each chapter are designed to give students practice in exam questions and to foster the development of *Problem-solving Skills* (selection of relevant information, presentation of information, processing of information, planning experimental procedures, drawing valid conclusions and making predictions). These questions are especially useful as extensions to class work and as homework.

Cell Biology

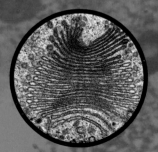

1 Cell variety in relation to function

At first glance a random selection of structures from living things, such as a chicken bone, a human nerve, a frog kidney, an octopus eye, a rose petal, a moss leaf and a dandelion root, have little in common with respect to their outward appearance. However microscopic examination reveals that they all share one essential feature – they are made up of **cells**.

The cell is the basic unit of life. It is the smallest structure that is able to lead an **independent life** and show all the **characteristics** of living things. A cell is like a chemical factory. It possesses a controlling centre, the **nucleus**, which governs the cell 'machinery'. The latter consists of a variety of subcellular structures which are illustrated in Appendix 2. Their functions will be discussed in later chapters.

Unicellular organisms

Amongst living things there exists a variety of tiny organisms each of which consists of only one cell. To survive, such an organism must be capable of manufacturing all the necessary chemicals (**metabolites**) and performing all the functions essential for the continuation of life, **within one cell**. This is made possible by the presence of a variety of structures each of which performs a specific role. Such **unicellular** level of organisation is illustrated by *Pleurococcus* (see Figures 1.1 and 1.2) which is a tiny

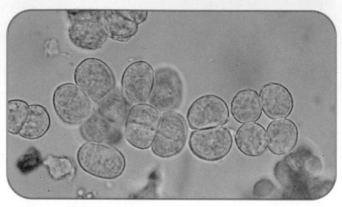

Figure 1.2 *Pleurococcus*

plant and *Paramecium* (see Figures 1.3 and 1.4) which is a tiny animal. Some unicellular organisms (see Question 5 on page 6) cannot be so easily classified since they possess plant- and animal-like characteristics.

Multicellular organisms

Advanced plants and animals consist of an enormous number (often millions) of cells. It would be inefficient for every one of these cells to perform every function essential for the maintenance of life. Instead the cells are arranged into **tissues**. A tissue is a group of cells specialised to perform a particular function (or functions). This **multicellular** level of organisation is therefore said to show a **division of labour**. Figures 1.5

structure	function	details
cell wall	support and protection	composed of cellulose fibres which combine strength with a degree of elasticity
nucleus	control of cell activities	contains genetic material which controls day to day running of cell and is passed on to daughter cells formed by asexual reproduction
chloroplast	photo-synthesis	a massive lobed structure which produces all of the cell's carbohydrate food

Figure 1.1 Unicellular level of organisation in *Pleurococcus*

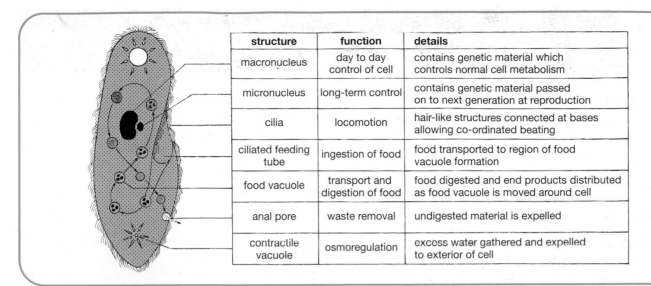

structure	function	details
macronucleus	day to day control of cell	contains genetic material which controls normal cell metabolism
micronucleus	long-term control	contains genetic material passed on to next generation at reproduction
cilia	locomotion	hair-like structures connected at bases allowing co-ordinated beating
ciliated feeding tube	ingestion of food	food transported to region of food vacuole formation
food vacuole	transport and digestion of food	food digested and end products distributed as food vacuole is moved around cell
anal pore	waste removal	undigested material is expelled
contractile vacuole	osmoregulation	excess water gathered and expelled to exterior of cell

Figure 1.3 Unicellular level of organisation in *Paramecium*

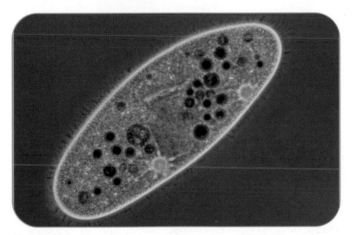

Figure 1.4 *Paramecium*

and 1.6 illustrate a selection of tissues (and their component cells) from two multicellular organisms.

Cell variety

Variation within one tissue

Although the cells that make up a tissue may all be of the same type, smooth muscle for example consists of spindle-shaped cells only, a tissue often possesses a **variety** of cell types. Blood contains red blood cells white blood cells and platelets. Phloem tissue consists of sieve tubes and companion cells.

Variation between different tissues

When one tissue is compared with another, for example blood compared with ciliated epithelium or phloem compared with xylem, then **variation** in cell structure becomes even more apparent.

Testing Your Knowledge

1 Why is the cell described as the basic unit of life? (1)

2 Explain the meaning of the terms *unicellular* and *multicellular* organism. (1)

3 Describe the roles played by a named unicellular plant's nucleus, chloroplast and cell wall. (4)

4 a) Make a simple diagram of a named unicellular animal which includes its nucleus, one food vacuole and one contractile vacuole. (2)

   b) Describe the role played by each of the two types of vacuole. (2)

Structure in relation to function

Structural variation in relation to function exists because each cell's structure is exactly tailored to suit its function. Once a new cell has been formed by cell division, it becomes specialised by undergoing physical and chemical changes until its size, shape and chemical 'machinery' are perfectly suited to its future role in a particular tissue.

Tables 1.1 and 1.2 describe how the structure of each of the cell types shown in Figures 1.5 and 1.6 is related to the cell's function.

Advantage of specialisation

With each specialised tissue performing its role so effectively, a multicellular organism is able to function at a much more advanced level than a unicellular orga

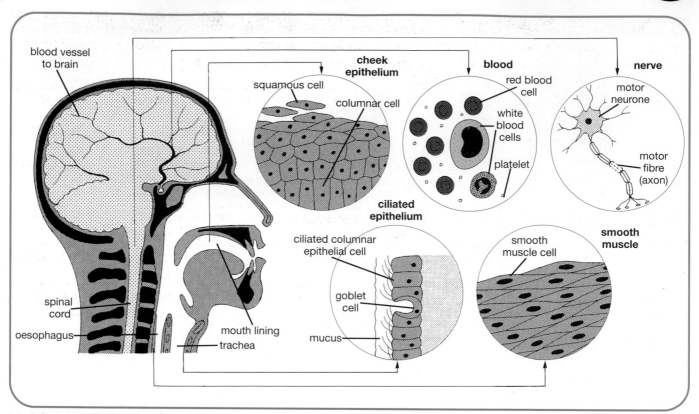

Figure 1.5 Human tissues and cells (not drawn to scale)

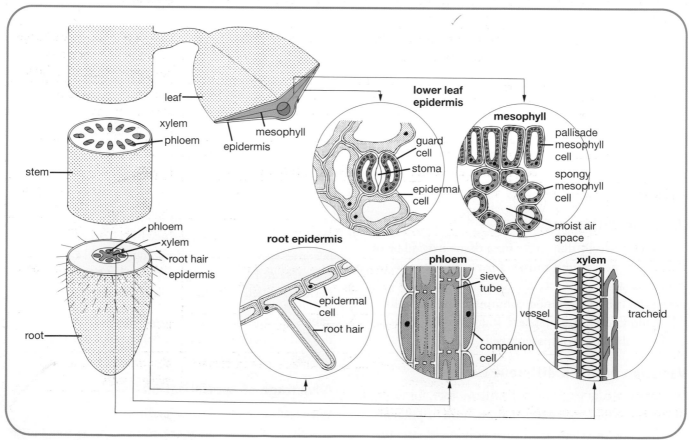

Figure 1.6 Plant tissues and cells (not drawn to scale)

Tissue	cell type	specialised structural features	function
cheek epithelium	epithelial cell	flat irregular shape (allowing cells to form a loose covering layer constantly replaced from below during wear and tear)	protection of mouth lining
blood	red blood cell	small size and biconcave shape present large surface area; rich supply of haemoglobin present	uptake and transport of oxygen to living cells
	white blood cell	able to change shape; sacs of microbe-digesting enzymes present in some types	destruction of invading pathogens
nerve	motor neurone	axon (long insulated extension of cytoplasm)	transmission of nerve impulses
ciliated epithelium	goblet cell	cup shape; able to produce mucus	secretion of mucus which traps dirt and germs
	ciliated epithelial cell	hair-like cilia which beat upwards	sweeping of dirty mucus up away from lungs
smooth muscle	smooth muscle cell	spindle shape (allowing cells to form sheets capable of contraction)	movement of food down gullet by peristalsis

Table 1.1 Structure of animal cells in relation to function

Tissue	cell type	specialised structural features	function
lower leaf epidermis	epidermal cell	irregular shape (allowing cells to fit like a jigsaw into a strong layer)	protection
	guard cell	sausage shape; thick inner cell wall facing stoma; chloroplasts present	control of gaseous exchange by changing shape and opening or closing stomata
mesophyll	palisade mesophyll cell	chloroplasts present; columnar shape (allows densely-packed green layer to be presented to light)	primary region of light absorption and photosynthesis
	spongy mesophyll cell	'round' shape allows loose arrangement in contact with moist air spaces for absorption of carbon dioxide	secondary region of photosynthesis
phloem	sieve tube	sieve plates and continuous system of cytoplasmic strands	transport (translocation) of soluble carbohydrates
	companion cell	large nucleus in relation to cell size	control of sieve tube
xylem	vessel	hollow tube; walls strengthened by lignin; lignin deposited as rings or spirals allowing expansion and contraction	support and water transport
root	epidermal cell	box-like shape allowing cells to fit together like a brick wall	protection
	root hair	long extension presenting large surface area in contact with soil solution	absorption of water and mineral salts

Table 1.2 Structure of plant cells in relation to function

Testing Your Knowledge

1 Describe a structural feature possessed by some white blood cells that enables them to destroy disease-causing bacteria. (1)

2 **a)** State the function of cheek epithelial cells.

   b) Explain how these cells are suited to perform this function. (2)

3 What role is played by the axon of a motor neurone? (1)

4 **a)** Make a simple diagram of a small sample of the tissue that covers the inner surface of the human windpipe as it would be seen in a longitudinal section under the microscope. (2)

   b) Name both types of cell that you have drawn and describe how each is ideally suited to the function that it performs. (2)

5 **a)** State TWO functions of xylem tissue. (2)

   b) Describe how a xylem vessel is structurally suited to perform each of these functions. (2)

6 **a)** Identify THREE types of cell capable of photosynthesis in a leaf of a normal dicotyledonous land plant. (3)

   b) In what way are a leaf's epidermal cells well suited to their role of protection? (1)

Applying Your Knowledge

1 'A red blood corpuscle is an example of a cell whose structure can easily be related to its function.' Justify this statement. (2)

2 Figure 1.7 shows a transverse section of a young root.

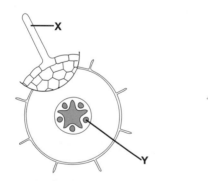

Figure 1.7

   a) Identify structure X and state its function. (2)

   b) Name TWO types of cell that would be present in tissue Y and compare them with respect to their structure and function. (4)

   c) Does a division of labour exist in a root? Explain your answer. (1)

3 Figure 1.8 shows the epithelial tissue which makes up the surface layer of a villus in the human small intestine.

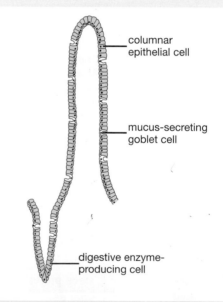

columnar epithelial cell

mucus-secreting goblet cell

digestive enzyme-producing cell

Figure 1.8

State TWO ways in which this tissue **i)** resembles; **ii)** differs from, the epithelium that lines the windpipe (see Figure 1.5). (4)

4 **a)** Identify the type of sex cell shown in Figure 1.9. (1)

   b) Give TWO reasons why this cell is well suited to its function. (2)

5 The unicellular organism shown in Figure 1.10 is called *Euglena*.

   a) Suggest how it obtains its food. (1)

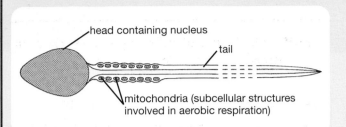

head containing nucleus

tail

mitochondria (subcellular structures involved in aerobic respiration)

Figure 1.9

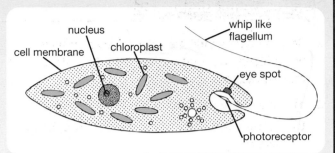

nucleus

cell membrane

chloroplast

whip like flagellum

eye spot

photoreceptor

Figure 1.10

b) Suggest how it **i)** senses light; **ii)** moves towards light. (2)

c) Briefly explain why *Euglena* cannot be satisfactorily classified as either a plant or an animal. (2)

6 'A division of labour amongst its cells enables a multicellular organism to function at a more advanced level than a unicellular organism.' Discuss the truth of this statement, giving named examples in your answer. (10)

2 Absorption and secretion of materials

Absorption means the uptake of materials by a cell from its external environment. **Secretion** means the discharge of useful intracellular molecules into the surrounding medium by a cell. The movement of small molecules or ions (tiny electrically-charged particles) into or out of a cell normally occurs as a result of **diffusion**, **osmosis** or **active transport** depending upon the nature of the substance involved.

Cell boundaries

All living cells are surrounded by a **cell membrane** (plasma membrane). In addition, plant cells possess a cell wall. The structure of each of these boundaries is closely related to the role that it plays in the movement of materials into and out of the cell.

Structure of the cell wall

The cell wall is a non-living layer composed mainly of cellulose. This complex carbohydrate consists of

unbranched chains of glucose molecules grouped together as **fibres**. The fibres are closely packed into layers and run in different well-defined directions (see Figures 2.1 and 2.2).

This criss-cross arrangement makes the cell wall strong, fairly rigid and yet slightly elastic. It is able to stretch slightly when the cell absorbs water. However when it reaches the limit of its elasticity, it resists further uptake of water by the cell and prevents the latter from bursting (see also page 13).

Many spaces are present between the cellulose fibres of the cell wall. Since cellulose is **hydrophilic** ('water-loving'), these spaces are normally water-filled.

Adjoining cell walls and intercellular spaces (see Figure 2.1) provide a **continuous water-conducting** pathway throughout the plant. This allows water to move easily from tissue to tissue without having to enter and leave every living cell along the way.

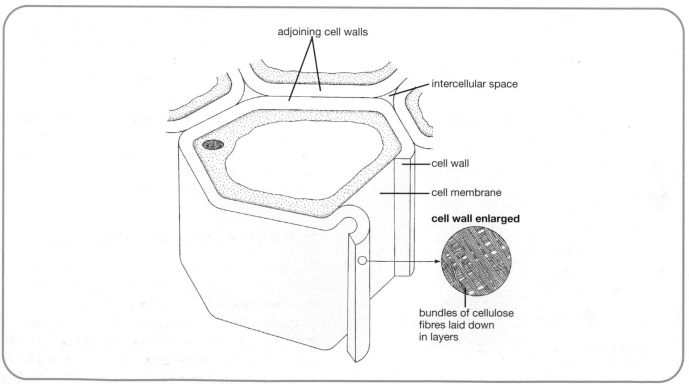

adjoining cell walls

intercellular space

cell wall

cell membrane

cell wall enlarged

bundles of cellulose fibres laid down in layers

Figure 2.1 Structure of a plant cell

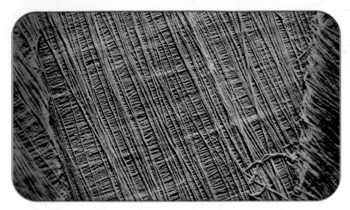

Figure 2.2 Fibrous nature of cell wall

Investigating the chemical nature of the cell membrane

The cell sap present in the central vacuole of a beetroot cell (see Figure 2.3) contains red pigment. 'Bleeding' (the escape of this red cell sap from a cell) indicates that the cell's plasma and vacuolar membranes have been damaged.

In the experiment shown in Figure 2.4, four identical cylinders of fresh beetroot are prepared using a cork

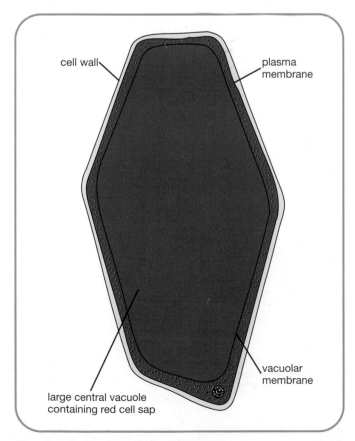

Figure 2.3 Beetroot cell

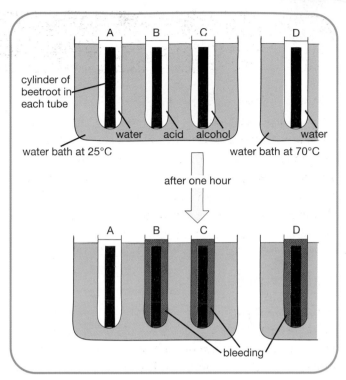

Figure 2.4 Investigating the chemical nature of the cell membrane

borer. The cylinders are thoroughly washed in distilled water to remove traces of red cell sap from outer damaged cells. The figure shows the results of subjecting the cylinders to various conditions.

Bleeding is found to occur in B, C and D showing that the membranes have been destroyed. Molecules of protein are known to become denatured when exposed to acid or high temperatures. Molecules of lipid are known to be soluble in alcohol. It is therefore concluded that the cell membrane contains **protein** (as indicated by B and D whose denatured protein has allowed red cell sap to leak out) and **lipid** (as indicated by C whose lipid molecules have dissolved in alcohol permitting the pigment to escape).

Structure of plasma membrane

The plasma membrane is now known to consist of **protein** and **phospholipid** molecules. Although the precise arrangement of these molecules is still unknown, most evidence supports the **fluid mosaic model** of cell membrane structure (see Figure 2.5). This proposes that the plasma membrane consists of a **fluid bilayer** of constantly moving phospholipid molecules containing a **patchy mosaic** of protein molecules.

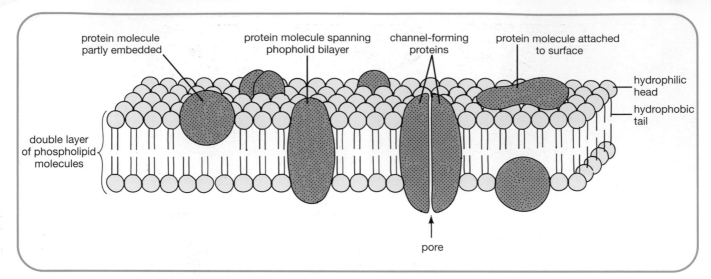

Figure 2.5 Fluid mosaic model of plasma membrane

Phospholipid bilayer

One end of a phospholipid molecule is **hydrophilic** ('water-loving') and the other end is **hydrophobic** ('water-hating'). In the company of other similar molecules, phospholipid molecules arrange themselves into a bilayer.

The **water-soluble** hydrophilic heads (see Figure 2.5) make up the two outer surfaces of the bilayer where they form hydrogen bonds with water molecules. One layer of heads forms the outside of the cell membrane in contact with extracellular fluid and the other layer forms the inside of the membrane in contact with the intracellular fluid.

The **water-insoluble** hydrophobic tails point inwards to the centre of the bilayer since they are attracted to those in the opposite layer.

This arrangement of phospholipid molecules is **fluid** yet at the same time it forms a **stable** and effective **boundary** round the cell. It allows tiny molecules such as water to pass through it rapidly. Larger molecules such as glucose depend on the membrane's protein molecules for entry to or exit from the cell.

Proteins

The protein molecules in the plasma membrane vary in size and structure. Some extend partly into the phospholipid layer while others extend across it from one side to the other. Some of these enclose narrow channels making the membrane porous. The fluid

nature of the membrane allows some movement of the proteins (and pores).

The protein molecules also vary in function (see Figure 2.6) in that they

- provide structural **support**;
- contain **channels** allowing transport of small molecules through the membrane;
- act as **carriers** which actively 'pump' molecules across the membrane;
- serve as **enzymes** catalysing biochemical reactions in and on the membrane;
- act as **receptors** for hormones arriving at the outer surface of the membrane;
- serve as **antigenic markers** which identify the animal cell's blood or tissue type.

Constant cell environment

For a cell to function efficiently, its internal environment must remain fairly **constant** with respect to the concentration of water and soluble substances present in its cytoplasm. A constant environment is successfully maintained within the cell by the cell membrane **regulating** the entry and exit of materials as required. It does this by allowing **selective communication** between intracellular and extracellular environments. This involves the processes of diffusion, osmosis and active transport.

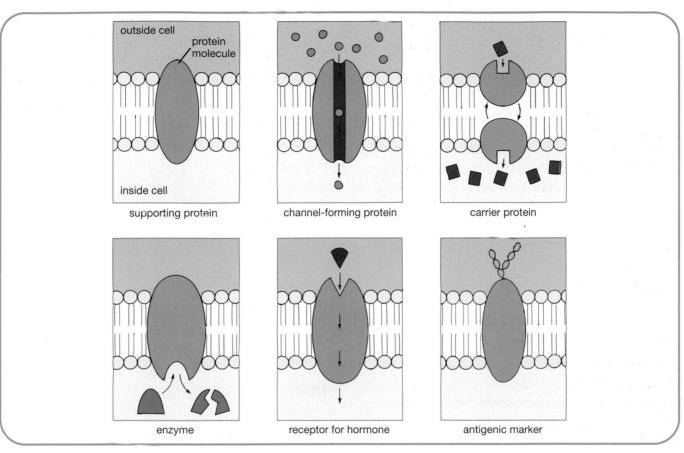

Figure 2.6 Functions of plasma membrane proteins

Diffusion

Diffusion is the net movement of molecules or ions from a region of high concentration to a region of low concentration of that type of molecule or ion. The difference that exists between two regions before diffusion occurs is called the **concentration gradient**. During diffusion, molecules and ions always move down a concentration gradient from high to low concentration.

Diffusion is a basic cell process. It is the means by which useful substances such as oxygen enter and waste materials such as carbon dioxide leave a cell. Diffusion is a **passive** process and does not require energy.

Effect of cell wall on diffusion

Many water-filled spaces and some large pores occur amongst the cellulose fibres in a cell wall. These make it **freely permeable** to all molecules in solution. The cell wall does not, therefore, act as a barrier to diffusion.

Effect of plasma membrane on diffusion

The plasma membrane is freely permeable to tiny molecules such as water, oxygen and carbon dioxide that are small enough to diffuse rapidly through the phospholipid bilayer. However the plasma membrane is **not** equally permeable to all substances. Figure 2.7 shows diffusion in two types of cell. Molecules of urea are able to diffuse slowly through the phospholipid bilayer. Larger molecules such as glucose and amino acids depend on certain protein molecules to **facilitate** (help) their movement across the membrane. These protein molecules contain channels which allow glucose and amino acid molecules to diffuse slowly into and out of a cell. Even larger molecules such as starch and protein are unable to pass through the membrane of a cell. The plasma membrane is said therefore to be **selectively permeable**.

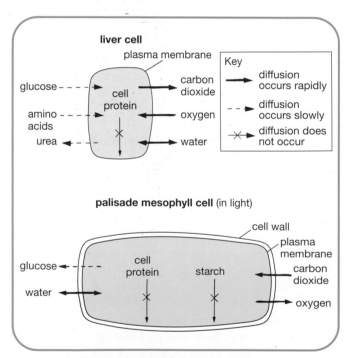

Figure 2.7 Diffusion into and out of cells

Osmosis

Osmosis is the net movement of water molecules from a region of higher water concentration (HWC) to a region of lower water concentration (LWC) through a selectively permeable membrane.

Investigating osmosis

In the experiment shown in Figure 2.8, visking tubing sausage A is found to gain weight after one hour whereas B remains unchanged.

It is concluded that A has gained weight because small water molecules have passed rapidly through the tiny pores in the selectively permeable membrane down a concentration gradient from a region of higher water concentration (HWC) to a region of lower water concentration (LWC). The solution with the HWC is said to be **hypotonic** to the solution with the LWC which is said to be **hypertonic**.

It is concluded that sausage B shows no change in weight because both solutions have an equal water concentration and neither makes a net gain of water molecules. Two solutions of equal water concentration are said to be **isotonic**.

Osmosis and cells

Movement of water by osmosis occurs in living things between neighbouring cells and between cells and their immediate extracellular environment. For example, in humans water passes continuously by osmosis from the gut cells into the bloodstream. Table 2.1 lists several examples of osmosis in a green plant. Osmosis is a basic cell process. It is **passive** and does not require energy.

Role of plasma membrane in osmosis

Whenever a cell is in contact with a solution (or another cell) of differing water concentration, osmosis

direction of movement of water	significance
soil solution → root hairs	water absorbed by plant from soil
xylem vessels → stem cells	water makes cells turgid giving support
xylem vessels → green leaf cells	water used as raw material for photosynthesis

Table 2.1 Movement of water by osmosis in plant cells

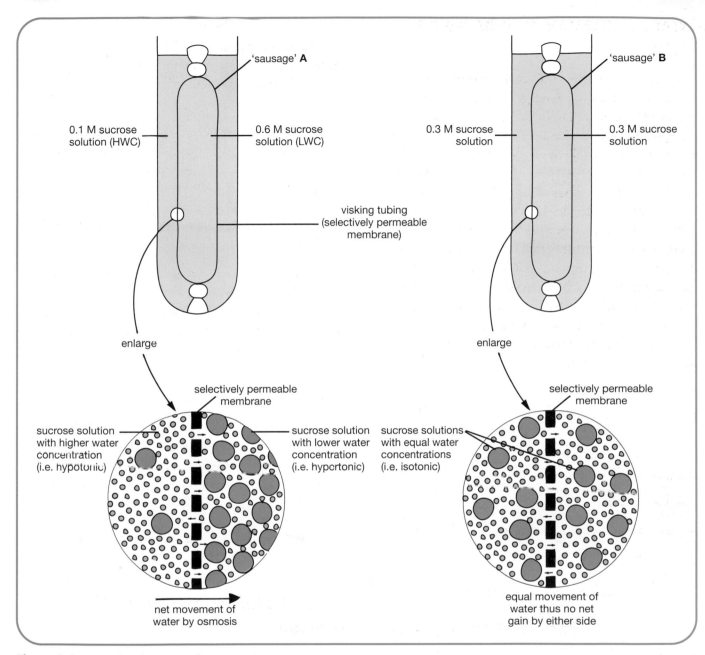

Figure 2.8 Investigating osmosis

occurs. This is made possible by the fact that the plasma membrane is selectively permeable. It allows the rapid movement of water molecules through it but only allows larger molecules to move across slowly or not at all.

The direction in which net movement of water molecules occurs depends upon the water concentration of the liquid in which the cell is immersed compared with that of the cell contents (see Figures 2.9, 2.10, 2.11, 2.12, 2.13 and 2.14).

Role of cell wall

Unlike an animal cell, a plant cell placed in water does not burst. As water enters a plant cell by osmosis, its central vacuole swells up and presses the cytoplasm and plasma membrane against the cell wall which stretches slightly. As this process continues, the wall presses back (i.e. exerts a **wall pressure**) on the cell contents. Eventually a point is reached where the wall pressure stops further water from entering and prevents the cell from bursting.

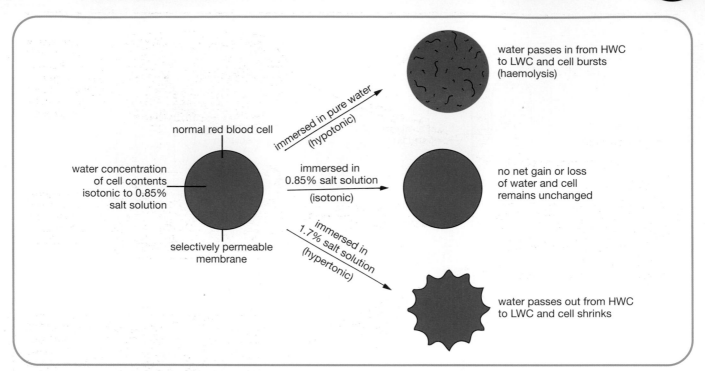

Figure 2.9 Osmosis in a red blood cell

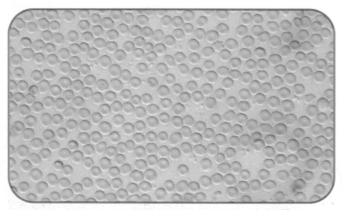

Figure 2.10 Red blood cells in isotonic salt solution

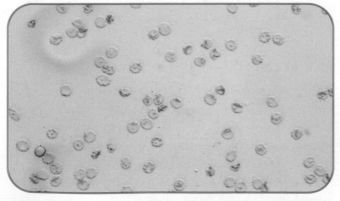

Figure 2.11 Red blood cells in hypertonic salt solution

No wall pressure exists in a **plasmolysed** cell (see Figures 2.12 and 2.15) where the contents have shrunk and pulled away from the cell wall.

Active transport

Active **transport** is the movement of molecules and ions across the plasma membrane from a low to a high concentration i.e. **against a concentration gradient.** Active transport works in the opposite direction to the passive process of diffusion and always requires **energy.** This energy is released during respiration (see chapter 4).

Consider the two situations shown in Figure 2.16 on page 17. Ion type A is being actively transported into the cell whereas ion type B is being actively transported out of the cell.

Active transport in a plant cell

From the experiment shown in Figure 2.17 on page 18, it is concluded that the marine plant, *Valonia* (see Figure 2.18), is able to **select** and **accumulate** potassium ions in its cell sap to a concentration greatly in excess of the concentration in the external environment. This **selective ion uptake** is brought about by active transport.

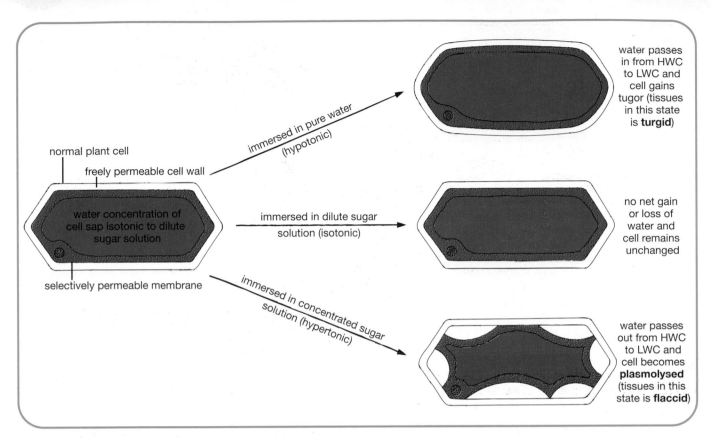

Figure 2.12 Osmosis in a plant cell

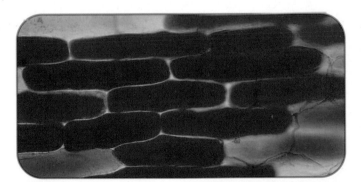

Figure 2.13 Red onion cells in isotonic sugar solution

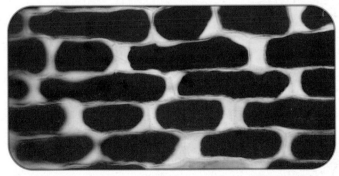

Figure 2.14 Red onion cells in hypertonic sugar solution

Active transport of sodium in the opposite direction accounts for the fact that the cell simultaneously maintains a low concentration of sodium ions in its cell sap despite the high concentration present in the surrounding sea water.

The plant constantly loses potassium ions and gains sodium ions by the passive process of diffusion. However by employing active transport, it is able to maintain the optimum level of these ions in its cell sap by continuously 'pumping' potassium in and sodium out against their respective concentration gradients.

Role of the plasma membrane in active transport

Certain protein molecules present in the plasma membrane act as carrier molecules. These protein

15

Practical Activity and Report

Measuring the water concentration of potato cell sap

Information

- The cell sap present in the central vacuole of a potato cell is a solution of sugar and mineral salts dissolved in water.

- In this experiment, cylinders of potato are going to be bathed in sucrose (sugar) solutions of varying water concentration. The higher the molar concentration of sucrose, the lower the water concentration of the solution.

- If a cylinder is found to have gained mass, water must have passed into it by osmosis showing that the sucrose solution has a higher water concentration than the potato cell sap.

- If a cylinder is found to have lost mass, water must have passed out of it by osmosis showing that the sucrose solution has a lower water concentration than the potato cell sap.

- If a cylinder is found to have neither gained nor lost mass, there must have been no net movement of water into or out of the cylinder by osmosis showing that the sucrose solution has a water concentration equal to that of potato cell sap.

You need

1 fresh potato

1 cork borer

1 ceramic tile

5 test tubes in a stand

5 labels or 1 marker pen

1 dropping bottle of each of 0.1, 0.2, 0.3, 0.4 and 0.5 molar (M) sucrose solution

1 safety razor

   paper towels

   access to electronic balance

What to do

1 Read all of the instructions in this section and prepare your results table before carrying out the experiment.

2 Label 5 test tubes 0.1, 0.2, 0.3, 0.4 and 0.5 M and add your initials.

3 Half-fill each test tube with its sucrose solution.

4 Using a cork borer, cut out 5 cylinders of potato tuber.

5 Trim the cylinders to equal length using a safety razor.

6 Roll each cylinder on a paper towel and blot its ends to remove surface liquid before weighing.

7 Weigh each cylinder and note the initial mass in your table.

8 Immerse each cylinder in its sucrose solution and leave at room temperature.

9 After a minimum of 1 hour, remove each cylinder from its test tube and dry it off as before.

10 Weigh each cylinder and note this final mass in your table.

11 If other students have carried out the same experiment, pool the results.

Reporting

Write up your report by doing the following:

1 Rewrite the title given at the start of this activity.

2 Put the subheading '**Aim**' and state the aim of your experiment.

3 a) Put the subheading '**Method**'.

   b) Draw a diagram of your apparatus set up at the start of the experiment after the cylinders have been weighed and immersed in the solutions.

   c) Briefly describe the experimental procedure that you followed using the impersonal passive voice. *Note:* The impersonal passive voice avoids the use of 'I' and 'we'. Instead it makes the apparatus the subject of the sentence. In this experiment, for example, you could begin your report by saying 'Five test tubes were labelled ... etc.' (not 'I labelled five test tubes ... etc.')

   d) Continuing in the impersonal passive voice, state how your results were obtained.

4 Put the subheading '**Results**' and draw a final version of your table of results.

5 a) Put a subheading '**Analysis and Presentation of Results**'.

b) Calculate the gain or loss and then the percentage gain or loss in mass for each cylinder and extend your table to include these values. (This conversion is necessary to standardise the results since the initial masses of the potato cylinders were probably not identical.)

c) Present your results as a line graph by plotting the five points and drawing a line of best fit.

d) From your graph read off the concentration of sucrose solution at which water was neither gained nor lost by the potato cells.

6 Put a subheading **'Conclusion'** and then state the water concentration of the cell sap of the potato.

7 Put a final subheading **'Evaluation of Experimental**

Procedure'. Give an evaluation of your experiment (keeping in mind that you may comment on any stage of the experiment that you wish).

Try to incorporate answers to the following questions in your evaluation. Make sure that at least one of your answers includes a supporting argument.

a) Why is the same cork borer used to cut all five cylinders?

b) Why were the cylinders dried off before each weighing?

c) Why were the changes in mass converted to percentages?

d) By what means could the reliability of the results from your experiment be checked?

Figure 2.15 Plasmolysis problem

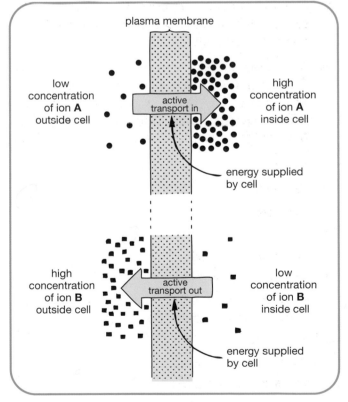

Figure 2.16 Active transport of two different ions

molecules 'recognise' specific ions and transfer them across the plasma membrane (see Figure 2.19). The energy required for this active process is supplied by ATP (adenosine triphosphate – see chapter 3) formed during respiration.

Sodium/potassium pump

Active transport carriers are often called **'pumps'**. Some carrier molecules have a dual role in that they **exchange** one type of ion for another. An example of this is the sodium/potassium pump. The same carrier

molecule actively pumps sodium ions out of the cell and potassium ions into the cell, each against its concentration gradient. The resulting difference in ionic concentrations maintained by the pump is specially important for the functioning of nerve cells.

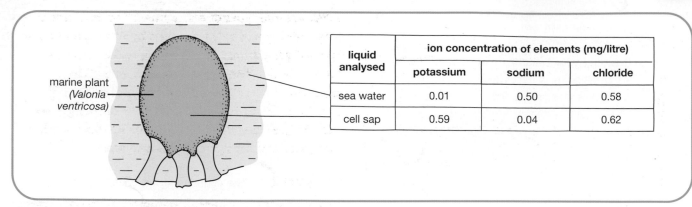

liquid analysed	ion concentration of elements (mg/litre)		
	potassium	sodium	chloride
sea water	0.01	0.50	0.58
cell sap	0.59	0.04	0.62

Figure 2.17 Investigating active transport in *Valonia*

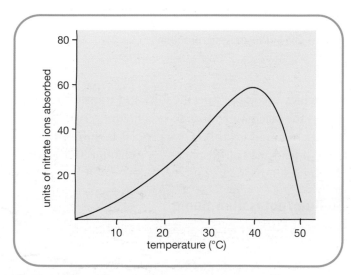

Figure 2.18 *Valonia*

Conditions required for active transport

Factors such as **temperature**, availability of **oxygen** and concentration of **respiratory substrate** (e.g. glucose) which directly affect a cell's respiration rate also affect the rate of active transport.

Figures 2.20 and 2.21 show the effects of increasing temperature and oxygen concentration on the rate of active uptake of nitrate ions by the cells of barley roots. Increase in temperature brings about an increase in ion uptake until at high temperatures the enzymes needed for respiration become denatured and the cell dies. Increase in oxygen concentration results in increased rate of ion uptake until some other factor affecting the process becomes limiting.

Gross movements of the cell membrane

Membrane transport systems involving proteins can cope adequately with the transport of relatively small

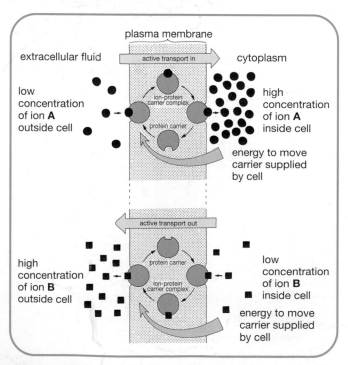

Figure 2.19 Role of protein carriers in active transport

Figure 2.20 Effect of temperature on ion uptake

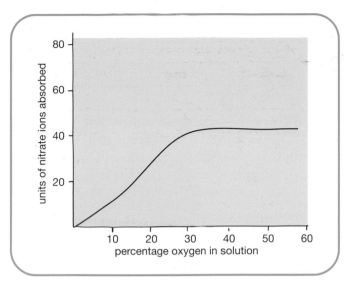

Figure 2.21 Effect of oxygen concentration on ion uptake

particles such as ions and molecules into and out of the cell. However there are some occasions when the cell needs to take in or pass out large particles. Such movements are beyond the scope of the protein transport systems; they require **gross movements** of the whole membrane.

Endocytosis

Endocytosis is the process by which a cell engulfs and takes in relatively large particles or quantities of material. This involves the plasma membrane folding inwards to form a 'pouch' (see Figure 2.22). When this becomes closed off and detached from the cell membrane, it is called an **intracellular vesicle**.

There are two types of endocytosis. The engulfing of large solid particles (e.g. bacteria by white blood cells) is called **phagocytosis** ('cell-eating'). The contents of the vesicle are then digested (see page 78). The formation of small liquid-filled vesicles by the cell membrane is called **pinocytosis** ('cell-drinking' – see page 351). Endocytosis is the means by which a cell often acquires hormones, lipids and proteins.

Exocytosis

Exocytosis is the reverse of endocytosis. Vesicles formed inside the cell fuse with the plasma membrane (see Figure 2.23) allowing their contents to be expelled from the cell. By this means intracellular products such as enzymes, glycoproteins and hormones can be **secreted** by the cell.

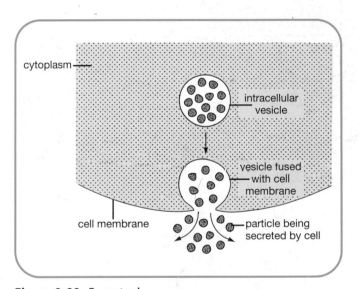

Figure 2.23 Exocytosis

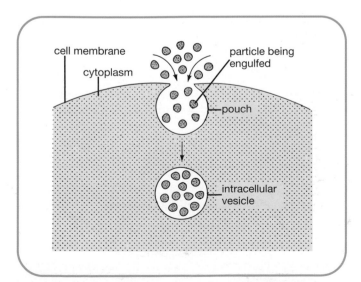

Figure 2.22 Endocytosis

19

Testing Your Knowledge

1 a) Define the term *osmosis*. (2)

   b) Rewrite the following sentences to include only the correct word(s) from each choice in brackets.

   Osmosis is a form of (diffusion/ion uptake). It is (an active/a passive) process which (requires/does not require) energy. During osmosis (oxygen/water) molecules move from a (high/low) water concentration to a (higher/lower) water concentration (down/against) a concentration gradient. (7)

2 a) With reference to the contents of a red blood cell and very concentrated salt solution, explain the meaning of the terms *hypotonic* and *hypertonic* solution. (2)

   b) Why does a red blood cell burst when placed in a hypotonic solution yet a red onion cell does not? (1)

   c) i)   Describe the effect of a hypertonic solution of sugar on red onion cells.

      ii) What name is given to cells in this state?

      iii) By what means could such cells be restored to their normal turgid condition? (4)

3 a) Define the term *active transport*. (2)

   b) Copy and complete Table 2.2 by answering the five questions that it contains relating to two processes by which particles pass through the plasma membrane. (5)

	diffusion	active uptake
1 What is an example of a substance that moves through the cell membrane by this process?		
2 Do the particles move from high to low or from low to high concentration?		
3 Do the particles move down or against a concentration gradient?		
4 Is energy required?		
5 Is the process passive?		

Table 2.2

Applying Your Knowledge

1 Figure 2.24 shows a possible arrangement of the molecules in a plasma membrane.

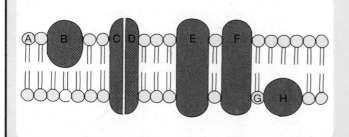

Figure 2.24

   a) What name is given to this model? (1)

   b) Identify molecule types A and B. (1)

   c) i) Which lettered structures enclose a narrow channel?

      ii) What is the function of such a channel? (2)

2 Decide whether each of the following statements is true or false and then use T or F to indicate your choice. Where a statement is false, give the word that should have been used in place of the word in bold print.

   a) Molecules of carbon dioxide move out of a respiring cell by **diffusion**.

   b) The carrier molecules that pump ions across a membrane are made of **phospholipid**.

   c) Red blood cells **shrink** when placed in concentrated salt solution.

   d) Phagocytosis and pinocytosis are two examples of **exocytosis**.

e) During active transport, energy is needed to move molecules **down** a concentration gradient from low to high concentration. (5)

3 a) Name a chemical molecule (other than water) that would be found diffusing out of a palisade leaf cell in **i)** light; **ii)** darkness. (2)

b) Give an example of a chemical molecule that is too large to diffuse out of a palisade leaf cell. (1)

4 Figure 2.25 shows the direction of movement of two different substances through the plasma membrane of an animal cell.

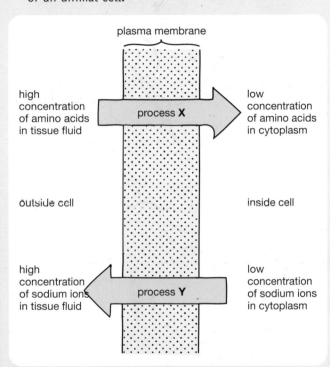

Figure 2.25

a) i) Name processes X and Y.
ii) Which of these processes requires energy?
iii) Which process will be unaffected by a decrease in oxygen concentration in the animal's environment? (4)

b) i) Predict what will happen to the rate of process Y if the temperature of the cell is reduced to 4°C for several hours.
ii) Give a reason for your answer. (2)

5 In an experiment, groups of potato discs of known masses were immersed in sucrose solutions of different concentration for a few hours and then reweighed. The results were plotted as shown in Figure 2.26.

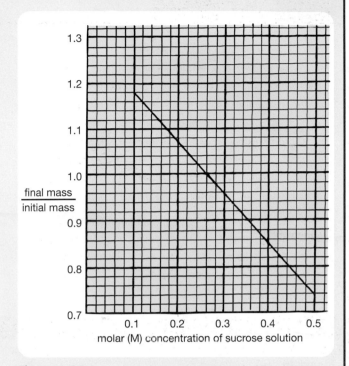

Figure 2.26

a) i) Was 0.2 M sucrose solution hypotonic or hypertonic to the potato cell sap?
ii) Explain your answer. (2)

b) i) Was 0.4 M sucrose solution hypotonic or hypertonic to the potato cell sap?
ii) Explain your answer. (2)

c) To which molarity of sucrose was the potato cell sap equal in water concentration? (1)

6 A sample of epidermal cells from red onion was mounted in a solution isotonic to the cell sap and examined to establish the normal appearance of the cells (see Figure 2.27).

The cells were then immersed for 10 minutes in solution 1 and re-examined. This procedure was repeated using different concentrations of sugar solution. The appearance of the cells at each stage is shown in the diagram.

a) Which of the five solutions was **i)** most hypertonic, **ii)** most hypotonic relative to the others? (2)

b) Identify the process occurring at each numbered immersion in Figure 2.27 using the appropriate letter from Table 2.3. (3)

c) What substance would be present in region R in Figure 2.27 after the cells had been immersed for 10 minutes? (1)

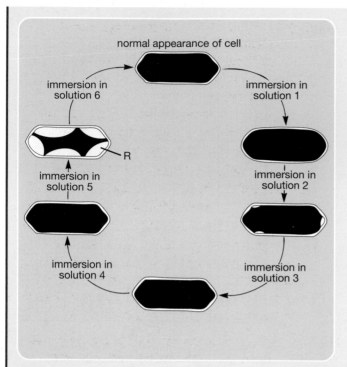

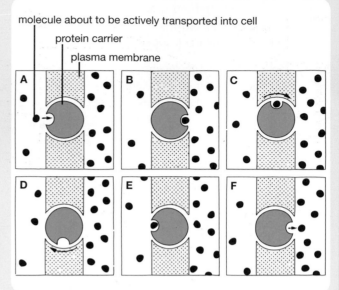

Figure 2.27

Figure 2.28

process	overall movement of water molecules
X	movement out of cell exceeds movement in
Y	movement into cell exceeds movement out
Z	movement into cell equals movement out

Table 2.3

7 It is possible that some protein carrier molecules actively transport materials against a concentration gradient by rotating within the plasma membrane. Rearrange the six stages in Figure 2.28 to give the correct sequence in which this would occur, starting with A. (1)

8 Table 2.4 refers to the concentrations of certain chemical ions in the cells of the alga, *Nitella*, and in the pond water in which this green plant lives.

   a) Make a generalisation about the concentration of

ions in the pond water compared with their concentration in the cell sap. (1)

b) Calculate the accumulation ratio for potassium and sodium. (1)

c) In what way do the data support the theory that a cell membrane is selective with respect to the process of ion uptake? (1)

d) i) Do the data support or dispute the suggestion that ion uptake occurs as a result of diffusion?

  ii) Explain your answer. (2)

9 In Figure 2.29, graph 1 presents the results of an experiment set up to investigate the effect of oxygen concentration on uptake of potassium ions and consumption of sugar by the cells of excised (cut off) roots of barley seedlings.

Graph 2 gives the results from an experiment set up to investigate the effect of temperature on potassium ion uptake by barley roots.

substance analysed	ion concentration of element (mg/l)				
	calcium (Ca^{2+})	chloride (Cl$^-$)	magnesium (Mg^{2+})	potassium (K$^+$)	sodium (Na$^+$)
cell sap	380	3750	260	2400	1988
pond water	26	35	36	2	28
accumulation ratio	14.1:1	107.1:1	7.2:1		

Table 2.4

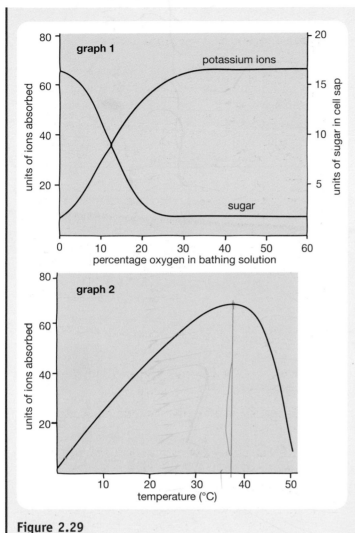

Figure 2.29

a) i) From graph 1, state the effect that an increase in oxygen concentration from 0 to 30% has on rate of ion uptake.

ii) Suggest why ion uptake levels off beyond 30% oxygen.

iii) What relationship exists between units of ion absorbed and units of sugar present in cell sap. Suggest why. (4)

b) i) State the temperature at which greatest uptake of ions occurs in graph 2.

ii) Account for the sudden decline in ion uptake shown by the graph. (2)

c) A farmer decided to drain a field which had become water-logged each year following heavy rainfall. With reference to ion uptake, give ONE possible reason why his crop of barley showed a greatly increased yield the next year. (1)

10 Give an account of the functions of phospholipids and proteins within the plasma membrane. (10)

11 Distinguish between *diffusion* and *osmosis*. Using TWO examples of each, describe how these processes play important roles in the life of a green plant. (10)

What You Should Know

(Chapters 1–2)
(See Table 2.5 for Word bank)

actively	freely	phospholipid
against	gradient	porous
bursting	membrane	protein
cell	mosaic	specialised
cellulose	multicellular	transport
diffusion	passive	unicellular
energy	permeable	variation

Table 2.5 Word bank for chapters 1–2

1 The _____ is the basic unit of life.

2 A _____ organism consists of one cell which possesses a variety of structures enabling it to perform all the functions necessary for the maintenance of life.

3 A _____ organism consists of more than one cell. In advanced animals and plants these are arranged into tissues giving a division of labour.

4 _____ in cell structure exists between cells of one type of tissue and cells of different tissues since cells are _____ to perform particular functions.

5 Cells absorb molecules in solution by _____, osmosis and active _____.

6 The cell wall surrounding plant cells is made of _____ and is _____ permeable to solutions. It prevents the cell from _____ when water is absorbed.

7 The plasma _____ surrounding the living contents of all cells is selectively _____. It consists of protein and _____ molecules thought to be arranged as in the fluid _____ model.

8 The plasma membrane is fluid and _____ in nature. It allows the _____ transport of small molecules by diffusion and osmosis down a concentration _____.

9 Other chemical substances such as ions are _____ transported across the plasma membrane _____ a concentration gradient by _____ carriers. This process requires _____.

3 ATP and energy release

Effect of adenosine triphosphate (ATP) on muscle fibre

In the experiment shown in Figure 3.1, only ATP is found to bring about contraction of the muscle fibres. It is therefore concluded that ATP is able to immediately provide the energy required for muscle contraction whereas glucose, despite being an energy-rich compound, is unable to do so.

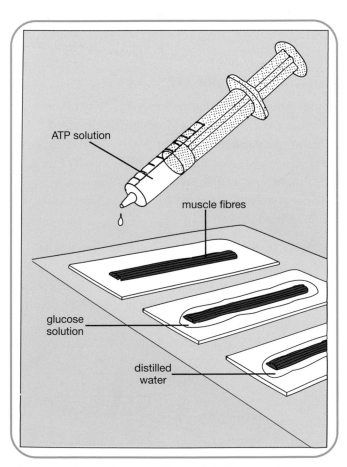

Figure 3.1 Investigating effect of ATP on muscle

Structure of ATP

A molecule of **adenosine triphosphate** (ATP) is composed of adenosine and three inorganic phosphate (Pi) groups as shown in Figure 3.2.

Energy stored in an ATP molecule is released when the bond attaching the terminal phosphate is broken by

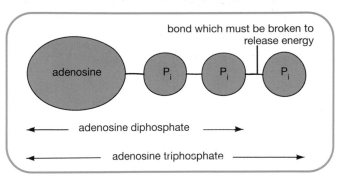

Figure 3.2 Structure of ATP

enzyme action. This results in the formation of **adenosine diphosphate** (ADP) and **inorganic phosphate (Pi)**. On the other hand, energy is required to regenerate ATP from ADP and inorganic phosphate by an enzyme-controlled process called **phosphorylation**. This reversible reaction is summarised by the following equation:

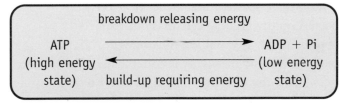

Production of ATP

When an energy-rich substance such as glucose is broken down (oxidised) in a living cell, it releases energy which is used to produce ATP.

When glucose is burned in a dish in the laboratory, its energy is released in one quick burst of heat and light. However in a living cell, the breakdown of glucose during cell respiration is a **gradual** process involving many enzyme-controlled steps (see chapter 4). This **orderly release** of energy is the ideal means by which the chemical energy needed to regenerate ATP from ADP + Pi is made available.

ATP is also regenerated and used during photosynthesis (see chapter 6).

Role of ATP

Many molecules of ATP are present in every living cell.

Since ATP can rapidly revert to ADP + Pi, it is able to make energy available for energy-requiring processes such as:

- muscular contraction
- synthesis of proteins and nucleic acids
- active transport of molecules
- transmission of nerve impulses.

ATP is important because it acts as the **link** between energy-releasing reactions and energy-consuming reactions. It provides the means by which chemical energy is **transferred** from one type of reaction to the other in a living cell as shown in Figure 3.3.

Since the breakdown of ATP is **coupled** with energy-demanding reactions, this promotes the transfer of energy to the new chemical bonds (e.g. those joining amino acids together – see Figure 3.4) and helps to reduce the amount of energy that is lost as heat.

Turnover of ATP molecules

It has been estimated that an active cell (e.g. a bacterium undergoing cell division) requires approximately two million molecules of ATP per second to satisfy its energy requirements. This is made possible by the fact that a **rapid turnover** of ATP molecules constantly occurs in a cell. At any given moment some ATP molecules are undergoing breakdown and releasing the energy needed for cellular processes while others are being **regenerated** from ADP + Pi using energy released during cell respiration.

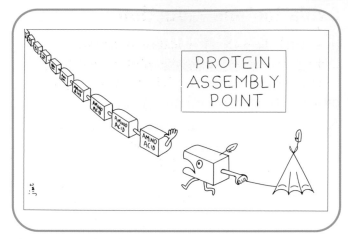

Figure 3.4 "What do you mean, I need **ATP**? I thought you said **a tepee**."

Constant quantity of ATP

In the human body a rise in level of activity by working cells (e.g. exercising muscle) leads to an increased demand being made on the supply of ATP molecules to break down and release energy. This is rapidly followed by an increase in rate of tissue respiration which produces the energy needed to regenerate the ATP that has been used up.

Since ATP is manufactured at the same time as it is used up, there is no need for the body to possess vast stores of ATP. The quantity of ATP present in the body is found to remain fairly constant at about 50 g despite the fact that the body may be using up and regenerating ATP at a rate of about 400 g/hour.

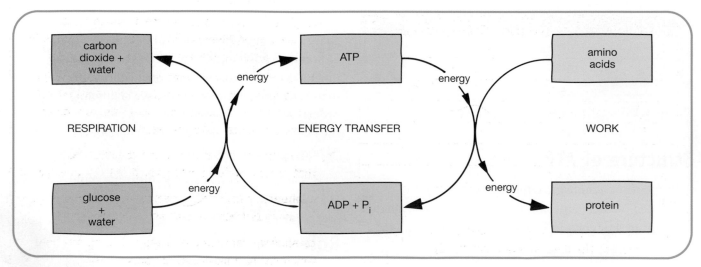

Figure 3.3 Transfer of chemical energy by ATP

Oxidation and reduction

In a metabolic pathway, **oxidation** occurs when hydrogen is removed from the substrate and energy is released as shown in Figure 3.5. Oxidation occurs during cell respiration.

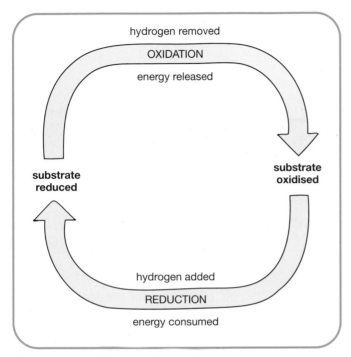

Figure 3.5 Oxidation and reduction

The process of **reduction**, on the other hand, involves the addition of hydrogen to the substrate and the consumption of energy. Reduction occurs during photosynthesis (see chapter 6).

The mnemonic OILRIG is useful for remembering the meanings of oxidation and reduction. Oxidation Is Loss of hydrogen. Reduction Is Gain of hydrogen.

Testing Your Knowledge

1 **a)** What compound is represented by the letters ATP? (1)

   b) Give a word equation to indicate how ATP is regenerated in a cell. (2)

2 Explain briefly the importance of ATP to a cell. (3)

3 The human body produces ATP at a rate of approximately 400 g/hour, yet at any given moment there is only about 50 g present in the body. Explain why. (2)

4 Give TWO differences between the processes of oxidation and reduction. (2)

Applying Your Knowledge

1 Table 3.1 shows the results of an experiment set up to investigate the effect of ATP solution on the length of a sample of skeletal muscle from a cow.

time (min)	length of muscle sample (mm)
0	50.0
1	48.0
2	46.5
3	44.8
4	42.3
5	40.7
6	40.7

Table 3.1

   a) Making the best use of a sheet of graph paper, present the data as a line graph. (2)

   b) i) Between which two times did a decrease in length occur at the fastest rate?

   ii) Between which two times did no decrease in length occur? (2)

   c) Give an equation to represent the chemical reaction involving ATP that occurs in the muscle cells enabling them to contract. (2)

   d) It could be argued that the muscle sample was going to decrease in length whether or not ATP was added. In what way should the experiment be altered to make the results valid? (1)

2 Metabolism is the sum of all the chemical changes that occur in a living organism. It falls into two parts:

   anabolism consisting of energy-requiring reactions involving synthesis of complex molecules and

   catabolism consisting of energy-yielding reactions in which complex molecules are broken down. →

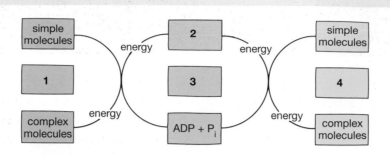

Figure 3.6

Transfer of energy from catabolic reactions to anabolic reactions is brought about by **ATP**. Figure 3.6 is a summary of the above information.

a) Copy the diagram and add four arrow heads to show the directions in which the two coupled reactions occur. (2)

b) Complete boxes 1–4 using each of the terms given in bold print in the passage. (2)

3 The graph in Figure 3.7 shows the volume of carbon dioxide released by yeast cells during the fermentation of glucose.

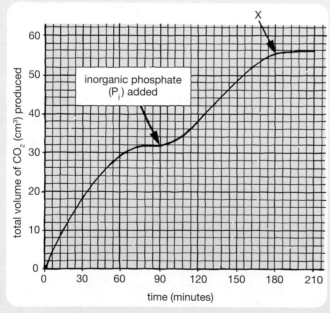

Figure 3.7

a) For which specific biochemical process does a yeast cell require a supply of inorganic phosphate (Pi)? (1)

b) Name the substance to which the inorganic phosphate added at 90 minutes became chemically combined. (1)

c) i) State what happened to the rate of CO_2 production following the addition of inorganic phosphate at 90 minutes.

ii) Suggest a reason for this change. (2)

d) Give a possible explanation for the decline in CO_2 production at point X on the graph. (1)

4 One mole of glucose releases 2880 kJ of energy. During aerobic respiration in living organisms, 44% of this is used to generate ATP. The rest is lost as heat.

a) What percentage of the energy generated during aerobic respiration is lost as heat? (1)

b) Out of a mole of glucose, how many kilojoules are used to generate ATP? (1)

c) Name TWO forms of cellular work (other than muscular contraction) that the energy held by ATP could be used to carry out. (2)

5 Give an account of the production of ATP and its role in cellular processes. (10)

4 Chemistry of respiration

Respiration is the process by which chemical energy is released from a foodstuff by **oxidation**. It occurs in every living cell and involves the regeneration of the high energy compound ATP by a complex series of chemical reactions.

Metabolism is the sum of all the chemical processes that occur in a living organism. Respiration is an example of a metabolic pathway.

Glycolysis

In the cytoplasm of a living cell, the process of cell respiration begins with a molecule of 6-carbon glucose being broken down by a series of enzyme-controlled steps to form two molecules of 3-carbon pyruvic acid (see Figure 4.1).

This process of 'glucose-splitting' is called glycolysis. It requires energy from two molecules of ATP to trigger it off but later in the process sufficient energy is released to form four molecules of ATP giving a **net gain** of two ATP.

During glycolysis, hydrogen released from the respiratory substrate becomes temporarily bound to a **coenzyme** molecule which acts as a hydrogen acceptor and carrier. The coenzyme involved is normally **NAD** (full name – nicotinamide adenine dinucleotide). For the sake of simplicity any coenzyme that acts as a hydrogen acceptor in this pathway will be referred to as 'NAD' and represented in its reduced state as '$NADH_2$'. At no point in the pathway does hydrogen exist as free atoms or molecules.

The process of glycolysis does not require oxygen but the hydrogen bound to reduced NAD only produces further molecules of ATP (at a later stage in the process) if oxygen is present. In the absence of oxygen, anaerobic respiration occurs (see page 33).

Mitochondria

When oxygen is present, **aerobic** respiration occurs in the cell's **mitochondria** (see Figures 4.2 and 4.3). Mitochondria are sausage-shaped organelles present in the cytoplasm of living cells. The inner membrane of each mitochondrion is folded into many plate-like extensions (**cristae**) which present a large surface area upon which respiratory processes can take place. The cristae project into the fluid-filled interior (**central matrix**) which contains enzymes.

Electron micrographs reveal that each crista bears many stalked particles. These are the site of ATP production. Cells requiring much energy such as sperm, liver, muscle and nerve cells contain numerous mitochondria which possess many cristae.

Testing Your Knowledge

1 Rewrite the following sentences including only the correct answer from each choice in brackets.

   During respiration, (chemical/light) energy is released from a foodstuff by (reduction/oxidation). This metabolic process occurs in (animal cells only/all living cells) and involves the (regeneration/degeneration) of the high energy compound (ADP/ATP). (2)

2 a) What substance results from the breakdown of a glucose molecule during glycolysis? (1)

   b) How many molecules of ATP are gained by the cell as a result? (1)

   c) How many carbon atoms are present in a molecule of:

      i) pyruvic acid; ii) glucose? (2)

3 a) Name the type of organelle responsible for aerobic respiration. (1)

   b) What is the fluid-filled inner cavity of this organelle called? (1)

   c) What name is given to the folded extensions of this organelle's inner membrane which present a large surface area? (1)

Fate of pyruvic acid

Each molecule of 3-carbon pyruvic acid produced during glycolysis diffuses into the central matrix of a

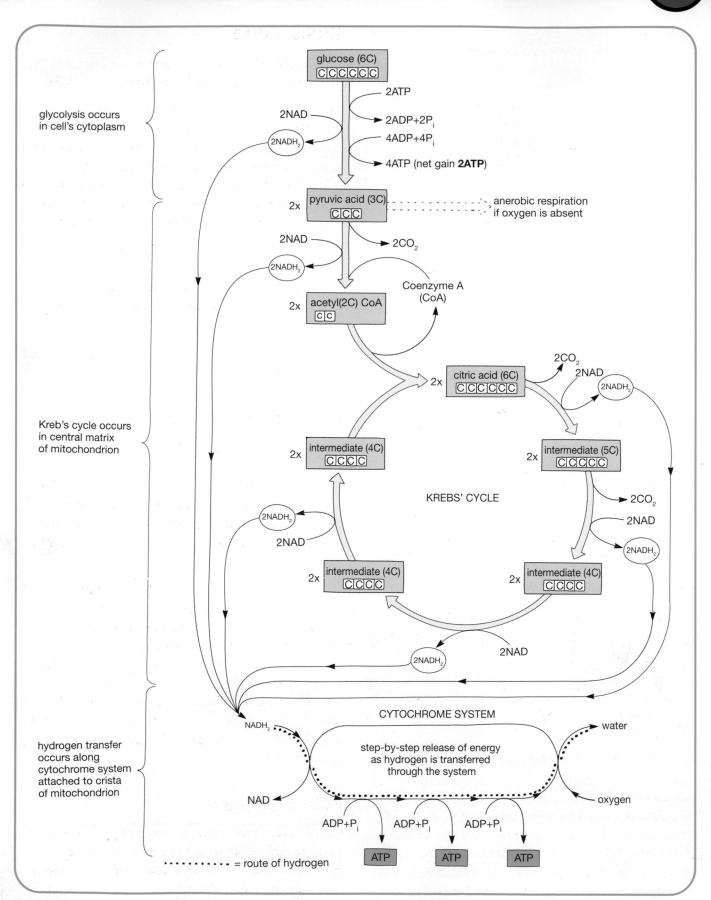

Figure 4.1 Chemistry of respiration

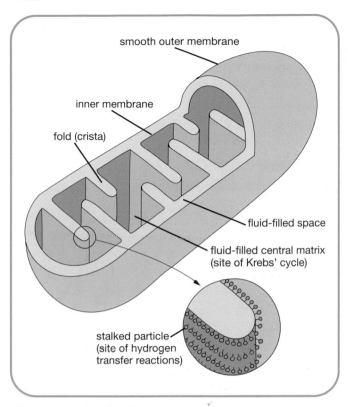

Figure 4.2 Mitochondrion

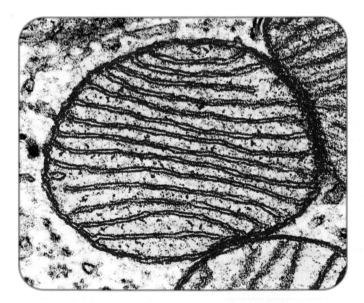

Figure 4.3 Electron micrograph of mitochondrion

mitochondrion where it is broken down into carbon dioxide and a 2-carbon fragment called an acetyl group. Each 2-carbon acetyl group becomes attached to coenzyme A (CoA) to form a molecule of **acetyl CoA**. This reaction is accompanied by the release of hydrogen which again becomes bound to NAD, the coenzyme acting as a **hydrogen acceptor**.

Krebs' Cycle

This aerobic phase of respiration is also known as the Citric Acid Cycle and the Tricarboxylic Acid Cycle.

Each molecule of 2-carbon acetyl CoA reacts with a molecule of a **4-carbon compound** present in the central matrix of the mitochondrion to form **6-carbon citric acid** (which is a tricarboxylic acid). This is gradually converted back to the 4-carbon compound by a cyclic series of enzyme-controlled reactions which bring about the removal of carbon and hydrogen from the respiratory substrate (see Figure 4.1).

Enzymes controlling the release of carbon to form carbon dioxide are called **decarboxylases**. The carbon dioxide formed diffuses out of the cell as a waste product.

Enzymes controlling the release of hydrogen are called **dehydrogenases**.

Cytochrome system

There are six points along the pathway where hydrogen is released and becomes temporarily bound to NAD, the coenzyme molecule. Reduced coenzyme ($NADH_2$) transfers hydrogen to a chain of hydrogen carriers called the **cytochrome system**. Every mitochondrion possesses many of these systems attached to each of its cristae.

Transfer of hydrogen from each $NADH_2$ along the cytochrome system releases sufficient energy to produce *three ATP*. This process is called **oxidative phosphorylation**. The complete oxidation of one glucose molecule yields a total of 38 ATP (two ATP during glycolysis + 36 ATP during oxidative phosphorylation). Since the hydrogen carrier (cytochrome) system generates most of the 38 ATP, it is responsible for releasing most of the energy during respiration.

Role of oxygen

Oxygen is the final hydrogen acceptor. Hydrogen and oxygen combine under the action of the enzyme cytochrome oxidase to form water. Although oxygen only plays its part at the very end of the pathway, its presence is essential for hydrogen to pass along the

31

cytochrome system. In the absence of oxygen, the oxidation process cannot proceed beyond glycolysis.

Alternative respiratory substrate

Fatty acids are converted to acetyl CoA in the central matrix of a mitochondrion. This compound then enters Krebs' Cycle and produces ATP as before. Fat liberates more than **double** the energy released by the same mass of carbohydrate.

Amino acids from protein can also act as respiratory substrates. Alanine, for example, can be converted to pyruvic acid allowing it to enter the pathway and release energy. A certain amount of energy is always derived from excess **dietary** protein but tissue protein is only used as a source of energy during prolonged starvation.

Function of aerobic respiration

Aerobic respiration (summarised in Figure 4.4) is a metabolic pathway consisting of a series of enzyme-controlled reactions during which a respiratory substrate such as 6-carbon glucose is oxidised to form

carbon dioxide accompanied by the production of ATP from ADP + Pi.

This **regeneration** of ATP for use in other cellular processes is the key function of respiration. (ATP is also regenerated during photosynthesis by **photophosphorylation** – see page 46.)

photophosphorylation – see page 46.

Testing Your Knowledge

1   Pyruvic acid is converted to acetyl CoA on entering a mitochondrion.

   a) How many carbon atoms are present in one molecule of acetyl CoA? (1)

   b) i) What TWO substances are released during the breakdown of pyruvic acid?
   ii) Which of these becomes bound to a coenzyme molecule? (3)

2   **a)** How many carbon atoms are present in a molecule of citric acid? (1)

   b) In which region of a mitochondrion is citric acid formed? (1)

   c) i) What name is given to the cycle of reactions by which citric acid is gradually converted back to a 4-carbon compound?
   ii) What happens to the hydrogen released during this cycle? (2)

3   **a)** What is the function of the cytochrome system? (1)

   b) Where is the cytochrome system located in a mitochondrion? (1)

   c) Name the high energy compound whose synthesis depends on energy released when hydrogen passes along the cytochrome system. (1)

   d) Identify the final hydrogen acceptor in the aerobic phase of respiration. (1)

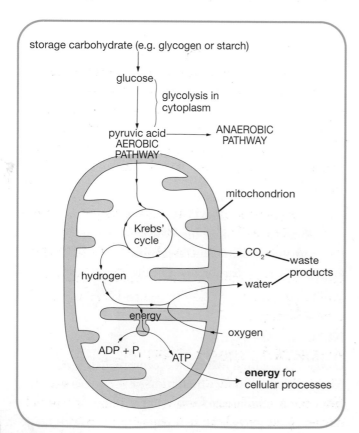

Figure 4.4 Summary of chemistry of respiration

Investigating the activity of dehydrogenase enzyme in yeast

- During respiration glucose is gradually broken down (oxidised) and hydrogen is released at various stages along the pathway. Each of these stages is controlled by an enzyme called a **dehydrogenase**.

- Yeast cells contain small quantities of stored food which can be used as a respiratory substrate.

- Resazurin dye is a chemical which changes colour upon becoming reduced (i.e. gaining hydrogen) as follows:

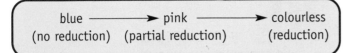

blue ⟶ pink ⟶ colourless
(no reduction) (partial reduction) (reduction)

- Before setting up the experiment shown in Figure 4.5, dried yeast is added to water and aerated for an hour at 35°C to ensure that the yeast is in an active state.

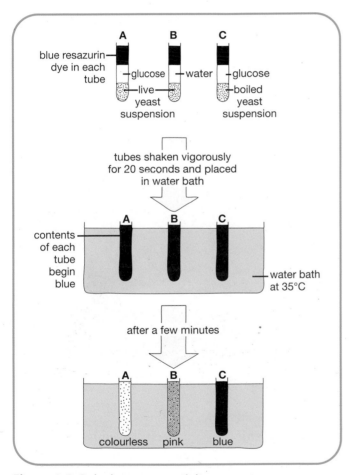

Figure 4.5 Dehydrogenase activity

Once the experiment has been set up, the contents of tube A are found to change from blue via pink to colourless much faster than those in tube B. Tube C, the control, remains unchanged.

It is concluded that in tube A hydrogen has been rapidly released and has reduced the resazurin dye. For this to be possible, dehydrogenase enzymes present in the yeast cells must have acted on glucose, the respiratory substrate, and oxidised it.

In tube B, the reaction was slower since no glucose was added and the dehydrogenases could only act on any small amount of respiratory substrate already present in the yeast cells.

In tube C, boiling has killed the cells and denatured the dehydrogenase enzymes.

Measuring rate of respiration

Figure 4.6 shows a simple respirometer set up to measure an earthworm's rate of respiration (as volume of oxygen consumed per unit time). Carbon dioxide given out by the animal is absorbed by the sodium hydroxide. Oxygen taken in by the animal causes a decrease in volume of the enclosed gas, therefore the coloured liquid rises up the tube.

After a known length of time (e.g. one hour), the syringe is used to find out the volume of air which must be injected to return the coloured liquid to its initial level. If, for example, 0.4 ml is needed then the worm's respiratory rate under these conditions = 0.4 ml oxygen/hour.

Design features and precautions

The control is set up without an earthworm. Since it could be argued that the observed differences were simply due to the fact that some space was occupied in the respirometer and not necessarily caused by the worm respiring, the best control includes some inert material (such as glass beads) equal in volume to the animal.

The experiment is carried out with the respirometer and control tubes in a large container of water at room temperature to ensure constant temperature throughout.

Handling of the apparatus during the experiment is avoided to prevent heat from the hands making the results invalid.

Anaerobic respiration

This is the process by which a little energy is derived from the partial breakdown of sugar in the absence of oxygen. Since oxygen is unavailable to the cell, the hydrogen transfer system and Krebs' Cycle cannot

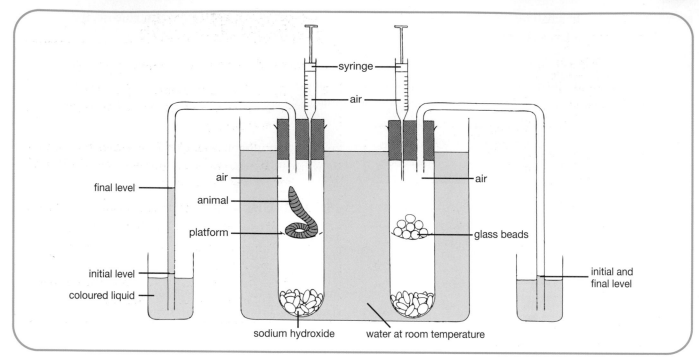

Figure 4.6 Simple respirometer and control

operate in any of the cell's mitochondria. Only glycolysis can occur. Each glucose molecule undergoes partial breakdown to pyruvic acid and yields only two molecules of ATP. The hydrogen released cannot go on to make ATP in the absence of oxygen.

An alternative metabolic pathway takes place in the cell's cytoplasm. The form that this takes depends on the type of organism involved.

Anaerobic respiration in plants

The equation below summarises this process in plant cells such as yeast deprived of oxygen, and cells of roots in water-logged soil:

glucose $\longrightarrow$ pyruvic acid $\longrightarrow$ ethanol + CO_2
(6C)   (2 × 3C)   (2 × 2C)

Anaerobic respiration in animals

The equation below summarises this process in animal cells such as skeletal muscle tissue:

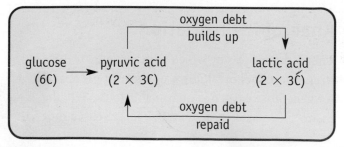

oxygen debt builds up

glucose $\longrightarrow$ pyruvic acid $\longrightarrow$ lactic acid
(6C)   (2 × 3C)   (2 × 3C)

oxygen debt repaid

Figure 4.7 Repayment of oxygen debt

During lactic acid formation, the body accumulates an **oxygen debt**. This is repaid when oxygen becomes available (see Figure 4.7) and lactic acid is converted back to pyruvic acid which then enters the aerobic pathway. Fortunately the biochemical conversion suggested in Figure 4.8 never really takes place.

Anaerobic respiration is a less efficient process since it produces only two ATP per molecule of glucose compared with 38 ATP formed by aerobic respiration. The majority of living cells thrive in oxygen and respire aerobically. They only resort to anaerobic respiration to obtain a little energy for survival while oxygen is absent.

Figure 4.8 Anaerobic nightmare

1 What is the purpose of a respirometer? (1)

2 a) With reference to oxygen, explain the difference between aerobic and anaerobic respiration. (1)

   b) State the number of ATP molecules formed from the breakdown of one glucose molecule during each type of respiration. (1)

3 a) Give the word equation of anaerobic respiration in i) a plant cell; ii) an animal cell. (4)

   b) i) Which of these forms of respiration is reversible?

      ii) Explain your answer with reference to the metabolic products formed. (2)

Applying Your Knowledge

1 a) State ONE way in which the structure of a mitochondrion is suited to its function. (1)

   b) Explain why a nerve cell contains a relatively higher number of mitochondria than a cheek epithelial cell. (1)

2 Table 4.1 refers to the process of cell respiration.

   a) Copy the table and complete the blanks indicated by brackets. (5)

   b) During which process in the table do the products of fat digestion enter the pathway and begin to undergo oxidation? (1)

   c) State which process(es) would fail to occur in the absence of oxygen. (2)

3 Give TWO reasons why a drop in pH occurs in the skeletal muscle tissue of a human during intensive physical training. (2)

4 Figure 4.9 shows a detailed version of the series of coupled reactions that occur along the hydrogen carrier system at the end of the aerobic respiration pathway.

   a) i) Name the final hydrogen carrier.
      ii) What form of respiration proceeds in the absence of this substance? (2)

   b) i) By what other name is the hydrogen carrier system known?
      ii) Exactly where in a cell would this series of carrier molecules be found? (2)

   c) i) Give the full names of substances X, Y and Z.
      ii) Which of these contains energy which can readily be made available to the cell when required? (4)

5 Figure 4.10 shows a set of apparatus about to be used to investigate the effect of temperature on an earthworm's rate of respiration.

process	site where process occurs in cell	reaction(s) involved in process	products
A glycolysis	[_____]	splitting of [_____] into pyruvic acid	ATP and $NADH_2$ and [_____]
B [_____] cycle	central matrix of [_____]	removal of [_____] from carbon compounds by oxidation; release of carbon dioxide	[_____] and $NADH_2$
C hydrogen transfer	[_____] of mitochondrion	release of energy from hydrogen	[_____] and [_____]

Table 4.1

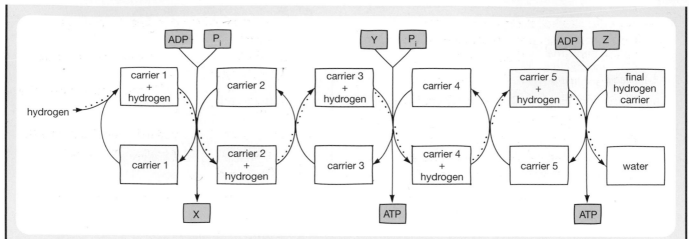

Figure 4.9

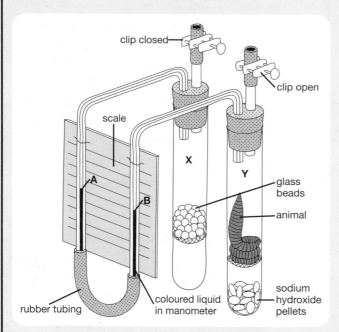

Figure 4.10

a) State TWO ways in which the apparatus would have to be altered before beginning the experiment. (2)

b) Suggest how a temperature of 25°C could be obtained simultaneously in tubes X and Y. (1)

c) i) Assume that the experiment is correctly set up and running. Predict the direction in which levels A and B will now move.
   ii) Explain why. (2)

d) i) In what way will the movement of the liquid levels differ when the experiment is repeated at 5°C for the same length of time?
   ii) Explain why. (2)

e) Why should the same earthworm be used each time? (1)

f) What is the purpose of the glass beads in tube X? (1)

g) How could the apparatus be redesigned to measure the actual volume of oxygen consumed per hour by the earthworm? (2)

6 Figure 4.11 shows a simplified version of the chemistry of respiration in an animal cell.

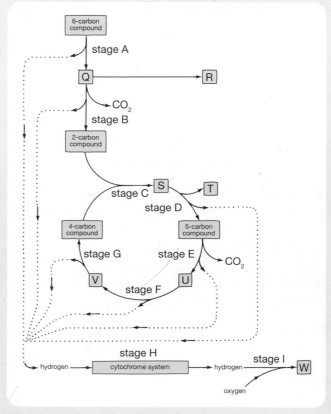

Figure 4.11

a) State the number of carbon atoms present in a molecule of each of the substances represented by boxes Q, R, S, T, U and V. (3)

b) Identify substances Q, R, S, T and W. (5)

c) The various stages in the pathway are labelled using the letters A–I.

    i)   At which of these stages is most ATP synthesised per molecule of glucose?

    ii) Identify another stage at which ATP is also synthesised (though in a much smaller quantity).

    iii) Which of the lettered stages occurs in the cell's cytoplasm?

    iv) Which stages occur in the central matrix of a mitochondrion? (4)

7 Give an account of ATP production resulting from the complete oxidation of one glucose molecule in a living cell. (10)

What You Should Know

(Chapters 3–4)
(See Table 4.2 for Word bank)

ADP	glycolysis	oxidation
aerobic	hydrogen	oxygen
anaerobic	Krebs'	phosphorylation
ATP	lactic	pyruvic
carbon dioxide	matrix	reduction
cytochrome	mitochondria	transfer
ethanol	NAD	water

Table 4.2 Word bank for chapters 3–4

1 _____ is a high energy compound able to release and _____ energy when it is required for cellular processes.

2 ATP is regenerated from _____ and inorganic phosphate by the process of _____ using energy released during respiration.

3 _____ involves the removal of hydrogen from a substrate and the release of energy; _____ involves the addition of hydrogen to a substrate and the consumption of energy.

4 _____ is a biochemical pathway common to aerobic and _____ respiration. It involves the breakdown of glucose to _____ acid in the cytoplasm of a cell with the net gain of two ATP.

5 In the presence of _____, aerobic respiration occurs in the central _____ of mitochondria where the respiratory substrate is oxidised during _____ Cycle and _____ is released.

6 This hydrogen becomes temporarily bound to _____, a coenzyme which transfers it to the _____ system on the cristae of _____ where energy is released and used to form ATP.

7 As a result of _____ respiration, one molecule of glucose yields 38 ATP. _____ and CO_2 are the final metabolic products.

8 In the absence of oxygen, anaerobic respiration occurs and one molecule of glucose yields two ATP. The final metabolic products are _____ and _____ in plant cells and _____ acid in animal cells (and some bacteria).

5 Role of photosynthetic pigments

Absorption, reflection and transmission

Although some of the light striking a leaf (see Figure 5.1) is reflected and some is transmitted, most of it is absorbed. Of this absorbed light, only a small part is used in photosynthesis. The rest is converted to heat and lost (e.g. by radiation).

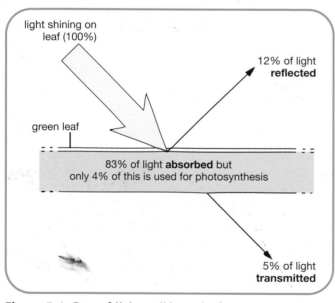

Figure 5.1 Fate of light striking a leaf

Extraction of leaf pigments

Fresh leaves are finely chopped and ground in a mortar containing propanone and a little fine sand as shown in Figure 5.2. The extract of soluble pigments is then separated from the cell debris by filtration.

Chromatography

Chromatography is a technique used to separate the components of a mixture which differ in their degree of solubility in a solvent.

During ascending paper chromatography, the mixture is applied to absorbent paper. As the solvent passes up through the paper, it carries the components of the

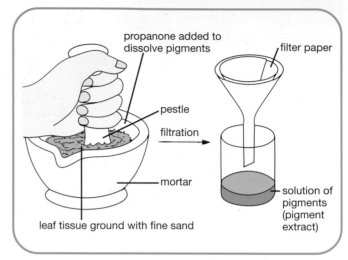

Figure 5.2 Extraction of leaf pigments

mixture up to different levels depending on the degree of their solubility in the solvent (and the extent to which they are absorbed by the paper).

Separation of leaf pigments using paper chromatography

A strip of chromatography paper is prepared as shown in Figure 5.3. This diagram illustrates the procedure followed during the spotting of the extract and the use of the chromatography solvent.

Spotting and drying of the extract is repeated many times and then the end of the paper is dipped into the solvent. The chromatogram is allowed to run until the solvent has almost reached the top of the paper. Table 5.1 gives the reasons for adopting certain techniques and precautions during this experiment.

Chromatogram

Figure 5.4 shows the chromatogram formed as a result of ascending paper chromatography using a solvent containing nine parts of petroleum ether to one part of propanone.

The solvent carries the most soluble pigment (carotene) to the highest position and so on down the paper to chlorophyll b, the least soluble. This is carried the

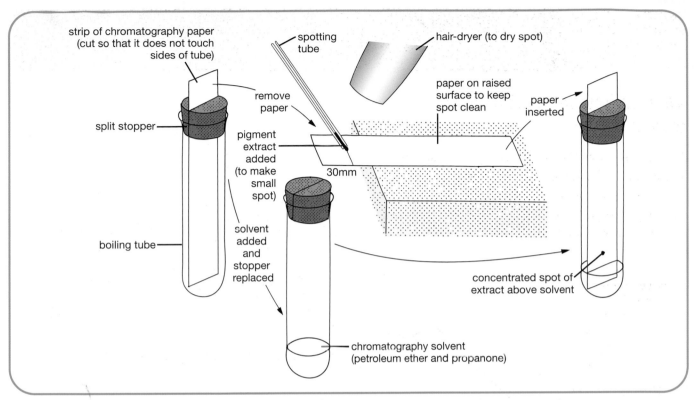

Figure 5.3 Separation of leaf pigments

design feature or precaution	reason
plant tissue ground in fine sand	to rupture cells allowing release of contents
chromatography paper cut so that it does not touch sides of tube	to ensure that solvent rises uniformly through paper rather than more rapidly up its edges
chromatography paper placed on raised surface and spotting done on overlap	to stop spot spreading and to prevent it from becoming contaminated with dirt or any other chemical
spotting and drying repeated many times	to obtain a concentrated spot of pigments
paper positioned in tube so that pigment spot is above solvent level at the start	to prevent extract dissolving in main bulk of solvent at bottom of tube
naked flames extinguished before starting experiment	to prevent fire risk since propanone and petroleum ether are highly flammable

Table 5.1 Design techniques

shortest distance. (A grey-coloured breakdown product of chlorophyll sometimes appears on the chromatogram but plays no part in photosynthesis.)

Separation of leaf pigments using thin layer chromatography

The pattern of pigments on a chromatogram depends on the absorbent material and the solvent used during the separation. The chromatogram in Figure 5.5 shows the result of **thin layer chromatography** using a thin layer strip and a solvent containing two parts of petroleum ether to one part of propanone.

Absorption spectra

When a beam of light is passed through a glass prism (or spectroscope), the **spectrum** of white (visible) light is produced (see Figure 5.6).

When a beam of white light is first passed through a sample of leaf pigments placed at X in Figure 5.7 and then passed through a glass prism, an **absorption spectrum** is produced.

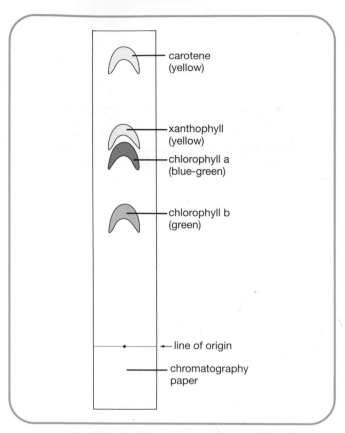

Figure 5.4 Paper chromatogram of leaf pigments

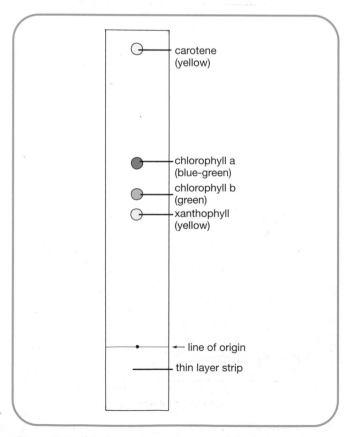

Figure 5.5 Thin layer chromatogram of leaf pigments

1 Some of the light that strikes a leaf is transmitted.

   a) What does this mean? (1)

   b) State TWO possible fates of the light that is not transmitted. (2)

2 **a) i)** Explain the purpose of the sand in the experiment shown in Figure 5.2.

     ii) Why do the contents of the mortar need to be filtered? (2)

   b) i) Name the technique in Figure 5.3 used to separate leaf pigments.

     ii) Briefly explain how it works.

     iii) Identify TWO chemicals suitable for use in the solvent mixture. (5)

3 What safety precaution must be strictly observed when carrying out the experiments shown in Figures 5.2 and 5.3. (1)

Each black band is a region of the spectrum where light energy has been absorbed by the leaf pigments and has therefore failed to pass through the prism and onto the screen. Each coloured band (e.g. green) is a region where light has not been absorbed by the extract.

Since the wavelengths of light not absorbed by a pigment are transmitted or reflected, chlorophyll appears green to the eye.

Graphs of absorption and action spectra

The **degree of absorption** at each wavelength of visible light by each pigment can be measured using a **spectrometer**. The data obtained allow a detailed graph of each pigment's absorption spectrum to be plotted (see Figure 5.8).

An **action spectrum** (see Figure 5.9) charts the effectiveness of different wavelengths of light at bringing about the process of photosynthesis.

Comparison of Figures 5.8 and 5.9 shows that a close correlation exists between the overall absorption spectrum for the pigments and the action spectrum. It is therefore concluded that the absorption of certain wavelengths of light for use in photosynthesis is the crucial role played by these pigments.

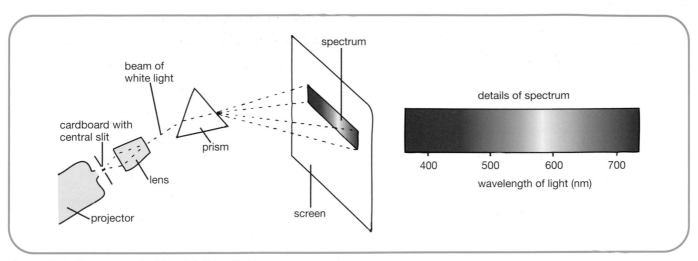

Figure 5.6 Spectrum of white (visible) light

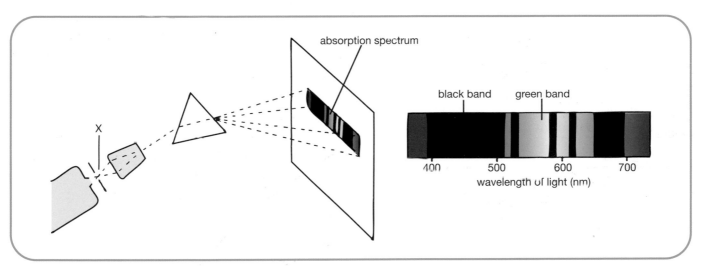

Figure 5.7 Absorption spectrum of leaf pigments

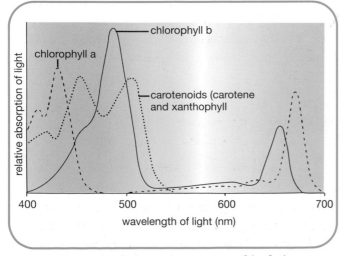

Figure 5.8 Graph of absorption spectra of leaf pigments

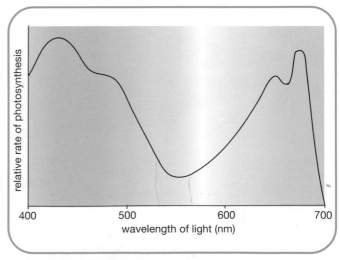

Figure 5.9 Action spectrum

41

Selective absorption

Chlorophyll a and b are structurally complex molecules containing **magnesium**. They absorb light primarily in the **red** and **blue** regions of the spectrum.

The accessory pigments (carotenoids) absorb light from other regions (e.g. the green part of the spectrum) and pass the energy on to chlorophyll.

Advantage to plant

Since each pigment absorbs different wavelengths of light, the total quantity of light absorbed is greater than it would be if only one pigment was involved.

Chloroplast

Chloroplasts (Figures 5.10 and 5.11) are relatively large, discus-shaped organelles situated in the cytoplasm of green plant cells. Each is bounded by a double membrane and possesses distinct internal structures made of grana and lamellae.

Each **granum** consists of a coin-like stack of flattened sacs containing photosynthetic pigments. Grana are the site of the **light-dependent stage** of photosynthesis (see page 46). Lamellae are tubular extensions which form an interconnecting network between grana but do not contain chlorophyll.

Figure 5.11 Electron micrograph of chloroplast

The widespread arrangement of grana throughout the chloroplast ensures that a large surface area of photosynthetic pigments is presented for absorption of light energy.

The colourless background material in a chloroplast is called **stroma**. It is the site of the **carbon fixation stage** of photosynthesis (see page 46). It lacks chlorophyll but contains important enzymes and starch grains (which act as temporary stores of photosynthetic products).

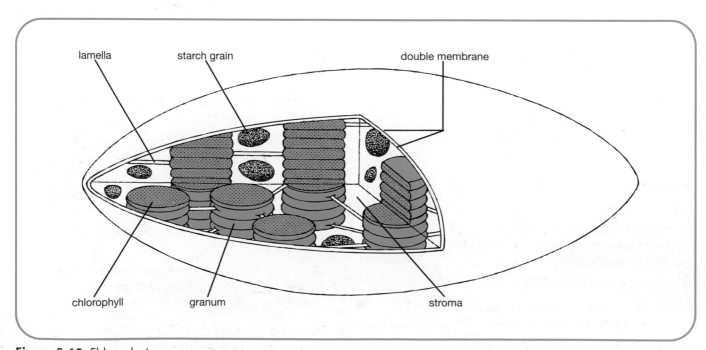

lamella   starch grain   double membrane

chlorophyll   granum   stroma

Figure 5.10 Chloroplast

Testing Your Knowledge

1 **a)** By what means can white light be split up into the spectrum of white (visible) light? (1)

   b) Which colour in the spectrum has light with the **i)** shortest; **ii)** longest wavelength? (2)

2 Explain the difference between an absorption spectrum and an action spectrum. (2)

3 Which TWO colours of light do chlorophylls a and b absorb mainly? (2)

4 Why does chlorophyll b appear green in colour? (2)

5 **a)** Give TWO structural differences between the grana and the stroma found in a chloroplast. (2)

   b) Identify the stage of photosynthesis that occurs in each of these regions. (2)

Applying Your Knowledge

1 Each pigment separated by chromatography has an **Rf value.** This is the ratio of the distance moved by the pigment front to the distance moved by the solvent front. It is usually expressed as a decimal fraction. For example substance X in Figure 5.12 has an Rf of 50/100 = 0.5. This means that X has moved half the distance moved by the solvent.

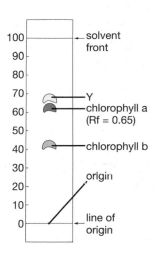

Figure 5.13

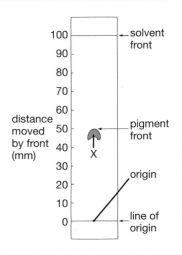

Figure 5.12

Figure 5.13 shows an incomplete paper chromatogram of the four photosynthetic pigments from a green leaf.

a) Calculate the Rf of chlorophyll b. (1)

b) Copy or trace Figure 5.13 and draw a spot to represent carotene which has an Rf of 0.95. (1)

c) Identify pigment Y and calculate its Rf. (1)

d) Suggest a method of separating the pigment extract from the cell debris in Figure 5.2 other than by filtration. (1)

e) Why must repeated spotting and drying of pigment extract be carried out when preparing the origin of pigment extract? (1)

f) Why must the origin be kept above the solvent level in the tube when the end of the paper is dipped into the solvent to run the chromatogram? (1)

g) Name a solvent that could be used to remove 'grass' stains from a garment at room temperature. (1)

2 The leaves of an oak tree change from green to brown in autumn prior to leaf fall. Outline the procedure that you would follow to investigate →

whether the pigment content of autumn leaves differs from that of green summer leaves. (5)

3 Rate of photosynthesis can be measured by counting the number of oxygen bubbles released per minute by the waterweed *Elodea*. In an experiment, a series of coloured filters were used in turn by inserting each between *Elodea* and the source of white light. Each coloured filter only allows one colour of light to pass through. The results are shown in Table 5.2.

   a) What was the one variable factor investigated in this experiment? (1)

   b) If a coloured filter only allows one colour of light to pass through, what happens to the other colours present in white light? (1)

   c) Explain why the average number of bubbles was calculated each time? (1)

   d) The experimenter allows a short space of time to elapse after removing one filter and before inserting the next. Suggest why. (1)

   e) Present the results as a bar chart. (2)

   f) Draw a conclusion from the results. (1)

   g) A nanometre (nm) is one thousandth of a micrometre (µm) which is one thousandth of a millimetre (mm) which is one thousandth of a metre (m). Draw a table to summarise this information and include a column which expresses each unit as a fraction of a metre using negative indices. (3)

4 a) Which pair of leaf pigments combined would give the absorption spectrum shown in Figure 5.14? (2)

   b) i) Which pair of pigments absorb most light energy in region Z in the graph?
      ii) Is this energy used for photosynthesis? Explain your answer. (4)

5 Figure 5.15 shows the result of placing a strand of alga in a liquid containing motile aerobic bacteria

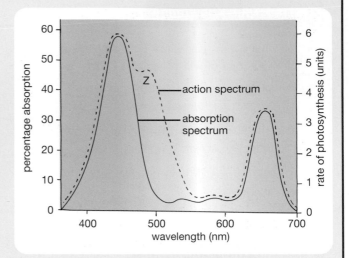

Figure 5.14

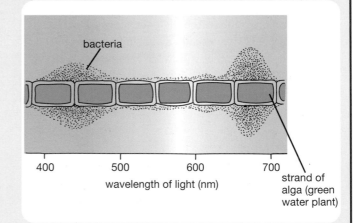

Figure 5.15

and illuminating the strand with a tiny spectrum of light.

a) In which colours of light did most bacteria congregate? (2)

b) Account for this distribution of the bacteria. (2)

6 Describe the structure of a chloroplast and then give an account of the role played by chlorophyll in

colour of light allowed through by filter	wavelength of light (nm)	average number of bubbles of oxygen released per minute
blue	430	15
green	550	1
yellow	600	4
red	640	12

Table 5.2

photosynthesis. Refer to action and absorption spectra in your answer. (10)

7 The information in Table 5.3 refers to photosynthetic pigments. The data in Table 5.4 refer to the amount of light absorbed at various wavelengths of light by two of these pigments each extracted from a different seaweed. All seaweeds belong to a large group of plants called algae.

a) Name ONE principal and THREE accessory pigments found in brown seaweeds. (1)

b) Name a type of plant that would lack all three classes of pigment listed in Table 5.3. (1)

c) Plot the data given in Table 5.4 as two line graphs (curves) to show the absorption spectra of pigments 1 and 2. (3)

d) Refer back to Figure 5.6 on page 41 which shows the spectrum of white (visible) light and devise a way of adding the six colours to one of the axes in your graph. (1)

e) Pigments 1 and 2 are known as the phycobilins.

i) One of them absorbs orange and red light and appears blue-green to the eye. Identify it by its number and its proper name.

ii) One of them absorbs green and yellow light and appears red to the eye. Identify it by its number and its proper name. (2)

f) Imagine that a sample of each of the phycobilin pigments were placed in turn at point X in the experiment shown in Figure 5.7 on page 41. Make a diagram of the absorption spectrum that you would expect to result in each case. (2)

g) Phycoerythrin is commonly found in seaweeds that live in deep sea water or in dimly lit rock pools. Suggest how the presence of this pigment helps the plant to survive. (1)

wavelength of light (nm)	amount of light absorbed (arbitrary units)	
	pigment 1	pigment 2
400	0	0
420	0	0
440	0	0
460	0.1	0
480	0.3	0.05
500	1.0	0.1
520	3.2	0.2
540	4.9	0.3
560	5.0	0.5
580	2.7	0.9
600	0.1	1.7
620	0.05	2.6
640	0	2.8
660	0	2.2
680	0	0.3
700	0	0

Table 5.4

class of pigment	name	principal (P) or accessory (A) pigment	location
chlorophyll	a	P	all photosynthetic plants
	b	A	higher plants and green algae
	c	A	brown algae
	d	A	some red algae
carotenoid	xanthophyll	A	all photosynthetic plants
	carotene	A	all photosynthetic plants
phycobilin	phycocyanin	A	main phycobilin in blue-green algae
	phycoerythrin	A	main phycobilin in red algae

Table 5.3

6 Chemistry of photosynthesis

Photosynthesis is the process by which organic compounds are synthesised by the **reduction** of carbon dioxide. The energy required for this process comes from light energy. The light energy is absorbed by photosynthetic pigments.

Photosynthesis consists of two separate parts: a **light-dependent** (photochemical) stage and a **temperature-dependent** (thermochemical) stage called **carbon fixation**.

Light-dependent stage

This light-dependent stage occurs in the grana of chloroplasts. **Solar** (light) energy is trapped by chlorophyll (and accessory pigments) and converted into **chemical** energy. The process involves several important events which are summarised in Figure 6.1.

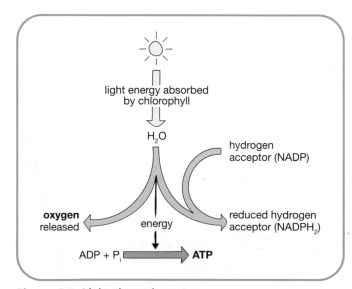

Figure 6.1 Light-dependent stage

Light energy is used to split molecules of water into hydrogen and oxygen. This is called **photolysis of water**. The oxygen is released as a by-product. The hydrogen combines with a hydrogen acceptor called **NADP** (full name – nicotinamide adenine dinucleotide phosphate) to form reduced hydrogen acceptor $NADPH_2$.

In addition, chlorophyll makes energy available for the regeneration of ATP from ADP and inorganic

phosphate. This process is called **photophosphorylation**.

The hydrogen held by $NADPH_2$ and the energy held by ATP at the end of the light-dependent stage are essential for use in **carbon fixation**, the second stage of photosynthesis.

Carbon fixation

This thermochemical stage occurs in the stroma of chloroplasts. It consists of several enzyme-controlled chemical reactions which take the form of a cycle (often referred to as the **Calvin Cycle** after the scientist who discovered it). It was formerly known as the 'dark' stage since it is not light-dependent.

Figure 6.2 summarises the cycle and indicates the number of carbon atoms present in the molecules of the metabolites involved.

On entering a chloroplast by diffusion, a molecule of carbon dioxide combines with a molecule of 5-carbon **ribulose bisphosphate (RuBP)** in the stroma to form a

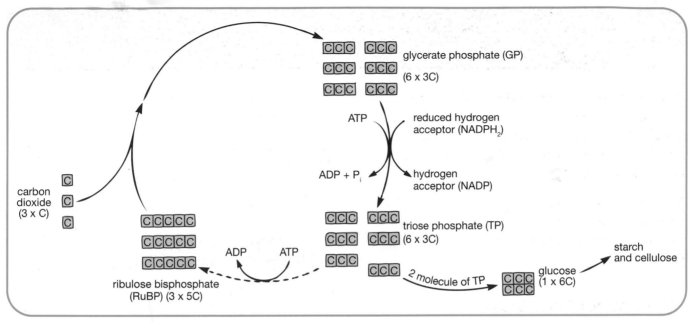

Figure 6.2 Calvin cycle (carbon fixation)

6-carbon molecule. This molecule is unstable and rapidly splits into two molecules of **3-carbon glycerate-3-phosphate (GP)**.

During the next stage in the cycle, GP is converted to a **3-carbon sugar (triose phosphate)** using the hydrogen temporarily bound to the reduced hydrogen acceptor (NADPH$_2$) and some of the energy held in ATP. Both of these are provided by the light-dependent stage.

Each molecule of glucose (a 6-carbon/hexose sugar) is synthesised from a pair of these 3-carbon sugar molecules which combine together in an enzyme-controlled sequence of reactions. Molecules of glucose may then be built up into starch and cellulose.

Simple products of photosynthesis are also converted by complex enzyme-controlled biochemical pathways into major biological molecules such as proteins, fats and nucleic acids.

Regeneration of RuBP

The 3-carbon triose phosphate molecules are not all used to make complex products. Some are needed to regenerate **RuBP**, the carbon dioxide acceptor. This conversion of triose phosphate (5 × 3C) to RuBP (3 × 5C) also requires energy which is derived from ATP.

Summary

The process of photosynthesis involves the **fixation of energy** and is summarised in Figure 6.3.

Reduction is the process by which hydrogen is added to a substrate. Photosynthesis is therefore a reduction process since carbohydrate is formed by the reduction of carbon dioxide.

Rate of photosynthesis

The rate of photosynthesis can be calculated by measuring one of the following:

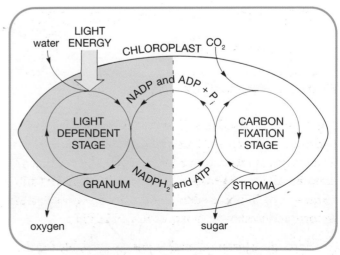

Figure 6.3 Summary of photosynthesis

- evolution of oxygen per unit time

- uptake of carbon dioxide per unit time

- production of carbohydrate (as increase in dry mass) per unit time.

Limiting factors

Rate of photosynthesis is affected by several environmental factors. These include **temperature, light intensity** and **carbon dioxide concentration.**

The rate at which photosynthesis proceeds is limited by whichever one of these factors is in short supply. For example light intensity would probably be the factor limiting photosynthesis on a dull wet summer's day.

The principle of **limiting factors** can be investigated using a water plant such as *Elodea* (see Figures 6.4 and 6.5.) The carbon dioxide concentration can be varied by adding a chemical to the water. The light intensity can be varied using a lamp with a dimmer switch. Photosynthetic rate is measured by counting the number of oxygen bubbles released per minute. Graph

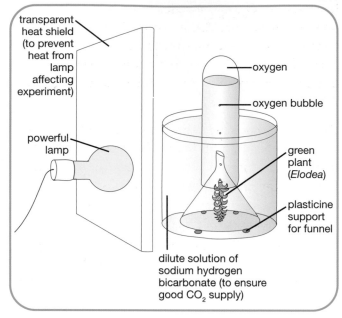

Figure 6.5 *Elodea* bubbler experiment

ABC in Figure 6.6 refers to a water plant kept in conditions of constant low light intensity.

When the plant is supplied with a CO_2 concentration of only one unit, photosynthetic rate is limited by this low concentration of CO_2 to three oxygen bubbles/min.

When CO_2 concentration is increased to two units, photosynthetic rate increases to six oxygen bubbles/min but no further since CO_2 concentration becomes limiting again.

A further increase in CO_2 concentration to three units brings about a further increase in photosynthetic rate. Beyond this point, the graph levels out and any further

Figure 6.4 Release of oxygen bubbles from *Elodea*

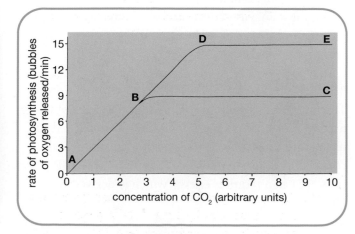

Figure 6.6 Limiting factors

increases in CO_2 concentration fail to affect photosynthetic rate. This is because light (which has been at constant low intensity throughout) has now become the factor limiting the process. Previously CO_2 concentration had been the limiting factor (i.e. on the AB part of the graph).

Graph ADE represents a further experiment using the same plant kept in conditions of constant high light intensity. This time an increase in CO_2 concentration to four and five units brings about a corresponding increase in photosynthetic rate in each case because, at these concentrations, CO_2 concentration is still the limiting factor when light intensity is high. However beyond five units of CO_2, the graph levels off again since light intensity (or some other factor) has become limiting.

Testing Your Knowledge

1 a) In which region of a chloroplast would carbon fixation be found to occur? (1)

   b) By what other name is the thermochemical stage of photosynthesis known? (1)

2 a) Identify the raw material which plays no part in the light-dependent stage yet is essential for the thermochemical stage of photosynthesis to proceed. (1)

   b) Name the substance that accepts this raw material into the cycle by combining with it. (1)

   c) Name the 3-carbon compound that is the first stable molecule formed during the process. (1)

   d) Give TWO possible fates of a molecule of triose phosphate. (2)

3 a) Give TWO methods by which a plant's rate of photosynthesis can be measured. (2)

   b) Name TWO environmental factors that can limit a plant's rate of photosynthesis. (2)

Applying Your Knowledge

1 Figure 6.7 represents the light-dependent stage of photosynthesis.

   a) Give the colour and the name of the substance responsible for trapping light energy and making it available for use in the reaction shown in the diagram. (2)

   b) i)   Identify substances X, Y and Z.
      ii) Which of these is a by-product of the process and diffuses out of the cell?
      iii) Which of these go on to play important roles at the thermochemical stage in photosynthesis? (6)

2 Present the process of photophosphorylation as a simple equation. (2)

3 Figure 6.8 shows the carbon fixation stage of photosynthesis.

   a) Copy the diagram and complete the blank boxes. (2)

   b) On your diagram, insert the number of carbon atoms possessed by a molecule of glycerate phosphate and triose phosphate. (2)

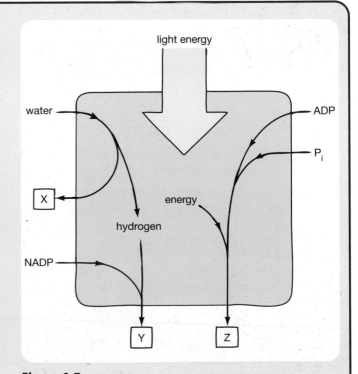

Figure 6.7

49

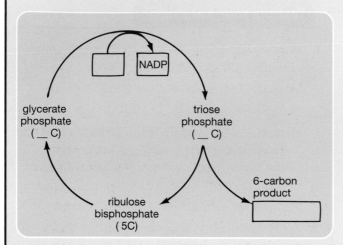

Figure 6.8

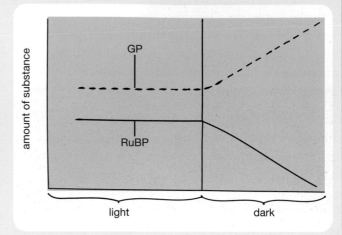

Figure 6.9

c) i) Which chemical acts as the carbon dioxide acceptor?

ii) Add an arrow and the symbol CO_2 to your diagram to show where CO_2 enters the cycle. (2)

d) i) What is the full name of ATP?

ii) Mark X on your diagram at TWO points at which ATP is needed for the reaction to proceed.

iii) Why is ATP necessary at these points? (3)

e) i) Which substance would accumulate if the plant were placed in darkness?

ii) Explain why. (2)

f) i) Which substance would accumulate if the plant were deprived of carbon dioxide?

ii) Explain why. (2)

4 a) Name the substances represented by the letters GP and RuBP in Figure 6.9. (2)

b) Describe the relationship that is thought to exist between GP and RuBP when light is present. (2)

c) Briefly explain how the results shown in the graph supply evidence that this relationship does exist. (2)

5 Figure 6.10 shows the effect of increasing light intensity on the rates of photosynthesis of a plant kept at two different concentrations of carbon dioxide.

a) What factor was limiting photosynthetic rate at region A on the graph? (1)

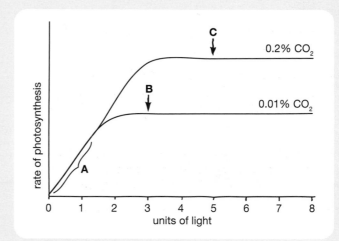

Figure 6.10

b) What factor was limiting photosynthetic rate at point B? (1)

c) By how many times did CO_2 concentration differ between the two experiments? (1)

d) What experimental factor not referred to in the graph could be limiting the rate of photosynthesis at point C? (1)

6 Write an essay on the biochemistry of photosynthesis under the following headings:

a) the light-dependent stage;

b) the Calvin Cycle. (10)

What You Should Know

Chapters 5–6
(See Table 6.1 for Word bank)

absorption	chromatography	photolysis
action	cycle	red
ATP	grana	reduction
blue	hydrogen	stroma
carbon dioxide	light-dependent	transmitted
carbon fixation	limited	yellow

Table 6.1 Word bank for chapters 5–6

1 Light is absorbed, reflected and _____ by a leaf.

2 The photosynthetic pigments from a leaf can be separated by _____.

3 Chlorophyll absorbs light primarily in the _____ and _____ regions of the spectrum of white light. _____ pigments absorb blue-green light.

4 The quantity of light absorbed by a pigment at different wavelengths of light can be presented as a graph called an _____ spectrum; the rate of photosynthesis that occurs in a plant at different wavelengths of light can be presented as a graph called an _____ spectrum.

5 Chloroplasts possess internal structures called _____ which contain photosynthetic pigments and are the site of the _____ stage of photosynthesis. The region between grana is called the _____. It is the site of the _____ stage of photosynthesis.

6 During the first stage of photosynthesis, light energy is used to split molecules of water into hydrogen and oxygen. This process is called _____. The light-dependent stage produces the energy (held in _____) and the hydrogen needed for the second stage (carbon fixation).

7 The second stage consists of a _____ of reactions which brings about the _____ of carbon dioxide using the ATP and _____ from the light-dependent stage to form carbohydrate.

8 Photosynthesis is affected by temperature, light intensity and _____ concentration. Its rate is therefore _____ by whichever one of these factors is in short supply.

7 DNA and its replication

Structure of DNA

Chromosomes (see Figure 7.1) are thread-like structures found inside the nucleus of a cell. They contain **deoxyribonucleic acid (DNA)**. DNA can be isolated from cells as shown in Figure 7.2.

Figure 7.1 Chromosomes

A molecule of DNA consists of two strands each made of repeating units called **nucleotides**. Each DNA nucleotide (see Figure 7.3) is made of a molecule of **deoxyribose sugar** joined to a **phosphate** group and a **base**. Since DNA possesses four different bases (**adenine, thymine, guanine** and **cytosine**) it has four different types of nucleotide.

A strong **chemical bond** forms between the phosphate group of one nucleotide and the deoxyribose sugar of another. These bonds are not easily broken and join neighbouring nucleotide units into a permanent strand as shown in Figure 7.4.

Two of these strands become joined together by weaker **hydrogen bonds** forming between their bases. However this union is temporary in that hydrogen bonds can be easily broken when it becomes necessary (e.g. during transcription of DNA into RNA – see page 59).

Each base can only join with one other type of base: adenine (A) always bonds with thymine (T), and guanine (G) always bonds with cytosine (C). A–T and G–C are called **base pairs**. Each member of a pair is complementary to its partner.

Double helix

The resultant double-stranded molecule is DNA and its two strands are arranged as shown in Figure 7.5. This twisted coil is called a **double helix**. It is like a spiral ladder in which the sugar–phosphate 'backbones' form the uprights and the base pairs form the rungs.

Testing Your Knowledge

1 a) What name is given to each of the repeating units that make up a strand of DNA? (1)

   b) How many strands are present in a molecule of DNA? (1)

2 a) Name the THREE parts of a DNA nucleotide. (3)

   b) i) Which type of bond joins nucleotides into a strand of DNA?

      ii) Rewrite the following sentence and complete the blanks.

      This type of bond forms between the _____ molecule of one nucleotide and the _____ molecule of the next nucleotide on the nucleic acid strand. (3)

3 a) i) How many different types of base molecule are found in DNA?

      ii) Name each type. (3)

   b) Which type of bond forms between the bases of adjacent strands of a DNA molecule? (1)

   c) Describe the base-pairing rule. (1)

4 a) What name is given to the twisted coil arrangement typical of DNA molecule? (1)

   b) If DNA is like a spiral ladder, which part of it corresponds to the ladder's i) rungs; ii) uprights? (2)

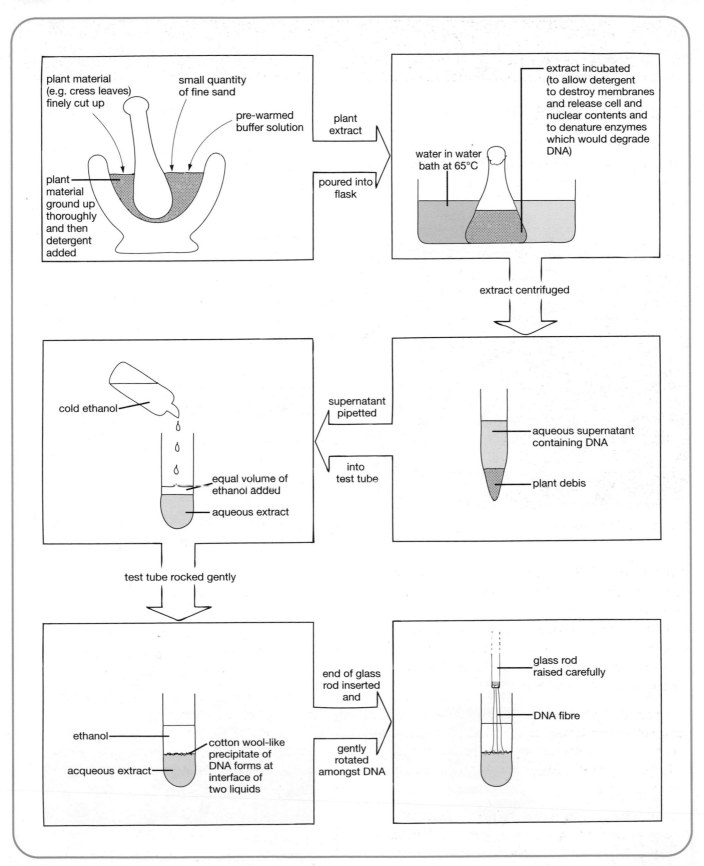

Figure 7.2 Isolation of DNA from plant tissue

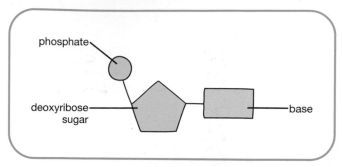

Figure 7.3 Structure of a nucleotide

Replication of DNA

DNA is a unique molecule because it is able to reproduce itself exactly. This process is called **replication**. It is illustrated in Figure 7.6.

Stage 1 shows a region of the original DNA molecule after it has just become unwound. Stage 2 is a little further ahead of stage 1 in the process. Here weak hydrogen bonds between two bases are breaking and causing the two component strands of DNA to separate ('unzip') and expose their bases.

At stage 3, pairing of two bases enables a free DNA nucleotide to find and align with its complementary nucleotide on the open chain. At stage 4, weak hydrogen bonds are forming between complementary base pairs.

At stage 5, slightly ahead of stage 4, a strong chemical bond is forming between the sugar of one nucleotide and the phosphate of the next one in the chain giving each strand its sugar–phosphate 'backbone'. This linking of nucleotides into a chain is controlled by an enzyme called **DNA polymerase**.

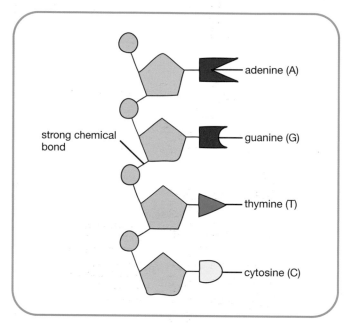

Figure 7.4 Strand of DNA nucleotides

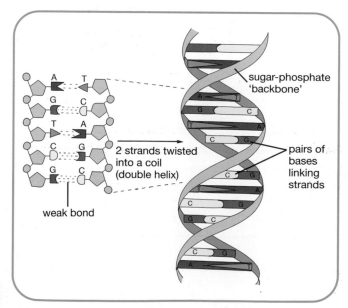

Figure 7.5 Structure of DNA

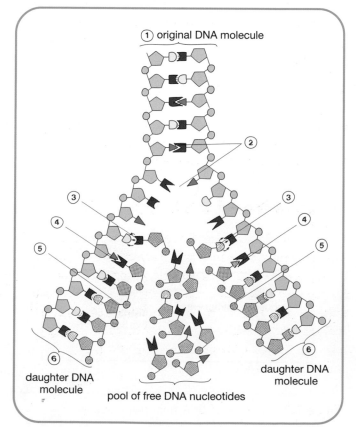

Figure 7.6 DNA replication

Stage 6 shows a newly formed daughter molecule of DNA about to wind up into a double helix. Daughter molecules have a base sequence identical to one another and to the original DNA molecule.

For DNA replication to occur the nucleus must contain:

- **DNA** (to act as a template for the new molecule);
- a supply of the four types of DNA **nucleotide**;
- the appropriate **enzymes** (e.g. DNA polymerase);
- a supply of **ATP** to provide energy.

Semi-conservative replication

DNA replication results in the formation of two new molecules, each of which receives one strand of the original parent molecule. It is therefore said to be semi-conservative (see Figure 7.7).

Maximum quantity of DNA

DNA replication is the means by which new genetic material is produced in the nucleus of a cell during **interphase**. The two identical daughter DNA molecules formed from each chromosome coil up and become **identical chromatids** held together by a **centromere** (see Figure 7.7).

The quantity of genetic material has doubled without changing the cell's chromosome number. It is at this time, following DNA replication but immediately before nuclear division (see chapter 12), that the cell's DNA content is at its maximum.

Importance of DNA replication

DNA replication ensures that an **exact copy** of the species' genetic information is passed from cell to cell during growth and from generation to generation during sexual reproduction. If DNA failed to replicate itself, the processes of mitosis (and cell growth) and meiosis (and gamete production) would be unable to take place. DNA is therefore essential for the continuation of life.

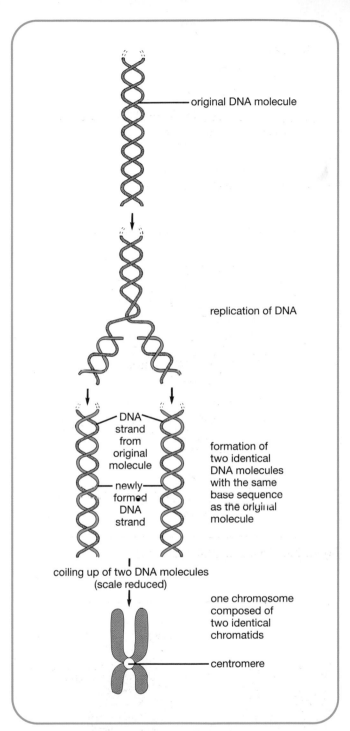

Figure 7.7 Semi-conservative replication and chromatid formation

Testing Your Knowledge

1 What is a molecule of DNA able to do that makes it unique compared to other chemical molecules? (1)

2 Study Figure 7.6 carefully and then answer the following questions.

a) At which numbered stage has the DNA molecule been involved in the replication process for the shortest time? (1)

b) i) What type of bond is breaking at stage 2?
ii) What effect does this have on the two component strands of the DNA molecule? (2)

c) At which stage is base-pairing seen to be occurring between the original DNA molecule and free DNA nucleotides? (1)

d) Name the type of bond formed and the types of molecule involved during **i)** stage 4; **ii)** stage 5. (2)

3 Name FOUR substances that must be present in a nucleus for DNA replication to occur. (4)

4 Why is DNA replication important? (2)

Applying Your Knowledge

1 Figure 7.8 shows part of a DNA strand.

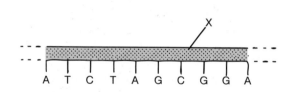

Figure 7.8

a) Of which types of molecule is region X composed? (2)

b) Draw the strand that would be complementary to the one shown. (1)

2 Calculate the percentage of thymine molecules present in a DNA molecule containing 1000 bases of which 200 are guanine. (1)

3 Arrange the following events that occur during DNA replication into the correct order:

a) bonds between opposite bases break,

b) bonds between opposite bases form,

c) DNA molecule uncoils from one end,

d) opposite strands separate,

e) daughter molecules coil into double helices,

f) sugar–phosphate bonds form. (1)

4 Table 7.1 gives the relative amounts of DNA present in some cell types from four different animals.

	sperm	red blood cell	kidney
carp	1.6	3.5	3.3
chicken	1.3	2.3	2.4
cow	3.3	0.0	6.4
human	3.3	0.0	6.6

Table 7.1

a) Account for the fact that the DNA content of human red blood cells is zero. (1)

b) Based on the data, suggest a structural difference that exists between the red blood cells of a cow and a chicken. (1)

c) Make a generalisation about the DNA content of sperm compared with kidney cells. (1)

d) When kidney cells are removed from a young animal and grown in a tissue culture, their DNA content is found at times to be higher than the values given in the table. Explain why. (2)

5 Figure 7.9 shows a cell's genetic material.

a) Name the parts enclosed in boxes 1, 2 and 3. (3)

b) Which of these boxed structures contains nucleic acid and consists of many different genes? (1)

c) Which of these structures is one of four basic units whose order determines the information held in a gene? (1)

d) In what way does region X in the diagram fail to represent one gene adequately? (1)

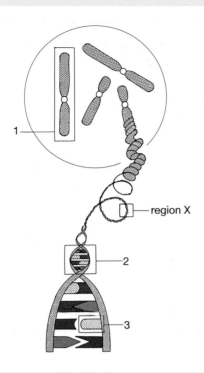

Figure 7.9

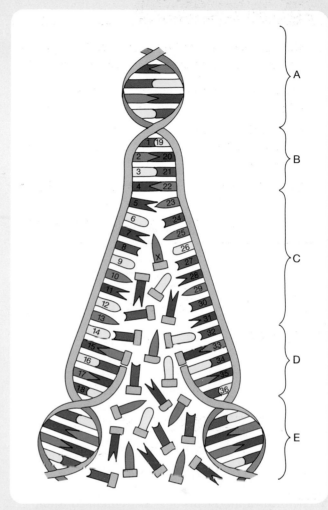

Figure 7.10

6 Figure 7.10 shows a molecule of DNA undergoing replication.

a) What process brings about the change in the DNA molecule's appearance between regions A and B in the diagram? (1)

b) What must happen to the DNA in region B before it can undergo replication? (1)

c) i) Name FOUR numbered base molecules to which nucleotide X could become attached as the diagram stands at present.

ii) Name TWO additional numbered bases to which nucleotide X could become attached if it were still free as the process of replication continued.

iii) How many of the free nucleotides in the pool are complementary to site 27? (7)

d) DNA polymerase is an enzyme which catalyses the assembly of nucleotides into DNA by promoting the formation of the sugar–phosphate 'backbone'. In which lettered region of the diagram would this process be taking place? (1)

e) Which lettered stage in the diagram shows the rewinding of daughter DNA molecules? (1)

7 Describe the main processes that occur during the replication of a molecule of DNA. (10)

8 RNA and protein synthesis

Structure of RNA

The second type of nucleic acid is called **ribonucleic acid** (RNA). RNA also consists of nucleotides (see Figure 8.1).

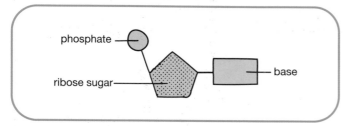

Figure 8.1 Structure of an RNA molecule

Although RNA's structure (see Figure 8.2) closely resembles that of DNA, a molecule of RNA differs from a molecule of DNA in three important ways as summarised in Table 8.1.

Sequence of DNA bases

The chemical components of DNA remain constant from species to species. However the DNA of one species differs from that of another in quantity and in the order in which the bases (A, T, G and C) occur along its length. It is this **sequence of bases** along the DNA strands which is unique to the organism. It contains the **genetic instructions** which control the organism's inherited characteristics.

Enzymes

Inherited characteristics are the result of many biochemical processes controlled by **enzymes**. In

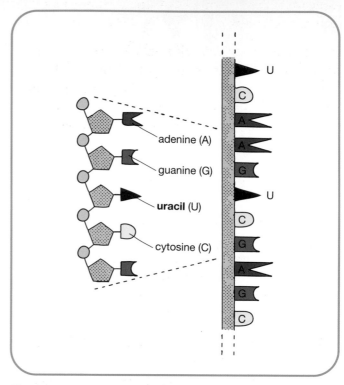

Figure 8.2 Structure of RNA

humans, for example, certain enzymes govern the biochemical pathways which lead to the formation of hair of a certain texture, eyes of a particular colour and so on.

Every enzyme is made of protein which is composed of amino acids. The protein's exact molecular structure, shape and ability to carry out its function all depend on the **sequence of its amino acids**. This critical order is determined by the sequence of the bases in the organism's DNA. By this means DNA controls the structure of enzymes and in doing so, determines the organism's **inherited characteristics**.

	RNA	DNA
number of nucleotide strands present in one molecule	one	two
complementary base partner of adenine	uracil	thymine
sugar present in a nucleotide molecule	ribose	deoxyribose (each molecule contains one fewer oxygen atom than ribose)

Table 8.1 Differences between RNA and DNA

Genetic code

The information present in DNA takes the form of a molecular code language called the **genetic code**. The sequence of bases along a DNA strand represents a sequence of 'codewords'.

DNA possesses only four different bases yet proteins contain about 20 different types of amino acid. The relationship cannot be one base coding for one amino acid since this would only allow four amino acids to be coded. Even two bases per amino acid would give only 16 (4^2) different 'codewords'.

Codon

However if the bases are taken in groups of three then this gives 64 (4^3) different combinations (see Appendix 3). It is now known that each amino acid is coded for by one (or more) of these 64 **triplets** of bases. Each triplet is called a **codon**. The codon is the basic unit of the genetic code (triplet code).

Thus a species' genetic information is encoded in its DNA with each strand bearing a series of base triplets arranged in a specific order for coding the particular protein needed by that species.

Protein synthesis
Transcription of DNA into mRNA

The genetic information carried on a section of DNA makes contact with structures responsible for protein synthesis in the cell's cytoplasm via a messenger. This go-between is called **messenger RNA (mRNA)** and it is

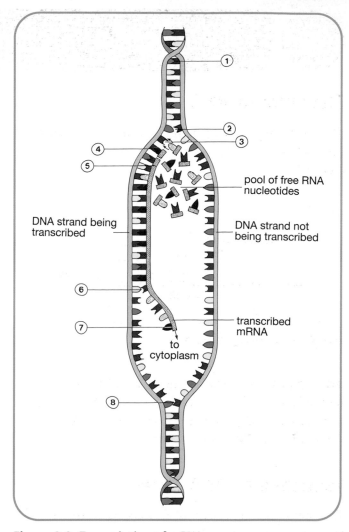

pool of free RNA nucleotides

DNA strand being transcribed

DNA strand not being transcribed

to cytoplasm

transcribed mRNA

Figure 8.3 Transcription of mRNA

formed (transcribed) from one of the DNA strands using free RNA nucleotides present in the cell's nucleus. The process of **transcription** is illustrated in Figure 8.3.

Testing Your Knowledge

1 State THREE ways in which RNA and DNA differ in structure and chemical composition. (3)

2 In what way does the DNA of one species differ from that of another which makes each species unique? (1)

3 a) Enzymes are made of protein. Name the sub-units of which proteins are composed. (1)

   b) Upon what do an enzyme's structure, shape and ability to carry out its function all depend? (1)

   c) What determines the sequence of amino acids in a protein? (1)

4 a) With reference to the relationship between the genetic code and the protein synthesised, why is it not possible that

   i) one base corresponds to one amino acid?
   ii) two bases correspond to one amino acid? (2)

   b) How many bases in the genetic code do correspond to one amino acid? (1)

   c) What name is given to the groups of bases that make up the genetic code? (1)

Stage 1 shows the DNA strands becoming unwound. Stage 2 is a little further ahead of stage 1. Here weak hydrogen bonds between two bases are breaking and causing the DNA strands to separate.

At stage 3, pairing of bases enables a free RNA nucleotide to find its complementary nucleotide on the DNA strand which is being transcribed. At stage 4, weak hydrogen bonds are forming between two complementary bases.

At stage 5, a little further ahead in the process, a strong chemical bond is forming between the sugar of one RNA nucleotide and the phosphate of the next one in the chain. This linking of nucleotides into a chain is controlled by an enzyme called **RNA polymerase**.

At stage 6, the weak hydrogen bonds between the DNA and RNA bases are breaking allowing the molecule of transcribed mRNA to become separated from the DNA template. Stage 7 shows transcribed mRNA ready to begin its journey out of the nucleus and into the cytoplasm.

At stage 8, weak hydrogen bonds between the two DNA strands reunite them and the molecule becomes wound up into a double helix once more. Although transcription of mRNA has been presented as a series of stages, in reality it is a continuous process.

mRNA

The completed molecule of **mRNA** leaves the nucleus through a pore in the nuclear membrane (see Figure 8.4) and enters the cytoplasm as shown in the

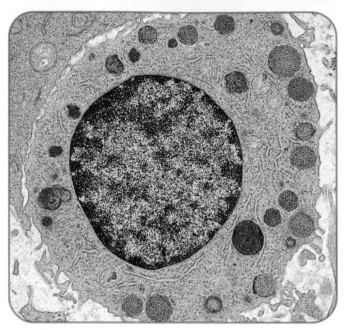

Figure 8.4 Porous nature of nuclear membrane

simplified version of the process given in Figure 8.5. Each triplet of bases on mRNA is called a **codon**.

tRNA

A second type of RNA is found in the cell's cytoplasm. This is called **transfer RNA (tRNA)** (see Figure 8.5). Each molecule of tRNA has only one of its triplets of bases exposed. This triplet, known as an **anticodon**, corresponds to a particular amino acid. Each tRNA molecule picks up the appropriate amino acid from the cytoplasm at its site of attachment. Many different types of tRNA are present in a cell, one or more for each type of amino acid.

Testing Your Knowledge

1 Study Figure 8.3 carefully and then answer the following questions.

   a) At which numbered stage is the DNA molecule becoming uncoiled in preparation for transcription? (1)

   b) i) What type of bond is breaking at stage 2?

   ii) What effect does this have on the component strands of the DNA molecule? (2)

   c) Name the type of bond formed and the types of molecule involved during **i)** stage 4;
   ii) stage 5. (4)

   d) What happens to a molecule of transcribed mRNA? (1)

   e) Describe the behaviour of the DNA strands once transcription has been completed. (1)

2 **a)** What name is given to a triplet of bases on a molecule of **i)** mRNA; **ii)** tRNA? (2)

   b) To what type of molecule does each tRNA triplet correspond? (1)

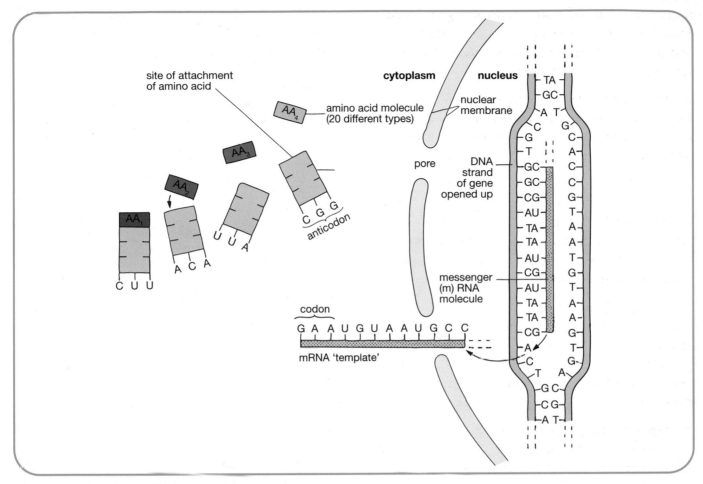

Figure 8.5 Two types of RNA

Ribosomes

Ribosomes are small, almost spherical structure found in all cells. Some occur freely in the cytoplasm, others are found attached to the **endoplasmic reticulum** (see Figure 8.7).

They are the site of the **translation** of mRNA into protein. Each ribosome contains enzymes essential for the process of protein formation. Large numbers of ribosomes are found in growing cells which need to produce large quantities of protein.

Translation of RNA into protein

A ribosome becomes attached to one end of the mRNA molecule about to be translated. Inside the ribosome there are sites for the attachment of tRNA molecules, two at a time (see Figure 8.6). This arrangement allows the anticodon of the first tRNA molecule to form weak hydrogen bonds with the complementary codon on the

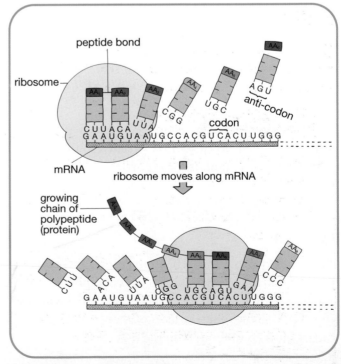

Figure 8.6 Translation of RNA into protein

mRNA. When the second tRNA molecule repeats this process, the first two amino acid molecules are brought into line with one another. They become joined together by a strong **peptide bond** whose formation is controlled by an enzyme present in the ribosome.

The first tRNA becomes disconnected from its amino acid and from mRNA and leaves the ribosome. The ribosome moves along the mRNA strand allowing the anticodon of the third tRNA to move into place and link with its complementary codon on mRNA. This allows the third amino acid to become bonded to the second one and so on along the chain.

Thus the process of translation brings about the alignment of amino acids in a certain order at the ribosome where they form a **polypeptide chain** (also see chapter 9). The completed polypeptide (normally consisting of very many amino acids) is then released into the cytoplasm.

Protein from polypeptides

Formation of a completed protein often involves folding or rearrangement of the polypeptide chain. Sometimes a number of polypeptide chains combine to form a protein molecule (see chapter 9).

Re-use of RNA molecules

Each tRNA molecule becomes attached to another molecule of its amino acid ready to repeat the process. The mRNA is often re-used to produce further molecules of the same polypeptide.

Protein synthesised in ribosomes is for use within the cell; protein made in ribosomes attached to the endoplasmic reticulum is for export.

Endoplasmic reticulum
Rough ER

The **rough endoplasmic reticulum (rough ER)** is illustrated in Figures 8.7 and 8.8. It is composed of a system of flattened sacs and tubules encrusted with ribosomes on the outer surfaces of their membranes. Rough ER is continuous with the outer nuclear membrane and provides a large surface area upon which chemical reactions can occur. It is present

throughout the cell and acts as a pathway for the transport of materials.

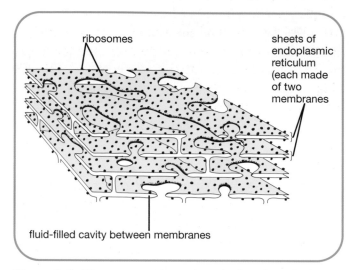

Figure 8.7 Ribosomes and rough endoplasmic reticulum

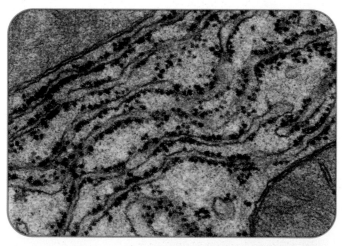

Figure 8.8 Electron micrograph of ribosomes and rough endoplasmic reticulum

The rough ER is the route taken by proteins that are to be secreted by a cell. When such a protein is synthesised in a ribosome on the surface of the rough ER, the emerging polypeptide chain is 'injected' into the ER. It then undergoes the coiling and folding processes involved in the formation of the protein's specific structure. Secretory proteins are passed on to the Golgi apparatus.

Golgi apparatus

The **Golgi apparatus** (or Golgi body) is composed of a group of flattened fluid-filled sacs (see Figures 8.9 and 8.10). **Vesicles** (tiny fluid-filled sacs) containing newly synthesised protein become pinched off from the rough ER. These vesicles fuse with the outermost sac of the Golgi apparatus and the protein is passed from sac to sac by means of vesicles. During this time the Golgi apparatus processes the protein, for example by adding a carbohydrate part to it to make it into a **glycoprotein**.

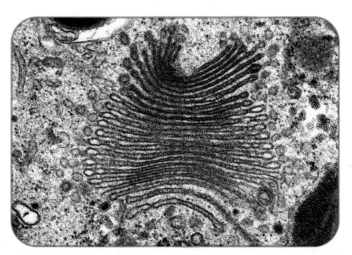

Figure 8.9 Electron micrograph of Golgi apparatus

Vesicles containing the finished product (e.g. mucus, digestive enzyme, hormone etc.) become pinched off at the ends of the Golgi apparatus. The diagram shows how these vesicles move towards and fuse with the cell membrane. The contents of a vesicle are then discharged to the outside. This **secretion** of intracellular products by the cell is called **exocytosis** (see page 19).

Golgi bodies are especially numerous and active in secretory cells such as the gastric gland cells in the stomach wall (which produce mucus and the enzyme pepsin) and certain cells in the pancreas (which produce the hormone insulin).

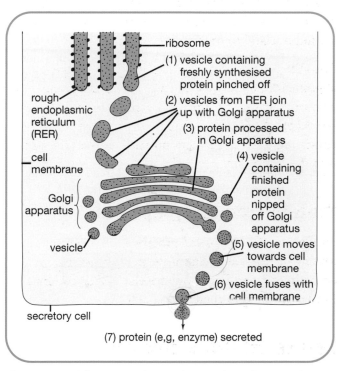

Figure 8.10 The processing and secretion of a protein

Testing Your Knowledge

1 a) What is a **ribosome**? (1)

  b) What name is given to the conversion of the genetic message (held by RNA) into protein? (1)

2 a) What type of bonds form between mRNA and tRNA anticodons brought together at a ribosome? (1)

  b) What type of bond forms between adjacent amino acids brought together by their tRNAs at a ribosome? (1)

  c) What is formed as the ribosome continues to move along the mRNA molecule? (1)

3 Is it valid to describe mRNA and tRNA as *re-useable* molecules? Justify your answer. (2)

4 a) Name the system of flattened sacs and tubules through which a molecule of newly synthesised protein passes on its way to the Golgi apparatus. (1)

  b) Describe ONE way in which a protein's structure could be altered during its time in the Golgi apparatus. (1)

  c) By what means does the protein become liberated from the Golgi apparatus and leave the cell? (2)

Applying Your Knowledge

1 a) The stages in the following list refer to the process of transcription of mRNA from DNA. Arrange them into the correct sequence starting with E.

A Bases along each DNA strand become exposed.

B Adjacent RNA nucleotides become joined to one another to form mRNA.

C DNA strands become separated as hydrogen bonds break between base pairs.

D Completed molecule of mRNA become detached from DNA template.

E A specific region of a DNA molecule unwinds.

F Hydrogen bonds form between bases of DNA strands which then rewind.

G Each base on the DNA strand attracts its complementary base on a free RNA nucleotide. (1)

b) Draw a diagram of the mRNA strand that would be transcribed from section X of the DNA molecule shown in Figure 8.11. (2)

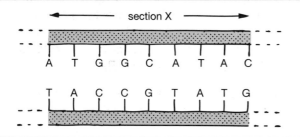

Figure 8.11

2 Figure 8.12 shows the method by which the genetic code is transmitted during protein synthesis. Table

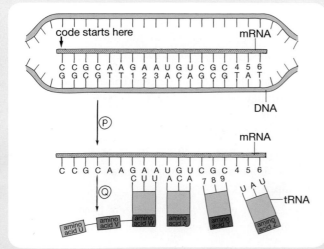

Figure 8.12

amino acid	codon	anticodon
alanine		CGC
arginine	CGC	
cysteine		ACA
glutamic acid	GAA	
glutamine		GUU
glycine	GGC	
isoleucine		UAU
leucine	CUU	
proline		GGC
threonine	ACA	
tyrosine		AUA
valine	GUU	

Table 8.2

8.2 gives some of the triplets which correspond to certain amino acids.

a) Identify bases 1–9. (2)

b) Name processes P and Q. (1)

c) Copy and complete Table 8.2. (2)

d) Give the triplet of bases that would be exposed on a molecule of tRNA to which valine would become attached. (1)

e) Use your table to identify amino acids U, V, W, X, Y and Z. (2)

f) i) Work out the mRNA code for part of a polypeptide chain with the amino acid sequence:

threonine–leucine–alanine–glycine.

ii) State the genetic code on the DNA strand from which this mRNA would be formed. (2)

3 Copy and complete Table 8.3. (3)

stage of synthesis	site in cell
formation of mRNA	
collection of amino acid by tRNA	
formation of codon–anticodon links	

Table 8.3

4 The information in Table 8.4 refers to the relative numbers of ribosomes present in the cells of a new leaf developing at a shoot tip.

age of new leaf (days)	relative numbers of ribosomes in cells
1	300
3	500
5	650
7	450
9	210
11	200
13	200
15	200

Table 8.4

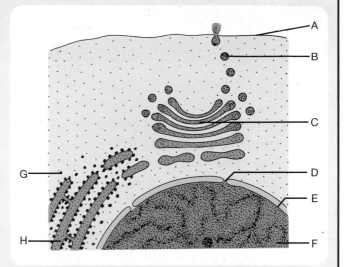

Figure 8.13

a) Construct a hypothesis to account for the trend shown by these data. (2)

b) Explain why the relative numbers of ribosomes in the cells of a fully grown leaf do not drop to zero. (1)

5 Figure 8.13 shows part of a secretory cell viewed under an electron microscope.

a) Name parts A–H. (4)

b) Name the type of chemical that would be present in B. (1)

c) State which letter indicates the site at which synthesis of this substance from its sub-units occurs. (1)

6 The chemical composition of the carbohydrate and fat components of the human body are common to everyone. Explain why it is the body's protein component that results in inherited differences existing between individuals, making each person different from everyone else. (2)

7 Give an account of protein synthesis under the headings: transcription and translation. (10)

9 Functional variety of proteins

Structure

Proteins are organic compounds. In addition to always containing the chemical elements **carbon** (C), **hydrogen** (H), **oxygen** (O) and **nitrogen** (N), they often contain **sulphur** (S).

Peptide bonds

Each protein is built up from a large number of sub-units called **amino acids** of which there are about 20 different types. These are joined together into chains by strong chemical links called **peptide bonds**. Each chain is called a **polypeptide** and it normally consists of hundreds of amino acid molecules linked together.

The amino acids are built into a particular genetically-determined sequence during the process of protein synthesis (see chapter 8). This sequence of amino acids determines the protein's structure and function.

Hydrogen bonds

Weak chemical links known as **hydrogen bonds** form between certain amino acids in a polypeptide chain causing the chain to become coiled into a spiral (helix) as shown in Figure 9.1.

Further linkages

Several polypeptide chains often become linked together by various types of cross-connection. These include bridges between sulphur atoms and additional hydrogen bonding.

The types of connection that occur at this stage are important since they determine the **final structure** of the protein enabling it to carry out its **specific function**. These further linkages lead to the formation of a fibrous or globular protein as shown in Figure 9.1.

Variety and functions of proteins

An enormous number of different proteins are found in living things. A human being possesses over 10 000 different proteins. Proteins can be classified as either fibrous or globular.

Fibrous proteins

A **fibrous** protein is formed by several spiral-shaped polypeptide molecules becoming linked together in parallel by cross-bridges forming between them. This gives the molecule of structural protein a rope-like structure (see Figure 9.1).

Collagen is an example of a fibrous protein. It is found in bone where its strong inelastic fibres provide rigid support.

Globular proteins

A molecule of **globular** protein consists of several polypeptide chains folded together into a roughly spherical shape like a tangled ball of string (see Figure 9.1). The exact form that the folding takes depends on the types of further linkage that form between amino acids on the same and adjacent polypeptide chains.

Globular proteins are vital components of all living cells and play a variety of roles.

Enzymes

All **enzymes** (biological catalysts) are made of globular protein (see Figure 9.2). Each is folded in a particular way to expose an active surface which readily combines with a specific substrate (see Figure 9.3). Since intracellular enzymes speed up the rate of biochemical

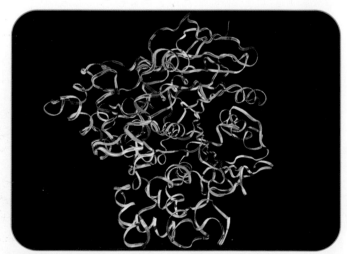

Figure 9.2 Computer-generated model of enzyme molecule

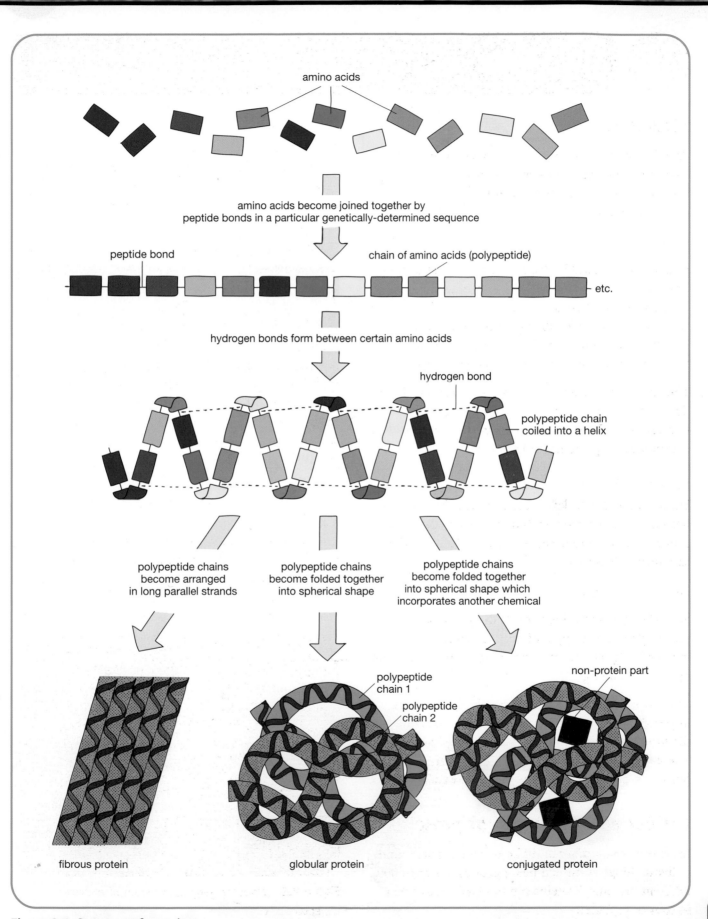

amino acids

amino acids become joined together by
peptide bonds in a particular genetically-determined sequence

peptide bond

chain of amino acids (polypeptide)

etc.

hydrogen bonds form between certain amino acids

hydrogen bond

polypeptide chain
coiled into a helix

polypeptide chains
become arranged
in long parallel strands

polypeptide chains
become folded together
into spherical shape

polypeptide chains
become folded together
into spherical shape which
incorporates another chemical

non-protein part

polypeptide
chain 1

polypeptide
chain 2

fibrous protein

globular protein

conjugated protein

Figure 9.1 Structure of proteins

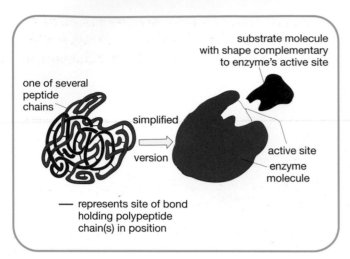

Figure 9.3 Enzyme structure

processes, such as photosynthesis, respiration and protein synthesis, they are essential for the maintenance of life.

Structural protein

Globular protein is one of the two components which make up the **membrane** surrounding a living cell (see page 9). Similarly it forms an essential part of all membranes possessed by subcellular structures within the cell, such as the nucleus. This type of protein therefore plays a vital structural role in every living cell.

Hormones

These are **chemical messengers** transported in an animal's blood to 'target' tissues where they exert a specific effect. Some hormones are made of globular protein and exert a regulatory effect on the animal's growth and metabolism. A few examples are given in Table 9.1.

Antibodies

Although Y-shaped rather than spherical, **antibodies** are also a type of globular protein (see Figure 9.4). They are made by white blood cells called **lymphocytes** and defend the body against antigens (see chapter 11).

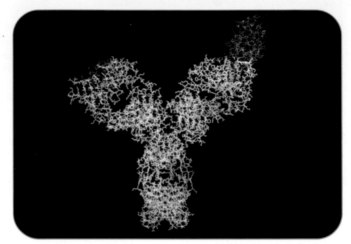

Figure 9.4 Antibody molecule

Conjugated proteins

Each **conjugated** protein consists of a globular protein associated with a non-protein chemical (see Figure 9.1).

A **glycoprotein** is composed of protein and carbohydrate. An example is mucus, the slimy viscous substance secreted by epithelial cells to lubricate or protect parts of the human body.

Haemoglobin is the oxygen-transporting pigment in blood. It is a conjugated protein consisting of the globular protein globin associated with **haem**, a non-protein part containing iron (see page 292). **Cytochrome**, which plays a key role in the cell's aerobic

hormone	secretory gland	role of hormone
somatotrophin	pituitary	promotes growth of long bones
insulin	pancreas	promotes conversion of glucose to glycogen
glucagon	pancreas	promotes conversion of glycogen to glucose
thyroxine	thyroid	controls rate of metabolic processes and growth

Table 9.1 Hormones composed of globular protein

respiratory pathway (see page 31), is also a conjugated protein which contains iron.

Proteins play a wide variety of roles. Some form part of the cell's structure, some defend the body, whilst others regulate biochemical reactions and metabolic processes.

Testing Your Knowledge

1 Name the chemical element always present in protein but absent from carbohydrates. (1)

2 a) Approximately how many different types of amino acid occur in living cells? (1)

 b) Copy the following sentences and complete the blanks.

   The amino acids in a protein are built into a particular order which is determined by the sequence of the _____ on a portion of DNA in a _____. This sequence of amino acids determines the protein's _____ and _____. (4)

3 Describe TWO ways in which polypeptide chains can become arranged to form a protein. (2)

4 Give TWO examples of a fibrous protein and for each state its function. (2)

5 Name THREE types of globular protein and briefly describe the role played by each one. (3)

Applying Your Knowledge

1 Copy and complete Table 9.2 (where G = globular and C = conjugated protein). (5)

name of protein	type of protein (G or C)	role of this protein
antibody		
catalase		
cytochrome		
haemoglobin		
insulin		

Table 9.2

2 Figure 9.5 is incomplete. It represents the action of an enzyme on its substrate. Draw a complete version of the diagram by replacing boxes P, Q and R with the omitted molecules. (3)

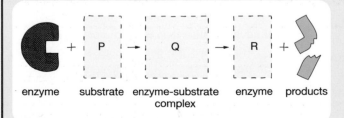

enzyme   substrate   enzyme-substrate   enzyme   products
                        complex

Figure 9.5

3 Figure 9.6 shows three different types of protein.

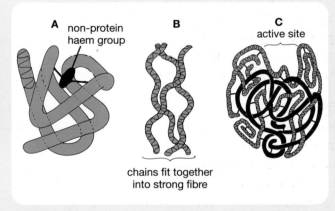

A non-protein haem group
B
C active site
chains fit together into strong fibre

Figure 9.6

a) Which of these is: i) a globular protein; ii) a conjugated protein; iii) a fibrous protein? (3)

b) i) Which diagram represents collagen?
   ii) Explain how you arrived at your answer.
   iii) State ONE way in which collagen's structure is suited to its function. (3)

c) i) Which diagram represents a molecule of enzyme?
   ii) Explain how you arrived at your answer. (2)

4 Some amino acids can be synthesised by the body from simple compounds; others cannot be

synthesised and must be supplied in the diet. The latter type are called the **essential amino acids.**

The graph in Figure 9.7 shows the results of an experiment using rats where group 1 was fed zein (maize protein), group 2 was fed casein (milk protein) and group 3 was fed a diet which was changed at day 6.

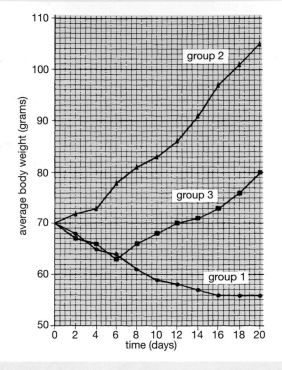

Figure 9.7

a) One of the proteins contains all of the essential amino acids whereas the other lacks two of them. Identify each protein and explain how you arrived at your answer. (4)

b) i) State which protein was given to the rats in group 3 during the first six days of the experiment.

   ii) Suggest TWO different ways in which their diet could have been altered from day 6 onwards to account for the results shown in the graph. (3)

c) By how many grams did the average body weight of the rats in group 2 increase over the 20-day period? (1)

d) Calculate the percentage decrease in average body weight shown by the rats in group 1 over the 20-day period. (1)

5 Write an essay on the variety and functions of proteins. (10)

What You Should Know

Chapters 7–9
(See Table 9.3 for Word bank)

adenine	endoplasmic	replication
amino acids	enzymes	ribose
antibodies	fibrous	ribosome
anticodons	globular	secreted
bonds	Golgi	thymine
cell	guanine	transcribed
code	helix	transfer
codons	nitrogen	triplet
cytosine	nucleotides	uracil
deoxyribose	polypeptide	vesicles

Table 9.3 Word bank for chapters 7–9

1 DNA consists of two strands twisted into a double _____. Each strand is composed of _____. Each nucleotide consists of _____ sugar, phosphate and one of four types of base (_____, thymine, _____ or cytosine).

2 Adenine always pairs with _____; guanine always pairs with _____.

3 DNA is unique because it is able to reproduce itself by _____. This allows the genetic message to be passed on from cell to _____ and generation to generation.

4 RNA consists of a single strand of nucleotides. _____ is found in place of thymine; _____ replaces deoxyribose.

5 The bases along a DNA strand take the form of a molecular language called the genetic _____. Each _____ of bases codes for a particular amino acid.

6 Messenger RNA (mRNA) is _____ from a strand of DNA and carries this genetic message from the nucleus out into the cytoplasm. At a _____, mRNA meets molecules of _____ RNA (tRNA) each carrying a specific amino acid.

7 Protein synthesis occurs in ribosomes; mRNA's triplets of bases, called _____, are 'read' and matched by tRNA's _____. This enables peptide _____ to form between adjacent amino acids.

8 Rough _____ reticulum bears ribosomes on its outer surface.

9 Freshly synthesised protein is transported via the endoplasmic reticulum to the _____ apparatus where it is processed and packaged in _____.

10 Some protein is _____ out of the cell by vesicles moving towards, and fusing with, the plasma membrane.

11 In addition to carbon, hydrogen and oxygen, proteins always contain _____.

12 A protein consists of sub-units called _____ (of which there are about 20 types) joined together by peptide bonds to form polypeptides.

13 A molecule of _____ protein consists of parallel _____ chains and has a structural function.

14 A molecule of _____ protein consists of polypeptide chains folded into a spherical shape. Some are structural (e.g. those in the plasma membrane); others act as _____, hormones or _____.

10 Viruses

Most disease-causing (pathogenic) organisms (e.g. fungi and bacteria) exhibit all the characteristics of living things such as nutrition, respiration, excretion, reproduction, growth and sensitivity.

Viruses (see Figure 10.1) are unusual micro-organisms in that **reproduction** is the only true characteristic of living things that they exhibit. Other features, such as the ability to form a crystalline structure, are typical of non-living substances.

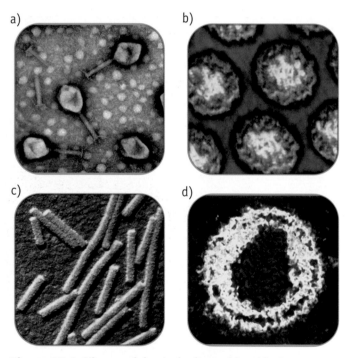

a) b)

c) d)

Figure 10.1 Viruses a) bacteriophage, b) cold virus, c) tobacco mosaic virus, d) HIV

Obligate parasites

Since viruses can only reproduce within the living cells of another organism (the **host**), they are described as **obligate parasites**. Animals, plants and bacteria are all susceptible to invasion by viruses. Since the host cell is subsequently damaged or destroyed, viruses are always associated with **disease**. These range from the common cold to deadly Lassa fever; further examples are given in Table 10.1.

Specificity

Viruses are often specific with respect to their chosen type of host cell. The poliomyelitis virus, for example, attacks nerve cells; the hepatitis virus targets liver cells.

Size

Viruses are much smaller than bacteria. They range in size from about 20 to 300 nanometres (nm). (1 nm = 1×10^{-9} m). Viruses can only be seen with the aid of an **electron microscope** (see page 349).

Spread of viruses

Coughing and sneezing spread the viruses that cause respiratory infections such as pneumonia. The poliomyelitis virus is excreted in faeces and can be passed to food by flies. The rabies virus is unusual in that it can affect several different mammals. Its usual mode of transmission to humans is via the bite of a

disease (or syndrome) caused by virus	host organism	effect
poliomyelitis	human	inflammation of spinal cord followed by paralysis
AIDS	human	destruction of immune system
hepatitis	human	inflammation of liver cells and jaundice
foot and mouth	farm animals	development of lesions in mouth
rabies	mammals	muscle spasms or paralysis
bushy stunt	tomato plant	stunted growth
leaf mosaic	potato plant	malformation of leaves

Table 10.1 Viral diseases

rabid dog. Human immunodeficiency virus (HIV) which causes acquired immune deficiency syndrome (AIDS) is transmitted by blood and sexual contact. Some viruses are able to spread when a host comes in contact with an object contaminated by a previous host (see Figure 10.2).

Figure 10.2 Spread of viruses

Viruses which attack plants are normally spread by insects such as greenfly. Seeds, bulbs and cuttings can also carry viruses.

Viral structure

A virus is not a cell. It consists of one type of nucleic acid (either DNA or RNA) surrounded by a protective coat (**capsid**) normally made of protein. Its genetic material carries the information necessary for viral multiplication but lacks the biochemical machinery to carry this out on its own.

A wide variety of viruses exist. Figure 10.3 shows a herpes virus which attacks human skin cells and causes cold sores on the lips (see Figure 10.4). Figure 10.3 also shows a type of virus called a bacteriophage which attacks bacteria. This virus' DNA thread is contained

inside a head from which a tail projects. Most viral particles do not have a tail.

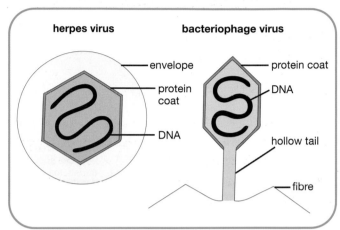

Figure 10.3 Structure of two viruses

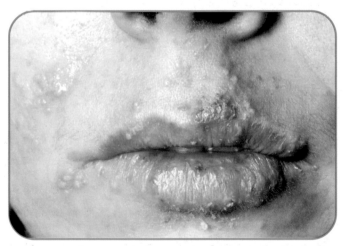

Figure 10.4 Cold sore

Invasion of a cell by a virus

A virus can lie dormant and lifeless for many years until it comes in contact with a suitable host cell which it then invades.

The virus becomes attached to the host cell by binding with certain molecules on the surface of the host cell. Different viruses employ different methods of introducing their nucleic acid into the host cell. In some cases, such as herpes virus, the whole virus enters the cell; in other cases, such as bacteriophage, the viral DNA is injected via the hollow tail. Figure 10.5 shows the invasion of a bacterium (the host) by a bacteriophage.

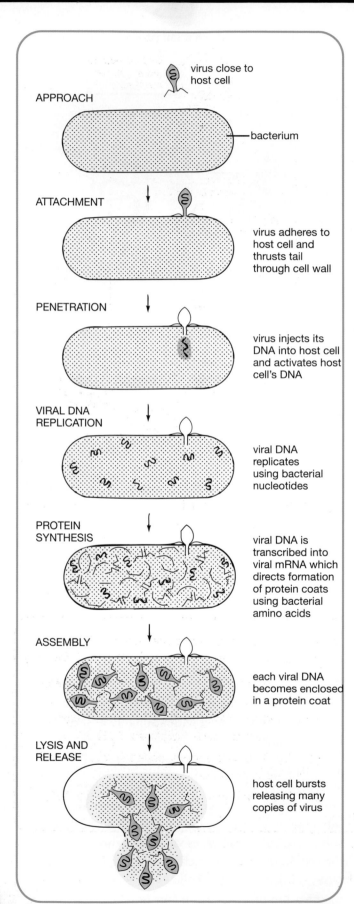

APPROACH — virus close to host cell

bacterium

ATTACHMENT — virus adheres to host cell and thrusts tail through cell wall

PENETRATION — virus injects its DNA into host cell and activates host cell's DNA

VIRAL DNA REPLICATION — viral DNA replicates using bacterial nucleotides

PROTEIN SYNTHESIS — viral DNA is transcribed into viral mRNA which directs formation of protein coats using bacterial amino acids

ASSEMBLY — each viral DNA becomes enclosed in a protein coat

LYSIS AND RELEASE — host cell bursts releasing many copies of virus

Figure 10.5 Multiplication of a virus

Alteration of cell instructions

Although completely inactive outside the host cell, once inside, the virus assumes control of the cell's biochemical machinery. It depends on the host cell for both energy (from ATP) and a supply of nucleotides and amino acids to build new viral particles.

The viral nucleic acid first suppresses the cell's normal nucleic acid replication and protein synthesis. It then uses the host's nucleotides and amino acids to produce many identical copies of viral nucleic acid and an appropriate number of protein coats.

New viral particles become assembled as shown in Figure 10.5. This is followed by the release of many copies of the virus capable of infecting new host cells and causing disease. In some viruses the release is achieved by viral nucleic acid directing the synthesis of enzymes that cause bursting (lysis) of the host cell membrane. In other cases, buds containing the virus are formed.

Some viruses infect a cell and remain dormant for an indefinite period before undergoing reproduction and release.

Retrovirus

If a virus contains RNA it is called a **retrovirus**. Since it lacks DNA to transcribe into mRNA, it adopts a different strategy. Along with its RNA it injects an enzyme called **reverse transcriptase** into the host cell. This enzyme brings about a reverse of normal transcription and produces viral DNA from viral RNA. Using this DNA, the virus is able to replicate itself many times. The new viral particles (containing viral RNA and reverse transcriptase) escape from the host cell and infect other cells.

Acquired immune deficiency syndrome (AIDS)

AIDS is caused by a retrovirus called **human immunodeficiency virus** (**HIV**). HIV attacks white blood cells called **helper T cells**. Glycoproteins on its surface become attached to specific receptors on the membrane of the helper T cell's surface (see Figure 10.6).

The envelope surrounding the HIV particle fuses with the membrane of the helper T cell and the virus enters the host cell. The events described above involving reverse transcriptase take place. Viral DNA becomes

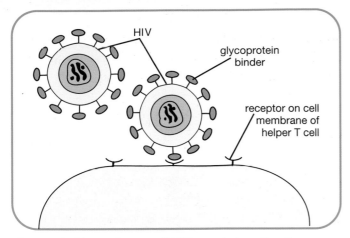

Figure 10.6 Attachment of HIV to a helper T cell

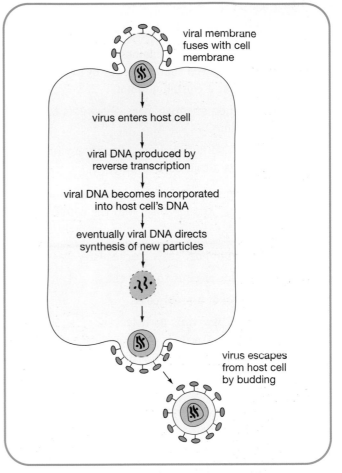

Figure 10.7 Life cycle of HIV

incorporated into the host cell's DNA where it can remain dormant for many years.

Eventually viral mRNA is transcribed and it then directs synthesis of new viral particles inside the host cells. These escape from the infected helper T cell by budding (see Figure 10.7) and move off to infect other cells. The original T cell's membrane becomes perforated during the process and this causes the destruction of the cell.

As the number of helper T cells gradually drops, the body's immunological activity decreases, leaving the person susceptible to serious opportunistic infections such as pneumonia and rare forms of cancer.

Testing Your Knowledge

1 a) Under what circumstances could a virus be
      i) active;   ii) inactive? (2)
   b) Why are viruses described as obligate parasites? (1)
   c) i) Name TWO viral diseases of plants and TWO viral diseases that affect human beings.
      ii) In addition to plants and animals, what third group of living things can play host to viruses? (5)

2 Viruses are too small to be seen under a light microscope. How do we know what they look like? (1)

3 a) i) Name TWO structural features possessed by all viruses.
      ii) Which of these enters the host cell and alters the cell's metabolism? (3)
   b) Name TWO substances supplied by the host cell during viral replication. (2)
   c) What is lysis and how is it brought about? (2)

Applying Your Knowledge

1 In the absence of a suitable host cell, a virus exhibits none of the characteristics of living things and may adopt a crystalline form.

   a) Why then do scientists consider viruses to be living things? (1)

   b) Scientists do not, however, classify a virus particle as a cell. Suggest why. (1)

2 The following steps occur during multiplication of a bacteriophage. Arrange them into the correct order.

   A assembly of completed viruses

   B alteration of host cell's biochemistry

   C attachment of virus to host cell

   D production of protein coats

   E release of new virus particles

   F injection of DNA into host cell

   G replication of viral DNA

   H transcription of viral mRNA from viral DNA (1)

3 Figure 10.8 shows the size of three viruses in relation to a red blood cell and a cell of the bacterium *Escherichia coli*.

   a) Construct a table which shows the given dimension of each of these life forms expressed in nanometres (nm), micrometres (µm), millimetres (mm) and metres (m). (4)

   b) i) By how many times is the length of a cell of *E. coli* greater than that of a polio virus?

   ii) By how many times is the diameter of a red blood cell greater than the length of a tobacco mosaic virus? (2)

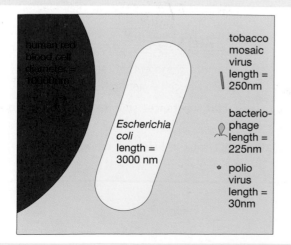

Figure 10.8

4 Figure 10.9 shows the replication cycle of the herpes virus which causes cold sores on the mouth.

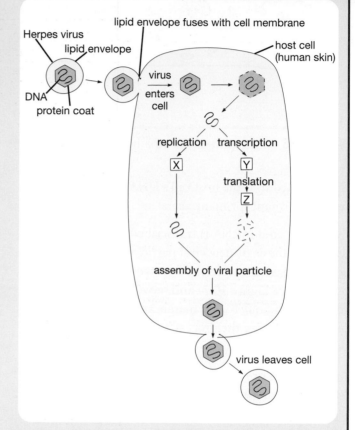

Figure 10.9

   a) Is herpes a retrovirus? Explain your answer. (1)

   b) Match boxes X, Y and Z with the terms viral RNA, viral protein and viral DNA. (3)

   c) i) What name is given to the method by which herpes leaves the host cell?

   ii) Predict the fate of the cell. (2)

   d) Compared with Figure 10.5, state ONE way in which Figure 10.9 is an oversimplification of the process of viral multiplication. (1)

5 The protein coat that surrounds the nucleic acid molecule of a virus is called a capsid. It consists of sub-units (capsomers) assembled into either a helical (spiral) or a polyhedral (many-sided) formation. Some capsids are in turn surrounded by a membranous envelope.

   Table 10.2 refers to eight different viruses.

type of nucleic acid	shape of capsid		state of capsid	example	host cell of example
DNA	helical		naked	coliphage virus	bacterium
			enveloped	smallpox virus	animal
	polyhedral		naked	bacteriophage virus	bacterium
			enveloped	herpes	animal
RNA	helical		naked	tobacco mosaic virus	plant
			enveloped	influenza virus	animal
	polyhedral		naked	bushy stunt virus	plant
			enveloped	human immunodeficiency virus	animal

Table 10.2

a) Name THREE structural features possessed by the virus that causes smallpox. (1)

b) i) Identify viruses X and Y shown in Figure 10.10.

ii) Which labelled component of virus X is responsible for making the host cell copy the viral RNA into viral DNA before multiplication can begin? (3)

c) Name the virus whose DNA strand is protected by an enveloped capsid composed of capsomers laid down in a polyhedral arrangement. (1)

d) Convert Table 10.2 into a key of paired statements. (4)

6 Give an account of the events that occur during the invasion of a host cell by a DNA virus. (10)

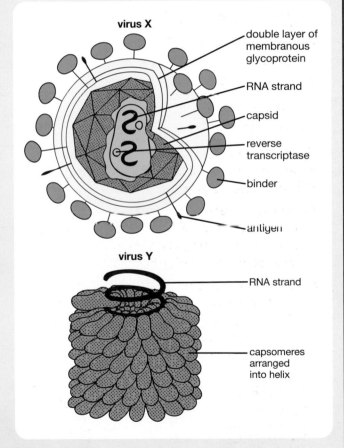

virus X

double layer of membranous glycoprotein

RNA strand

capsid

reverse transcriptase

binder

antigen

virus Y

RNA strand

capsomeres arranged into helix

Figure 10.10

11 Cellular defence mechanisms

Cellular defence mechanisms in animals

The warm, moist, food-rich conditions that exist inside the human body are ideal for the growth and multiplication of many types of disease-causing (pathogenic) micro-organisms. However these are normally prevented from causing disease by the body's defence mechanisms. Two of these mechanisms are brought about by **white blood cells** which recognise and respond to the presence of invading foreign particles. Some examples of white blood cells are shown in Figure 11.1.

Phagocytosis

Phagocytosis ('cell-eating') is the process by which foreign bodies such as bacteria are engulfed and destroyed. Cells capable of phagocytosis are called phagocytes (see Figure 11.2).

The process of phagocytosis is illustrated in Figure 11.3. A phagocyte detects chemicals released by a bacterium and moves up a concentration gradient towards it. The phagocyte adheres to the bacterium and engulfs it in a vacuole formed by an infolding of the cell membrane.

Figure 11.2 Phagocyte in action

A phagocyte's cytoplasm contains many specialised organelles called **lysosomes** which contain **digestive enzymes**. Some of these lysosomes fuse with the vacuole and release their enzymes into it. The bacterium becomes digested and the breakdown products are absorbed by the phagocyte.

During infection, hundreds of phagocytes migrate to the infected area and engulf many bacteria by phagocytosis. Dead bacteria and phagocytes often gather at a site of injury forming **pus**.

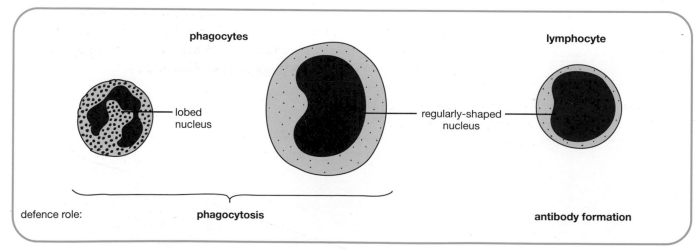

Figure 11.1 White blood cells

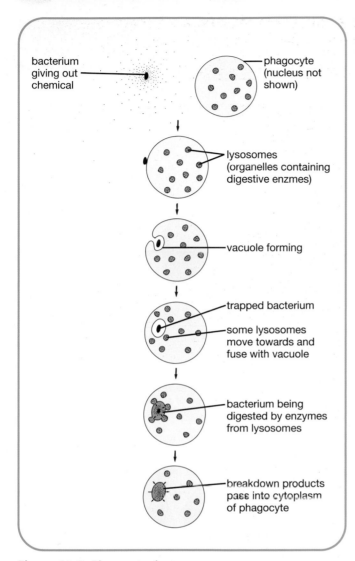

Figure 11.3 Phagocytosis

Immune response

Immunity is an organism's ability to resist infectious disease.

Non-specific immune response

Phagocytosis is an example of a non-specific immune response since it provides general protection against a wide range of micro-organisms.

Specific immune response (antibody production)

An antigen is a complex molecule, such as a protein, which is recognised as an alien by the body's lymphocytes. The antigen's presence stimulates lymphocytes to produce special protein molecules called antibodies.

An antibody is a Y-shaped molecule as shown in Figure 11.4. Each of its arms bears a receptor (binding) site which is specific to a particular antigen. The human body possesses thousands of different types of lymphocyte each capable of responding to one specific antigen and producing the appropriate antibody.

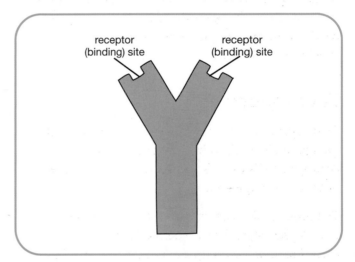

Figure 11.4 Antibody

A viral particle is surrounded by a coat of protein. This protein possesses many sites which act as antigens on the surface of the virus. When an antibody meets its complementary antigen, the two combine at their specific sites like a lock and key and the antigen is rendered harmless (see Figure 11.5).

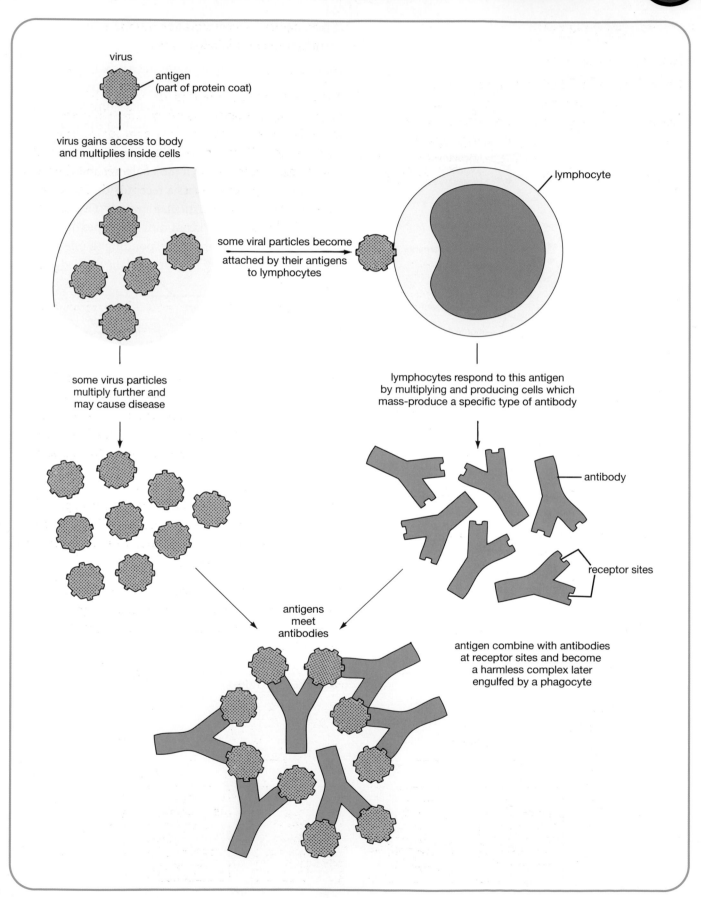

virus

antigen
(part of protein coat)

virus gains access to body
and multiplies inside cells

some viral particles become
attached by their antigens
to lymphocytes

lymphocyte

some virus particles
multiply further and
may cause disease

lymphocytes respond to this antigen
by multiplying and producing cells which
mass-produce a specific type of antibody

antibody

receptor sites

antigens
meet
antibodies

antigen combine with antibodies
at receptor sites and become
a harmless complex later
engulfed by a phagocyte

Figure 11.5 Action of antibodies

Immunological memory

Primary and secondary responses

When a person is infected by a disease-causing organism, their body responds by producing antibodies. This is called the **primary response** (see Figure 11.6). Due to a latent period elapsing before the appearance of the antibodies, this primary response is often unable to prevent the person from becoming ill.

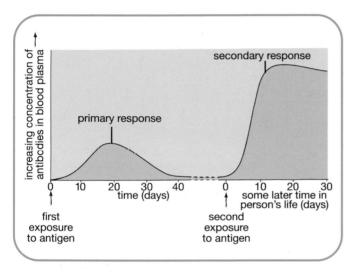

Figure 11.6 Primary and secondary response

If the person survives, exposure to the same antigen at a later date results in the **secondary response**. This time the disease is usually prevented because:

- antibody production is much **more rapid**;

- the concentration of antibodies produced reaches a **higher level**;

- the higher concentration of antibodies is maintained for a **longer time**.

Memory cells

The secondary response is made possible by the presence in the body of **memory cells**. These are lymphocytes specific to the antigen from the first exposure to it. When the body becomes exposed to the disease-causing micro-organism for a second time, the memory cells produce a **clone** of antibody-forming lymphocytes and fight it off. The person has acquired immunity by **natural** means.

Types of specific immunity

Active immunity

The organism produces its own antibodies in one of two possible ways.

Naturally acquired immunity

The person is exposed to the disease-causing antigen (e.g. chicken pox virus), suffers the disease, makes antibodies and continues to be able to do so for a long time (or even permanently) after recovery. The person is therefore **immune** to future attacks by that antigen.

Artificially acquired immunity (immunisation)

The person receives a small dose of vaccine (e.g. polio vaccine) containing antigens which have been treated in a certain way so that they induce antibody formation but do not cause the disease.

If the person is later exposed to the disease-causing antigen, their body is already prepared to fight the infection. The person has acquired immunity by **artificial** means.

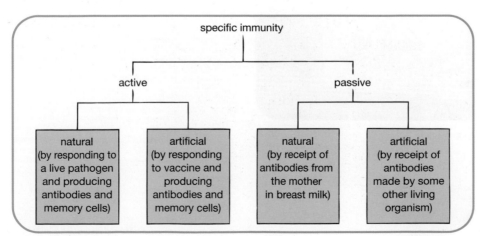

Figure 11.7 Types of specific immunity

Passive immunity

Instead of the person making antibodies in response to the antigen, **ready-made** antibodies are passed into their body.

Such passive immunity occurs **naturally** when antibodies cross the placenta from mother to foetus (and from mother's milk to suckling baby), giving the child protection for a short time until its own immune system develops.

Passive immunity is also brought about **artificially** by extracting antibodies that have been made by one mammal (e.g. horse) and injecting them into another (e.g. human). The effect is short-lived since the antibodies only persist for a short time. However they give the person protection until their own immune system has time to respond to the antigen. The types of specific immunity are summarised in Figure 11.7.

Rejection of transplanted tissues

When living tissue is transplanted from one individual to another (see Figure 11.8), some of the lymphocytes from the recipient's immune system regard the new tissue as a collection of foreign antigens and attempt to destroy them. This destruction process is called **tissue rejection** (see Figure 11.9) and always occurs unless donor and recipient are genetically identical twins.

"BOY! TALK ABOUT ORGAN REJECTION!"

Figure 11.9 Rejection of transplanted tissue

Successful transplants of tissues and organs (e.g. liver, kidney etc.) are made possible by choosing a donor and a recipient that are as genetically similar as possible (by tissue typing) and then administering **immunosuppressor drugs**. Unfortunately these drugs greatly inhibit the recipient's immune system and make the person susceptible to serious diseases such as pneumonia.

Drugs are now being developed which tend to eliminate the tissue rejection response while leaving part of the immune response intact.

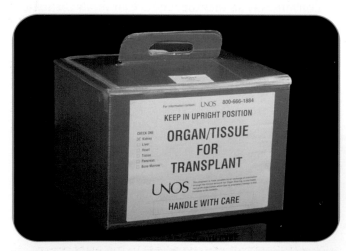

Figure 11.8 Organ transplant

Testing Your Knowledge

1 Name TWO types of white blood cell and for each state how it defends the human body. (4)

2 a) Arrange the following steps into the correct sequence in which they occur during phagocytosis:

    A bacterium digested by enzymes

    B bacterium engulfed

    C lysosomes fuse with vacuole

    D phagocyte meets bacterium. (1)

  b) What is pus? (2)

3 a) What is the difference between an antigen and an antibody? (2)

  b) Is antibody production a specific or a non-specific immune response? Explain. (2)

4 Distinguish between the following pairs:

  a) primary and secondary response; (4)

  b) active and passive immunity; (2)

  c) naturally and artificially acquired immunity. (2)

5 a) Why must immunosuppressor drugs normally be administered to patients undergoing kidney transplant surgery?

  b) i) What problem may result from their use?
     ii) Explain why. (2)

Cellular defence mechanisms in plants

Plants are constantly exposed to an immense variety of pathogenic (disease-causing) micro-organisms such as fungi (see Figure 11.10) which can penetrate plant cell walls by secreting digestive enzymes. In addition most plants find themselves the target of hungry herbivorous animals. Plants often defend themselves by employing one of the following methods.

Figure 11.10 Diseased sycamore leaf

Production of toxic compounds

Tannins

Tannins are a group of acidic substances widely distributed throughout the plant kingdom. They occur in the vacuoles, cytoplasm and walls of plant cells and are common in leaves (e.g. tea plant), unripe fruit (e.g. plum), seed coats (e.g. broad bean), insect galls (see page 85) and the bark of trees (see Figures 11.11 and 11.12).

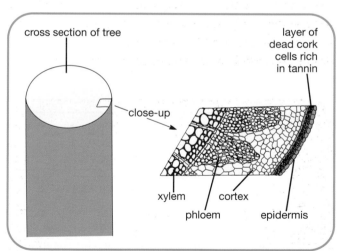

Figure 11.11 Tannin-rich cork cells

Tannins are toxic to many micro-organisms. They defend the plant by successfully preventing pathogens such as fungi from gaining access to the plant organ under attack. This protection is achieved by the tannins acting as enzyme inhibitors. They interfere with the invading pathogen's metabolism and render it harmless.

Cyanide

Hydrogen cyanide is a poison which acts by blocking an organism's cytochrome system. Some varieties of

Figure 11.12 Tannin-rich cork cells in bark

white clover *(Trifolium repens)* contain non-toxic glycoside which is hydrolysed by enzyme action to hydrogen cyanide when leaf tissue is damaged (e.g. during nibbling by a herbivore such as a slug). This production of cyanide is called **cyanogenesis**. It is summarised in the following equation.

$$
\text{glycoside} \xrightarrow{\text{enzyme}} \text{cyanide}
$$
$$
\text{(non-toxic)} \qquad\qquad \text{(toxic)}
$$

The ability to make glycoside and enzyme is determined genetically. A plant which possesses both of the necessary genes is said to be **cyanogenic**. Under normal circumstances the glycoside and the enzyme are kept apart in its leaf cells. Plants unable to make cyanide are said to be **acyanogenic**.

Table 11.1 shows the results from an experiment set up to investigate whether cyanogenic plants are better at resisting attack by herbivores than acyanogenic plants. Several members of the same species of slug were released into plastic boxes each containing equal numbers of leaves from cyanogenic and acyanogenic clover plants. The cyanogenic plants were found to suffer much less damage than the acyanogenic ones.

Cyanogenesis, however, does not always guarantee protection. For example the larvae of the common blue butterfly produce a further enzyme that converts poisonous cyanide to harmless thiocyanate.

Nicotine

Nicotine is a toxic chemical produced by the root cells of the tobacco plant and transported to its leaves. Since it is poisonous it protects the leaves against attack by herbivorous insects. For this reason nicotine can be extracted from tobacco plants and used by farmers as an insecticide.

Isolation of the problem

Insect galls

When a parasite such as an insect successfully penetrates the tissues of a host plant, the latter often produces a gall (see Figures 11.13 and 11.14) in response to a chemical stimulus made by the parasite.

A gall is an abnormal swelling of plant tissues resulting from active division of the cells at the site of injury. Many galls contain tannins which play a protective role.

This combination of extra layers of cells and rich deposits of tannin in a gall provides the plant with a **protective barrier** within which the parasite can be isolated (see Figure 11.15) and prevented from causing further damage.

type of clover	% number of leaves		
	undamaged	less than half of leaf eaten	half or more of leaf eaten
cyanogenic	80(+)	16(−)	4(−)
acyanogenic	42(−)	25(+)	33(+) ·

(+ and − indicate more or less than expected by chance)

Table 11.1 Damage to cyanogenic and acyanogenic clover leaves

Figure 11.13 Insect galls on stem

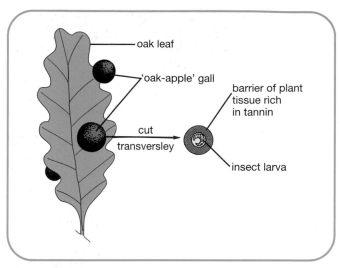

Figure 11.15 Inside an insect gall

Resin

Resin is a sticky acidic substance produced by many shrubs and trees especially conifers which possess numerous **resin canals** (see Figure 11.16). Each resin canal is lined with resin-secreting cells.

Figure 11.14 Insect galls on leaf

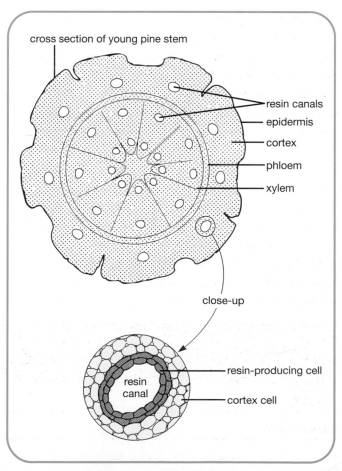

Figure 11.16 Resin canals in pine tree

When the plant becomes wounded or infected by a pathogen, the resin-secreting cells increase in activity. The resin produced flows through the canals to the site of injury. The resin is then secreted into the intercellular spaces around the area under attack and often into nearby vascular tissue causing local blockage.

The pathogen is therefore isolated since it can neither spread by attacking healthy cells near the wound nor by being transported in the phloem or xylem tissue to other parts of the plant.

Testing Your Knowledge

1 Describe the means by which tannins defend a plant against fungal attack. (2)

2 a) What is *cyanogenesis*? (2)

  b) How does it work as a defence mechanism in plants? (2)

3 What is nicotine and why do tobacco plants produce it? (2)

4 a) Name a type of plant which contains numerous resin canals in its stem. (1)

  b) Give TWO ways by which the secretion of resin prevents spread of an invading pathogen. (2)

Applying Your Knowledge

1 The 'split' cell in Figure 11.17 represents part of a phagocyte and part of a lymphocyte in action.

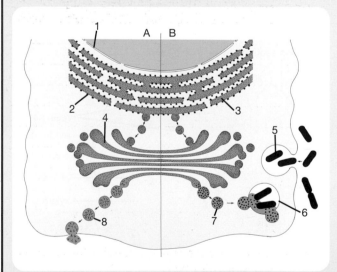

Figure 11.17

  a) Match letters A and B with the two types of white blood cell. (1)

  b) Name structures 1–8. (4)

  c) Compare the contents of structures 7 and 8. (2)

  d) Suggest how a phagocyte avoids self-digestion during the process of phagocytosis. (1)

2 Table 11.2 refers to the four types of immunity. The following list gives examples of ways in which immunity may be acquired:

	active	passive
natural	W	X
artificial	Y	Z

Table 11.2

  1) by receiving materials across the placenta;

  2) by suffering and surviving measles;

  3) by receiving an injection of antibodies made by a horse to prevent diphtheria;

  4) by being vaccinated against diphtheria.

  a) Match 1, 2, 3 and 4 with W, X, Y and Z in Table 11.2. (2)

  b) i) In the long-term, which is the better way to acquire immunity against diphtheria, method 3) or method 4)?

   ii) Explain why. (2)

3 Figure 11.18 shows a disease-causing virus (V) and an attenuated (weakened) version of it (W).

  a) Suggest why receiving a vaccine containing W gives immunity against attacks in the future by V. (2)

  b) Would such immunity be described as naturally or artificially acquired? (1)

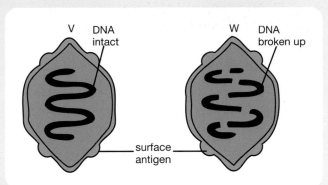

Figure 11.18

4 Table 11.3 refers to antibody proteins called immunoglobulins found in human blood.

The graph in Figure 11.19 refers to the sequence of events which occurs in response to two separate injections of a type of antigen into a small mammal.

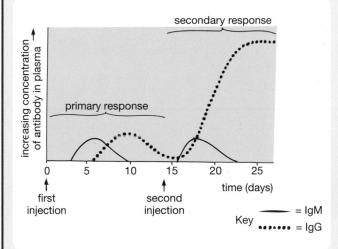

Figure 11.19

a) i) What immunoglobulin in the table would be found in the blood of an unborn baby?
   ii) Suggest why these antibodies are only needed by the baby for a few months after birth. (2)

b) Of the five types of immunoglobulin molecules, which is the **i)** largest; **ii)** rarest? (2)

c) State the normal serum concentration of IgA in mg/ml. (1)

d) With reference to IgM and IgG, state ONE feature common to both the primary and secondary response shown in Figure 11.19. (1)

e) With reference to IgG, state THREE differences between the primary and the secondary response. (3)

f) Antibodies such as IgG are now known to be produced by the activity of long-lived lymphocytes. With reference to the graph, suggest why the latter are called 'memory cells'. (1)

5 Give an account of the process of phagocytosis. (10)

6 a) Present the information in Table 11.4 as a graph by plotting the points and then drawing the best curve. (2)

age of bracken leaf (weeks)	average cyanide content of leaf (mg/100 g dry wt)
1	9.9
3	4.3
5	2.1
7	1.0
9	0.5
11	0.2
13	0.0

Table 11.4

b) 'State the relationship that exists between cyanide concentration and age of leaf. (1)

c) In what way could this relationship be of survival value to the plant? (1)

	immunoglobulin (Ig)				
	IgA	**IgD**	**IgE**	**IgG**	**IgM**
molecular weight	170 000	184 000	188 100	150 000	960 000
normal serum concentration	1.4–4.0 g/l	0.1–0.4 g/l	0.1–1.3 mg/l	8.0–16.0 g/l	0.5–2.0 g/l
ability to cross placenta	no	no	no	yes	no

Table 11.3

7 Figure 11.20 shows a close-up of part of the stem of a conifer tree under attack by a pathogen.

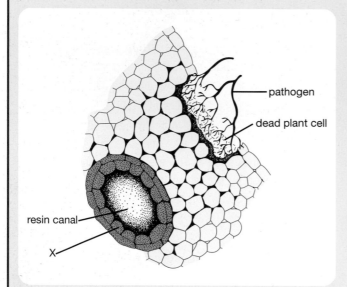

Figure 11.20

a) What is meant by the term *pathogen?* (1)

b) i) Is the pathogen in this case a bacterium or a fungus?

ii) By what means does this type of pathogen normally gain access to the host cells? (2)

c) What is the function of cell type X? (1)

d) Why is it unlikely that the pathogen in Figure 11.20 will successfully invade the host? (1)

e) Suggest why horse chestnut buds are covered with resin. (1)

8 A certain type of insect gall on willow trees only develops if the parasite's egg hatches into a larva. Eventually when the larva stops feeding and becomes a pupa, gall development stops.

a) From this relationship, identify the gall-maker. (1)

b) Suggest which stage in the life cycle of the gall-causer produces the chemical stimulus to which the gall-maker responds. (1)

c) What benefit is gained by a gall-maker in return for the materials and energy expended during gall formation? (1)

9 Write an essay on the defence mechanisms employed by plant cells. (10)

What You Should Know

Chapters 9–11
(See Table 11.5 for Word bank)

antibodies	nicotine	resin
antigen	nucleic	specific
defence	phagocytosis	tannins
host	protein	viruses
lymphocytes	non-specific	white

Table 11.5 Word bank for chapters 10–11

1 _____ exhibit living and non-living characteristics and can only reproduce within the living cells of another organism.

2 A virus consists of one type of _____ acid surrounded by a coat of _____.

3 Once inside a host cell, the virus alters the _____ cell's metabolism to produce many identical copies of itself.

4 The two main _____ mechanisms employed by the human body depend on the activities of _____ blood cells.

5 Phagocytes engulf and destroy microbes by _____ using lysosomes. This is a _____ immune response.

6 _____ recognise antigens on the surface of a microbe and produce_____. These possess receptors sites which bind to one particular type of _____ rendering it harmless. This is a _____ immune response.

7 Plants do not make antibodies. They defend themselves against invasion by producing toxic compounds such as _____, cyanide and _____.

8 Some plants can isolate an area of injury by secreting sticky _____.

Genetics and Adaptation

12 Meiosis

Role of sexual reproduction

Continuous and discontinuous variation exist amongst the members of a species. Much of this variation is determined by alleles of genes and is inherited. For a species to survive, reproduction must occur. During sexual reproduction, the nuclei of two sex cells (gametes) fuse to form a zygote. Since the two sex cells are genetically different from one another, the zygote formed is different from both of the parents and from all other members of the species. Thus sexual reproduction maintains and increases genetic variation within a population.

This chapter explores the means by which the gametes in a breeding population receive different alleles of genes and thereby provide the raw material for evolution to occur.

Haploid and diploid cells

Every species of plant and animal has a characteristic number of chromosomes (the chromosome complement) present in the nucleus of each of its cells. A diploid cell (e.g. a zygote as shown in Figure 12.1) has a double set of chromosomes (two of each type) which form pairs. A haploid cell (e.g. a sex cell as

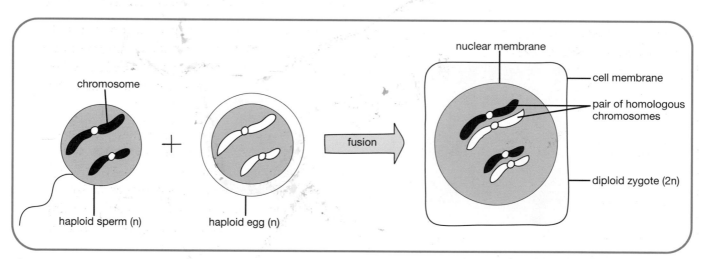

Figure 12.1 Haploid and diploid cells

organism	number of chromosomes present in haploid cell (n)	number of chromosomes present in diploid cell (2n)
fruit fly	4	8
buttercup	7	14
onion	8	16
rice	12	24
clover	16	32
cat	19	38
human	23	46
horse	33	66

Table 12.1 Chromosome complements

shown in Figure 12.1) has a single set of chromosomes (one of each type).

The chromosome complement of a haploid cell is represented by the letter n and that of a diploid cell by 2n. Some examples are shown in Table 12.1.

Since the members of each pair of chromosomes present in a diploid cell match one another gene for gene, they are said to be **homologous**. They may however possess different alleles (forms) of the same gene.

Mass of DNA

Chromosomes contain a cell's DNA. The mass of DNA present in a haploid cell (e.g. sperm) is half of that found in a diploid body cell (e.g. liver). Human red blood cells lack nuclei and have neither chromosomes nor DNA.

Testing Your Knowledge

1 What term is used to refer to the characteristic number of chromosomes possessed by a species? (1)

2 Distinguish clearly between the terms *haploid* and *diploid*. (2)

3 **a)** State ONE feature that the members of a pair of homologous chromosomes always have in common. (1)

   b) State ONE feature by which the members of a pair of homologous chromosomes may differ. (1)

4 The mass of DNA present in each of a mammal's sperm is equal to 50% of the DNA present in each of its kidney cells. Explain why. (2)

Meiosis

Meiosis is the form of nuclear division which results in the production of four haploid (n) gametes from one diploid (2n) gamete mother cell. It occurs at specific sites in living organisms as shown in Table 12.2.

Process of meiosis

Meiosis involves two consecutive nuclear (followed by cell) divisions. The gamete mother cell divides into two cells and these then divide again. Figure 12.2 refers to a gamete mother cell containing four chromosomes (two of 'paternal' and two of 'maternal' origin).

Before the process of meiosis begins, each chromosome replicates forming two identical **chromatids** (see page 55). Therefore when the nuclear material becomes visible (on staining), each chromosome is seen to consist of two chromatids attached at a centromere.

Homologous chromosomes pair up and come to lie alongside one another so that their centromeres and genes match exactly. While in this paired state, the chromosomes become even shorter and thicker by coiling up.

Next the members of each homologous pair begin to repel one another and move apart except at points called chiasmata (singular = chiasma) where they remain joined together. It is here that exchange of genetic material occurs by two chromatids twisting around one another and 'swapping' portions. This exchange is called **crossing over**.

Next the nuclear membrane disappears, spindle fibres form and homologous pairs become arranged on the equator of the cell. The arrangement of each pair relative to any other is random.

---

site of meiosis	diploid gamete mother cell	haploid gametes formed
testis of animal	sperm mother cell	sperm
ovary of animal	egg mother cell	eggs (ova)
anther of flowering plant	pollen mother cell	pollen
ovary of flowering plant	egg mother cell	eggs (ovules)

Table 12.2 Sites of meiosis

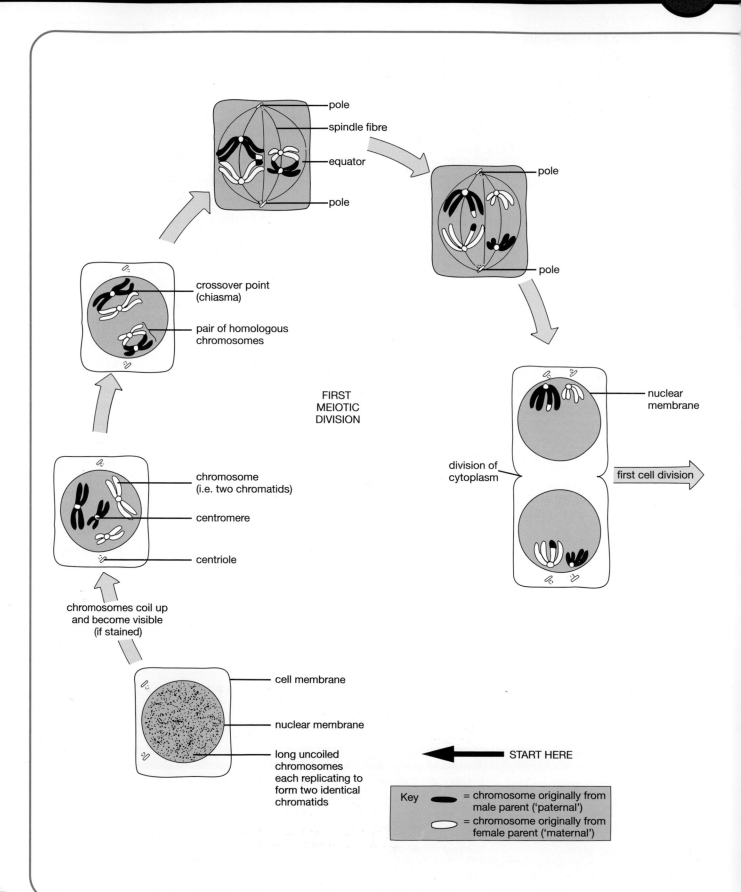

pole
spindle fibre
equator
pole

pole
pole

crossover point (chiasma)
pair of homologous chromosomes

nuclear membrane

division of cytoplasm

first cell division

FIRST MEIOTIC DIVISION

chromosome (i.e. two chromatids)
centromere
centriole

chromosomes coil up and become visible (if stained)

cell membrane
nuclear membrane
long uncoiled chromosomes each replicating to form two identical chromatids

START HERE

Key — = chromosome originally from male parent ('paternal')

○ = chromosome originally from female parent ('maternal')

Figure 12.2 Process of meiosis

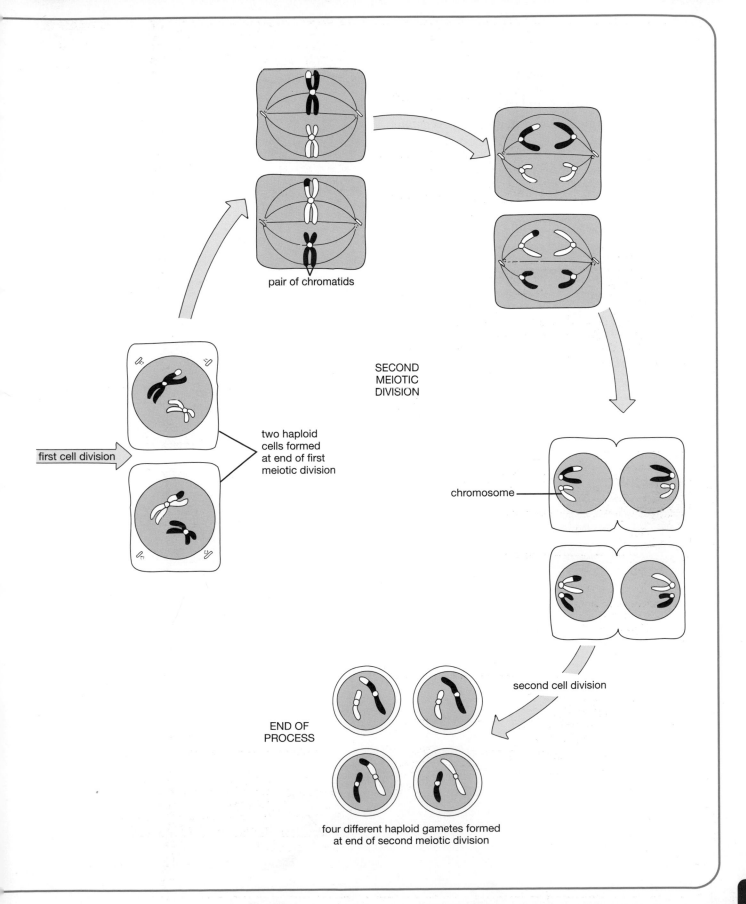

pair of chromatids

SECOND
MEIOTIC
DIVISION

two haploid
cells formed
at end of first
meiotic division

first cell division

chromosome

second cell division

END OF
PROCESS

four different haploid gametes formed
at end of second meiotic division

On contraction of the spindle fibres, one chromosome of each pair moves to one pole and its homologous partner moves to the opposite pole.

This is followed by a nuclear membrane forming round each group of chromosomes and then by division of the cytoplasm resulting in the formation of two haploid cells.

Each of these haploid cells now undergoes the second meiotic division as shown in Figure 12.2. Single chromosomes (each made of two chromatids) line up at each equator. On separation from its partner, each chromatid is regarded as a chromosome.

Each of the four gametes formed contains half the number of chromosomes present in the original gamete mother cell. Figure 12.3 shows the process of meiosis in the anther of a flower.

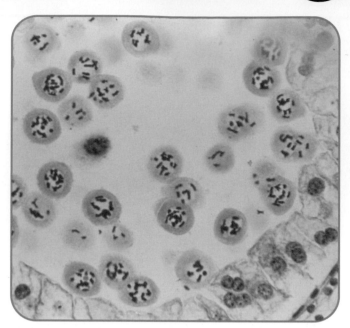

Figure 12.3 Meiosis in *Lilium* anther

Significance of meiosis

Prior to the mixing of one individual's genotype with that of another at fertilisation, meiosis provides the opportunity for new combinations of the existing alleles of genes to arise as follows.

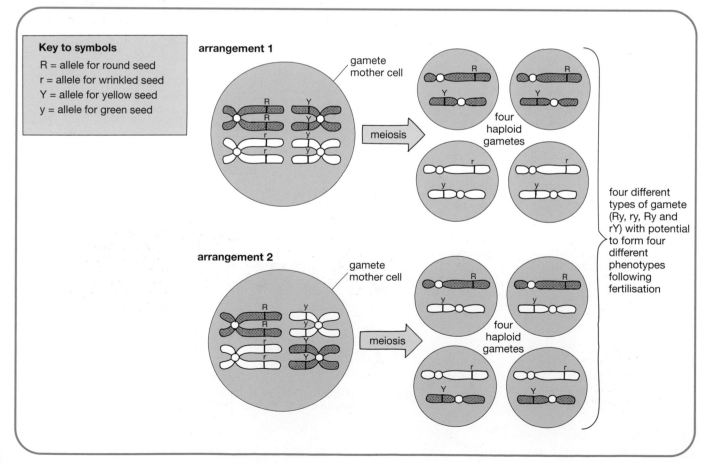

Figure 12.4 Independent assortment of chromosomes

Independent assortment of chromosomes

When homologous pairs of chromosomes line up at the equator during the first meiotic division, the final position of any one pair is random relative to any other pair.

Figure 12.4 shows a simple example of a gamete mother cell in a pea plant (where only two homologous pairs have been drawn). There are two ways in which the pairs can become arranged. Subsequent meiotic divisions bring about independent assortment of chromosomes. This gives rise to 2^2 (i.e. 4) different genetic combinations in the gametes. These may lead to the formation of new phenotypes in the next generation.

The larger the number of chromosomes present, the greater the number of possible combinations. For example, a human egg mother cell with 23 homologous pairs has the potential to produce 2^{23} (i.e. 8 388 608) different combinations.

Crossing over and separation of linked genes

Two genes situated on the same chromosome are said to be linked. When a chromosome (consisting of two chromatids) is aligned with its homologous partner prior to the first meiotic division, portions of chromatid may be exchanged at points called chiasmata (see Figure 12.2).

The process of crossing over involves chromatids becoming broken and the broken end of one joining with that of another. By this means, alleles of linked genes can become separated. This can result in the formation of new combinations of alleles and give four genetically different chromatids each of which ends up in a different gamete. This may lead to the formation of new phenotypes in the next generation.

A simple example is shown in Figure 12.5 which refers to one homologous pair of chromosomes in a gamete mother cell of a fruit fly. Consider a human gamete mother cell with 23 homologous pairs all capable of undergoing crossing over at one or more chiasmata. The potential for increased variation by this method is enormous.

Importance of variation

The processes of independent assortment of chromosomes and crossing over increase variation within a species by producing new combinations of genetic material in the species' gametes. Following

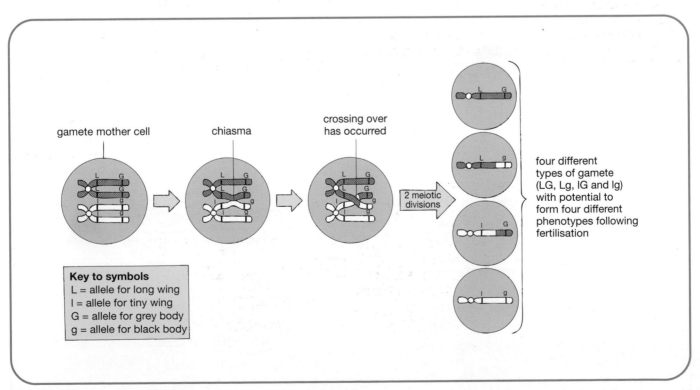

Figure 12.5 Crossing over

sexual reproduction, individuals are produced that are different from one another.

In the long term, such variation is of great importance because it helps a species to adapt to a changing environment. Imagine, for example, that a new disease appears. If great genetic variation exists amongst the members of a species, then there is a good chance that some of them will be resistant and survive. If they were all identical and susceptible to the disease, the whole species would be wiped out.

This 'survival of the fittest' occurs as a result of natural selection (see chapter 17).

Testing Your Knowledge

1 Briefly define the term *meiosis*. (2)

2 Name ONE site in an animal and ONE site in a plant where meiosis occurs. (2)

3 a) What name is given to the two identical structures which make up each chromosome prior to meiosis? (1)

   b) i) What name is given to points of direct contact between adjacent chromatids during the first meiotic division?

      ii) What may happen at these points of contact? (2)

4 a) Are the products of the first meiotic division haploid or diploid? (1)

   b) Are the products of the second meiotic division haploid or diploid? (1)

5 Compare the chromosomal number of the original gamete mother cell with that of each of the four gametes formed by meiosis. (1)

6 a) Identify the TWO ways in which new combinations of existing alleles may arise during meiosis. (2)

   b) Why is it important that variation exists amongst the members of a species? (2)

Applying Your Knowledge

1 Refer to Table 12.1 and then state how many chromosomes would be present in

   a) an egg from a fruit fly;

   b) a cheek epithelial cell from a human being;

   c) a sperm from a horse;

   d) a pollen mother cell from an onion flower;

   e) a root tip cell from a buttercup plant;

   f) an egg mother cell from a cat. (6)

2 The following statements refer to various stages that occur during meiosis. With reference only to their letters, arrange the statements into the correct sequence.

   A Homologous chromosomes become arranged on the equator of the cell.

   B Homologous chromosomes pair up and their centromeres and genes become aligned.

   C Homologous chromosomes become separated as each is pulled to an opposite pole of the cell.

   D Homologous chromosomes begin to repel one another but remain joined at points called chiasmata. (1)

3 With reference to the cell in Figure 12.6 state the number of

   a) chromatids present;

   b) centromeres present;

   c) chromosomes present;

   d) pairs of homologous chromosomes present;

   e) chromosomes that would be present in each gamete produced. (5)

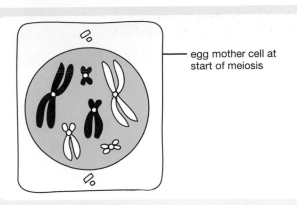

egg mother cell at start of meiosis

Figure 12.6

f) i) Draw diagrams to show TWO different ways in which the homologous pairs of chromosomes could become arranged at the equator during the first meiotic division.

ii) Give the gametes that would result from each arrangement (if no crossing over occurred). (4)

4 Figure 12.7 shows a pair of homologous chromosomes during meiosis.

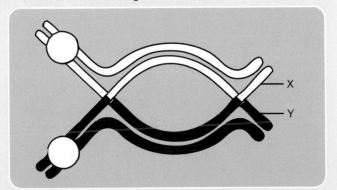

X

Y

Figure 12.7

a) How many chiasmata occur between chromatids X and Y? (1)

b) Draw a diagram showing the appearance of these chromosomes at the end of the first meiotic division. (2)

5 Figure 12.8 shows the human life cycle. Copy and complete the diagram using the terms *meiosis*, *mitosis* and *fertilisation* to complete the blank boxes. (2)

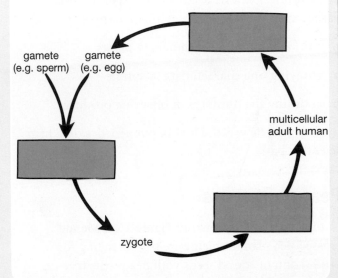

gamete (e.g. sperm)   gamete (e.g. egg)

multicellular adult human

zygote

Figure 12.8

6 Copy and complete Table 12.3 which compares the processes of mitosis and meiosis. (7)

	Mitosis	Meiosis
site of division	occurs all over body of growing animal	
pairing and movement of chromosomes		homologous chromosomes form pairs; chromosomes line up in pairs on equator
exchange of genetic material	chiasmata not formed and no crossing over occurs	
number of divisions		double division of nucleus
number and type of cells produced	following cell division, two identical daughter cells formed	
effect on chromosome number		chromosome number halved
effect on variation	does not increase variation within a population	

Table 12.3

13 Monohybrid cross

A cross between two true-breeding parents that possess different forms (alleles) of one particular gene is called a **monohybrid cross.**

Gregor Mendel (1822–84) performed early monohybrid crosses using varieties of pea plant which possessed characteristics showing discontinuous variation. He appreciated the importance of:

- working with **large numbers** of plants
- studying **one characteristic** at a time
- **counting the numbers** of offspring produced.

By doing so, he was the first to put genetics on a firm scientific basis.

Pea plant cross

In the experiment shown in Figure 13.1, Mendel crossed pea plants which were true-breeding for production of round seeds with pea plants true-breeding for wrinkled seeds.

All the seeds produced in the **first filial generation** (F1) were round. Once these seeds had grown into plants, they were self-pollinated. The resultant **second filial generation** (F2) consisted of 7324 seeds (5474 round and 1850 wrinkled). This is a ratio of 2.96:1.

When these figures are analysed statistically, they are found to represent a ratio of 3 round:1 wrinkled. An exact 3:1 ratio is rarely obtained because pollination and fertilisation are **random** processes both of which involve an element of **chance.**

The difference between the expected ratio and the observed ratio is known as the **sampling error.** The larger the sample, the smaller the error.

Since wrinkled seed, absent in the F_1, reappears in the F_2, 'something' has been transmitted undetected in the gametes from generation to generation. Mendel called this a **factor.** Today we call it a **gene.** In this case it is the gene for seed shape which has two forms (alleles) – round and wrinkled.

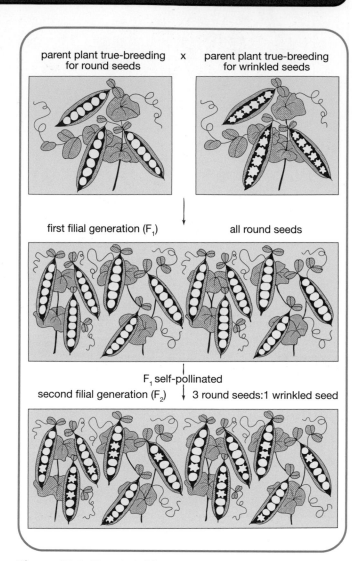

Figure 13.1 Monohybrid cross

Since the presence of the round allele masks the presence of the wrinkled allele, round is said to be **dominant** and wrinkled **recessive.**

Genotype and phenotype

An organism's **genotype** is its genetic constitution (i.e. alleles of its genes) that it has inherited from its parents. An organism's **phenotype** is its appearance resulting from this inherited information (see Figure 13.2).

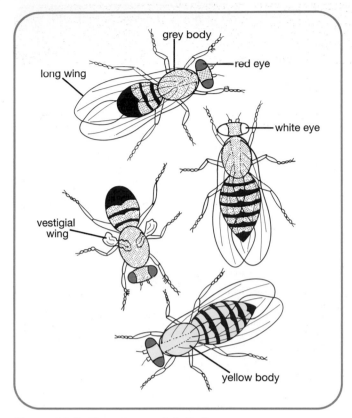

Figure 13.2 Phenotypes of fruit fly

Symbols

Using symbols R for round and r for wrinkled, the cross (shown in Figure 13.1) can be summarised as follows:

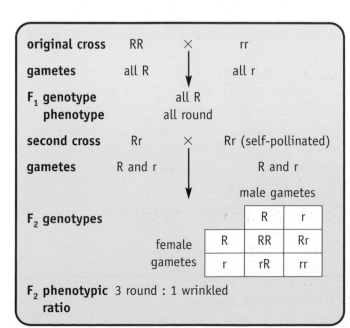

Mendel's first law

From the results of his many monohybrid crosses, Mendel formulated his first law, the **principle of segregation**. Expressed in modern terms it states that: the alleles of a gene exist in pairs but when gametes are formed, the members of each pair pass into different gametes. Thus each gamete contains only one allele of each gene.

Homozygous and heterozygous

When an individual possesses two similar alleles of a gene (e.g. R and R or r and r), its genotype is said to be **homozygous** (true-breeding) and all of its gametes are identical with respect to that characteristic.

When an individual possesses two different alleles of a gene (e.g. R and r), its genotype is said to be **heterozygous**. It produces two different types of gamete with respect to that characteristic.

Incomplete dominance

Sometimes one allele of a gene is not completely dominant over the other allele. This results in the formation of heterozygous offspring which exhibit a phenotype different from both of the homozygous parents.

Snapdragon

For example when true-breeding homozygous red snapdragon plants are crossed with true-breeding homozygous white (ivory) plants, the heterozygous F_1 are all pink. When the F_1 is self-pollinated, the F_2 occurs in the ratio 1 homozygous red:2 heterozygous pink:1 homozygous ivory.

This cross is shown in Figure 13.3 where R = allele for red flower and I = allele for ivory flower.

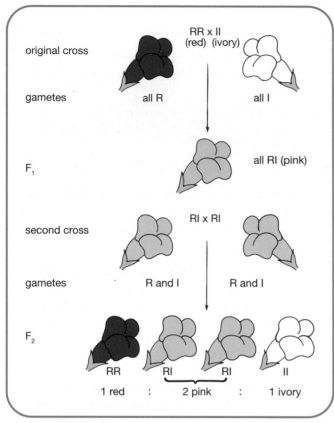

Figure 13.3 Incomplete dominance

Shorthorn cattle

In shorthorn cattle, red coat colour is incompletely dominant to white coat. Heterozygous offspring which possess a mixture of red and white hairs are described as 'roan' (see Figure 13.4).

Figure 13.4 Roan coat colour

Testing Your Knowledge

1 What term is used to describe a cross between two true-breeding parents which differ from one another in one way only? (1)

2 Give THREE examples of good scientific practice employed by Mendel in his experiments with pea plants. (3)

3 a) What term is used to refer to the different forms of a gene? (1)

   b) Which form of the gene for seed shape was masked in the F_1 in Mendel's experiment but reappeared in the F_2? (1)

   c) What term is used to apply to an allele that can be masked by another allele? (1)

   d) Which allele of the gene for seed shape is dominant? (1)

4 a) Using symbols, represent the genotype of:
   i) a plant which is true-breeding for production of round seeds;
   ii) a plant which is true-breeding for production of wrinkled seeds;
   iii) the F_1 offspring that result from crossing i) with ii). (3)

   b) State the phenotype of the F_1 offspring. (1)

   c) i) When these F_1 offspring are self-pollinated, what F_2 phenotypic ratio is produced?
   ii) Draw a Punnet square to show how this ratio arises. (3)

5 Distinguish between the following pairs:

   a) genotype and phenotype; (2)

   b) homozygous and heterozygous. (2)

Applying Your Knowledge

1 In guinea pigs, long hair is recessive to short hair. A long-haired female was crossed with a heterozygous male.

   a) Using symbols of your choice present this cross as a diagram. (2)

   b) What proportion of the offspring on average will be long-haired? (1)

2 Figure 13.5 shows a human family tree. The ability to roll the tongue is governed by the presence of the dominant allele R. The recessive allele is r.

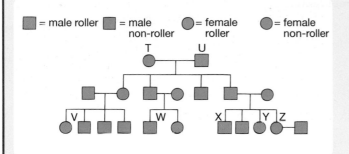

Figure 13.5

   a) How many male grandchildren do grandparents T and U have? (1)

   b) From the choice given in brackets, identify the person who could have the genotype RR. (V, W, X, Y). (1)

   c) Person Z marries a person of genotype Rr. The chance of each of their children being a tongue roller is:

   A 1 in 1 **B** 1 in 2 **C** 1 in 3 **D** 1 in 4 (1)

3 In rabbits, spotted coat is dominant to uniform coat. Show in a diagram the cross that should be set up to find out whether a rabbit with a spotted coat is homozygous or heterozygous for this gene. (4)

4 In snapdragon plants, the alleles of the gene for flower colour, red (R) and ivory (I), are incompletely dominant to one another. From the information in Table 13.1, work out the phenotypes of the two parents used in each of crosses b–f. (5)

5 **a)** Using suitable symbols, show in diagrammatic form the outcome of each of the following crosses which refer to coat colour in shorthorn cattle:

     i) red × red; **ii)** roan × roan;
     ii) roan × white; **iv)** red × white. (8)

   b) In each case state the phenotypes of the offspring and the ratio in which they occur. (4)

cross	phenotypes of two parents used	phenotypes and numbers of offspring produced
a)	red × white	158 pink
b)		41 red, 79 pink, 38 ivory
c)		161 red
d)		157 ivory
e)		81 red, 83 pink
f)		78 ivory, 82 pink

Table 13.1

14 Dihybrid cross and linkage

Dihybrid cross

A cross between two true-breeding parents that possess different forms (alleles) of **two** genes is called a **dihybrid** cross. In one of his experiments, Mendel crossed true-breeding pea plants which produced round yellow seeds with true-breeding plants which produced wrinkled green seeds.

This cross involves two genes: the gene for seed **shape** (alleles – round and wrinkled) and the gene for seed **colour** (alleles – yellow and green).

All the F_1 plants produced round yellow seeds. This shows that round is dominant to wrinkled and yellow is

dominant to green. The cross was followed through to the F_2 generation. A simplified version is shown in Figure 14.1. The actual observed F_2 consisted of 315 round yellow seeds, 108 round green, 101 wrinkled yellow and 32 wrinkled green.

Phenotypic ratio

When these figures are analysed statistically, they are found to represent a **9:3:3:1** ratio. Figure 14.2 shows how an F_2 with such a phenotypic ratio can arise. For the 16 possible combinations shown in the Punnett square to occur, each F_1 parent at the second cross must make four different types of gamete in equal numbers.

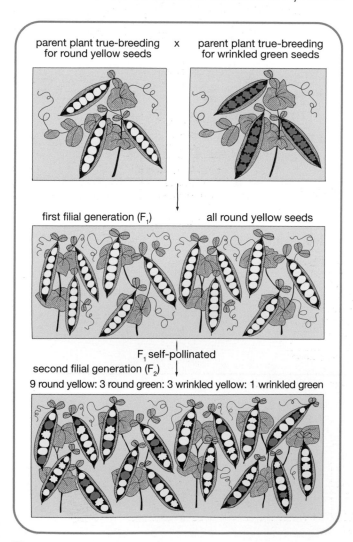

Figure 14.1 Dihybrid cross

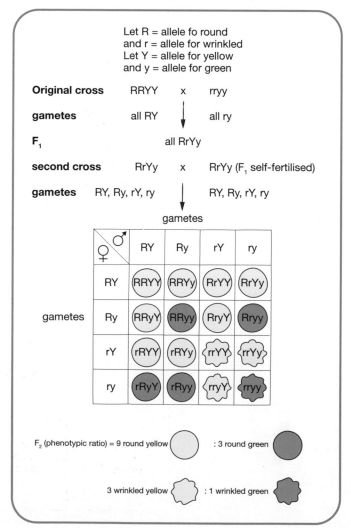

Figure 14.2 Dihybrid cross using symbols

How does this occur? If the genes for seed shape and colour are located on different chromosomes then these will be subjected to independent assortment during meiosis as shown in Figure 12.4 on page 94.

There are two possible ways in which homologous pairs can become arranged at the equator. As a result, approximately 50% of gamete mother cells produce RY and ry gametes and the other 50% produce Ry and rY gametes.

Mendel's second law

From the results of his dihybrid crosses, Mendel was the first to appreciate this idea and used it to formulate his second law, the **principle of independent assortment**. Expressed in modern terms it states that during gamete formation, the two alleles of a gene segregate into different gametes independently of the segregation of the two alleles of another gene.

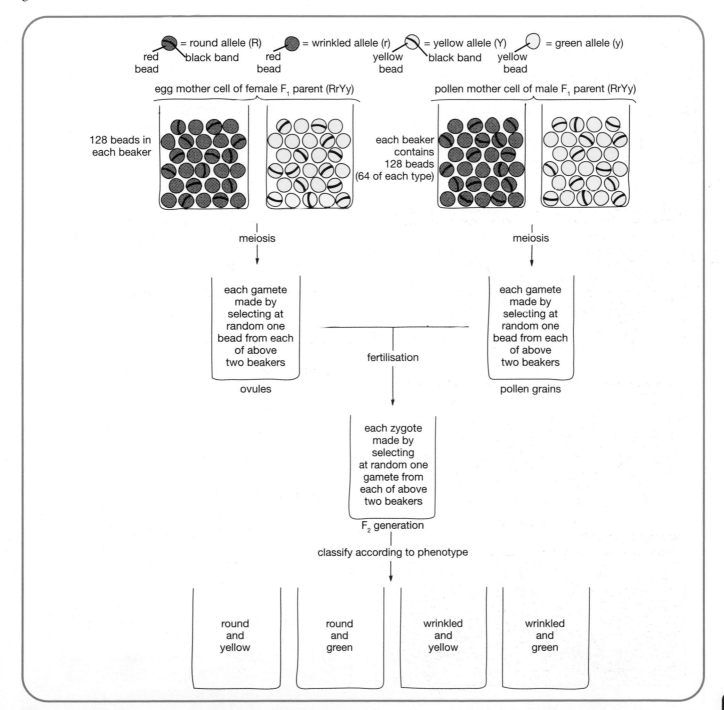

Figure 14.3 Bead model experiment of gamete formation

So in the above cross, there is just as much chance of R and y ending up together in a gamete as R and Y going together. This principle is demonstrated by carrying out the bead model experiment of gamete formation shown in Figure 14.3.

Recombination

In a dihybrid cross, two of the F_2 phenotypes resemble the original parents and two display new combinations of the characteristics. This process by which new combinations of parental characteristics arise is called recombination and the individuals possessing them are called recombinants (see Figure 14.4).

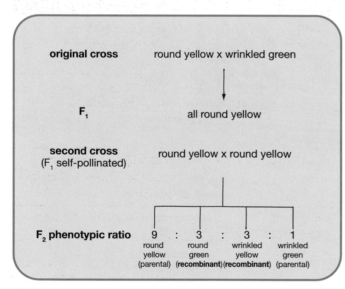

Figure 14.4 Recombinants

Use of fruit flies in breeding experiments

Table 14.1 lists some of the features that make the fruit fly (*Drosophila melanogaster*) suitable for use in genetics experiments.

feature	reason why feature is useful
short life cycle (about 10 days at 25°C)	allows the geneticist to study the transmission of an inherited characteristic through several generations in a short space of time
production of many offspring (female usually lays about 100 eggs)	allows the geneticist to make valid statistical analysis of results (if only a few offspring were produced they might be unusual)
adult male differs from female in several obvious ways (see Figure 14.5)	allows geneticist to easily identify and separate males from females before flies reach sexual maturity thus ensuring that females remain virgins until needed for next experiment

Table 14.1 Reasons for using fruit flies in genetics experiments

1 What is meant by the term *dihybrid cross*? (2)

2 **a)** Using symbols, represent the genotype of:
 i) a plant which is true-breeding for round yellow seeds;
 ii) a plant which is true-breeding for wrinkled green seeds;
 iii) the F_1 offspring resulting from a cross between **i)** and **ii)**. (3)

 b) i) When these F_1 offspring are self-pollinated, what phenotypic ratio is produced?
 ii) Draw a Punnett square to show how this ratio arises. (3)

3 **a)** What name is given to the process by which new combinations of characteristics arise? (1)

 b) What term is used to refer to the individuals possessing the new combinations of characteristics? (1)

4 **a)** What F_2 phenotypic ratio results from a dihybrid cross (involving two genes showing independent assortment) following self-fertilisation of the F_1? (1)

 b) What proportion of this F_2 generation could be described as showing **i)** both parental characteristics? **ii)** recombinant characteristics? (2)

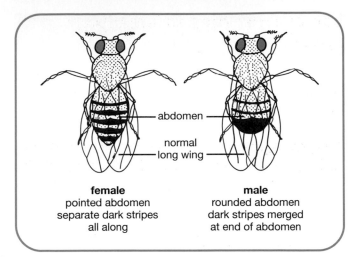

Figure 14.5 Comparison of male and female fruit flies

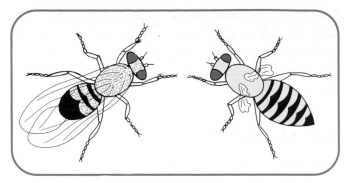

Figure 14.6 Parents for a *Drosophila* dihybrid cross

Wild type

Amongst the members of a species, the type that occurs most commonly in wild populations is called the wild type. In *Drosophila*, the wild-type has a grey body, long wings and red eyes. Each of these alleles is dominant to any other allele of the same gene. A wild type allele is often represented by the symbol '+'. However when the alleles of two genes are being studied simultaneously, confusion can arise.

To avoid this, the following dihybrid crosses adopt the normal policy of representing each dominant allele by the upper case of its first letter and each recessive allele by the lower case of the same letter. Thus grey body = G and yellow body = g; long wing = L and vestigial (tiny) wing = 1.

Dihybrid cross in *Drosophila*

In the following cross, true-breeding wild-type males are crossed with females possessing yellow bodies and vestigial wings (see Figure 14.6). The females are said to be **double recessive**. (The use of wild-type females and double recessive males would be a genetically identical cross.)

Table 14.2 lists the reasons for adopting certain techniques or precautions during this investigation.

design feature or precaution	reason
use of several culture tubes	to produce a large number of progeny and make the results statistically valid
use of several males and females in each tube	to allow for infertility or non-recovery from anaesthetic by individual flies
use of virgin females	to ensure that eggs have not been fertilised in advance by male flies of unknown genotype
removal of parents after egg-laying	to ensure that the parents are not confused with the next generation when they emerge
experiment repeated several times	to check that the results can be successfully repeated and are reliable and not the outcome of a lucky chance

Table 14.2 Design features

Results

Once the F_1 has emerged, its members are anaesthetised and examined. All of them are found to be wild type in appearance. Some F_1 males and females are then used as the parents of the second cross. When the F_2 generation emerges, it is anaesthetised and the number of each type of fly counted. Table 14.3 shows a typical set of results from this experiment.

	grey body, long wing	yellow body, long wing	grey body, vestigial wing	yellow body, vestigial wing
original cross	6♂	0	0	3♀
F$_1$	198	0	0	0
parents of F$_2$	6♂ and 3♀ from F$_1$	0	0	0
F$_2$	92	28	31	11

Table 14.3 Results of dihybrid cross

When the F$_2$ results are analysed statistically they are found to represent a 9:3:3:1 ratio. It is therefore concluded that the genes for body colour and wing length are on different chromosomes and their alleles show independent assortment during meiosis.

Using the symbols stated above, the cross can be represented as follows:

original cross	GGLL	×	ggll

				GL	Gl	gL	gl
gametes	all GL		all gl				
				GL	GGLL	GLl	GgLL	GgLl
F$_1$		all GgLl		Gl	GGLl	GGll	GgLl	Ggll
second cross	GgLl	×	GgLl	gL	GgLL	GgLl	ggLL	ggLl
gametes	GL, Gl, gL, gl		GL, Gl, gL, gl	gl	GgLl	Ggll	ggLl	ggll

female gametes

F$_2$ genotypes see Punnett square

F$_2$ phenotypic ratio 9 grey body, long wing:
3 grey body, vestigial wing:
3 yellow body, long wing:
1 yellow body, vestigial wing

Dihybrid backcross (testcross)

If the F_1 flies produced in the above cross are backcrossed (testcrossed) with double recessive flies (instead of being crossed with other F_1 flies), the following occurs:

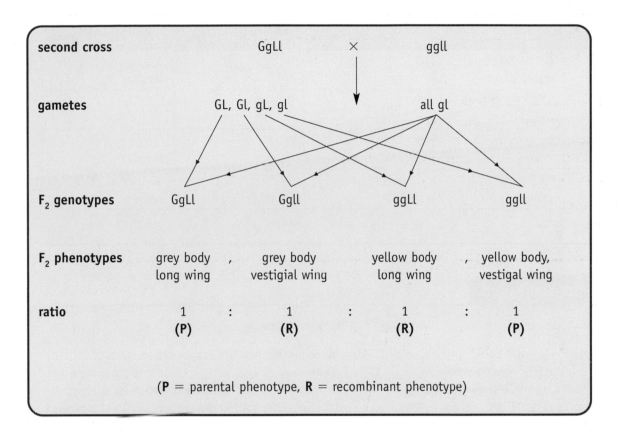

second cross	GgLl	×	ggll
gametes	GL, Gl, gL, gl		all gl

F_2 genotypes	GgLl	Ggll	ggLl	ggll
F_2 phenotypes	grey body long wing	grey body vestigial wing	yellow body long wing	yellow body, vestigal wing
ratio	1 (P)	1 (R)	1 (R)	1 (P)

(**P** = parental phenotype, **R** = recombinant phenotype)

This F_2 generation consisting of **50% parental phenotypes** and **50% recombinants** occurs because the two genes involved are on separate chromosomes and show independent assortment.

Testing Your Knowledge

1 a) What F_2 phenotypic ratio results from a dihybrid cross (involving two genes showing independent assortment) following a backcross of the F_1 to double recessive individuals? (1)

   b) What proportion of this F_2 generation could be described as showing
      i)   both parental characteristics?
      ii) recombinant characteristics? (2)

2 In fruit flies, grey body colour (G) is dominant to yellow (g) and long wing (L) is dominant to vestigial (l). Table 14.4 shows the results of an investigation carried out to examine the phenotypes arising from dihybrid crosses involving the body colour and wing length genes.

   a) Present the information in the table in the diagrammatic form used on page 106. (4)

   b) Why are no flies with yellow bodies or vestigial wings produced in the F_1 generation? (1)

   c) i)   What is the expected ratio of the members of the F_2 generation?
      ii) Why do the actual results vary slightly from the expected ones? (2)

   d) Explain why each of the following practices is adopted during this investigation:

i) several parental flies of each sex are used per culture tube;

ii) only virgin female flies are used;

iii) parental flies are removed a few days after the eggs have been laid. (3)

	grey body, long wing	yellow body, long wing	grey body, vestigial wing	yellow body, vestigial wing
parents of F_1	6 true-breeding males	0	0	6 true-breeding females
F_1	172 flies of both sexes	0	0	0
parents of F_2	6 males from F_1	0	0	6 true-breeding females
F_2	53 flies of both sexes	47 flies of both sexes	54 flies of both sexes	46 flies of both sexes

Table 14.4

Practical Activity and Report

Examining the phenotypes arising from dihybrid crosses in tomato plants

- The tomato seeds that you are going to plant belong to the F_1 and F_2 generations resulting from an original cross between two true-breeding strains of tomato plant. In one parental strain the seedlings always developed purple stems and 'cut' leaves; in the other strain the seedlings always developed green stems and 'potato' leaves (see Figure 14.7).

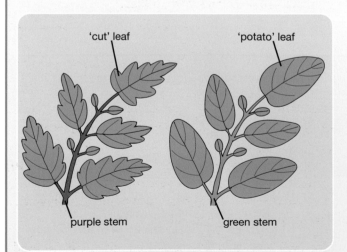

'cut' leaf   'potato' leaf

purple stem   green stem

Figure 14.7 Two strains of tomato plant

- Some members of the F_1 generation were **self-pollinated** to give an F_2 generation; other members of the F_1 were **backcrossed** to the double recessive to give an F_2. Your F_2 generation will be the product of one or other of these types of cross.

supply of tomato seeds (approximately 100) marked F_1

supply of tomato seeds (approximately 100) marked F_2

2 plant pots containing an equal mass of moist seed compost

2 labels

access to a seed propagator at 24°C

What to do

1 Read all of the instructions in this section and prepare your table of results before carrying out the experiment.

2 Label one of the pots 'F_1' and the other 'F_2' and add your initials.

3 Sprinkle the F_1 seeds as evenly as possible over the surface of the soil in pot 'F_1'.

4 Repeat step 3 for the F_2 seeds in pot 'F_2'.

5 Place the two pots in the seed propagator.

6 Inspect the pots regularly and remove them from the propagator when the seeds have germinated into seedlings (after about 14 days).

7 Count the number of plants in the F_1 and F_2 generations that possess **i)** a purple stem and 'cut' leaves; **ii)** a purple stem and 'potato' leaves; **iii)** a green stem and 'cut' leaves; **iv)** a green stem and 'potato' leaves. Record your results in a table.

Reporting

Write up your report by doing the following:

1 Rewrite the title given at the start of this activity.

2 Put the subheading **'Aim'** and state the aim of the experiment.

3 **a)** Put the subheading **'Method'**.

   b) Using the impersonal passive voice, briefly describe the experimental procedure that you followed and state how you obtained your results.

4 Put the subheading **'Results'** and draw a final version of your table.

5 Put a heading **'Analysis and Presentation of Results'**. Extend your results table to include a column giving your results expressed as a ratio.

6 Put a subheading **'Conclusions'** and answer the following questions:

   a) The genes being studied are stem colour and leaf shape.
   i) For each of these genes, which allele was dominant and which was recessive?
   ii) Explain how you arrived at your answer.

   b) To which whole number ratio are your F_2 results closest?

   c) Were the members of the F_2 generation that you examined the products of self-pollinating the F_1 or backcrossing the F_1 generation?

   d) Give the four alleles appropriate letter symbols and present the full cross through to the F_2 in diagrammatic form beginning with the true-breeding parents and including a Punnett square in your answer.

7 Put a final subheading **'Evaluation of Experimental Procedure'**. Give an evaluation of your experiment (keeping in mind that you may comment on any stage of the experiment that you wish).

Try to incorporate answers to the following questions in your evaluation. Make sure that at least one of your answers includes a supporting argument.

   a) Why is an equal mass of moist seed compost used in each plant pot?

   b) Why are the results for as many as 100 (or more) seedlings used when attempting to calculate the F_1 and F_2 proportions?

   c) By what means could the reliability of the results of your experiment be checked?

Linkage

If each chromosome consisted of just one gene then Mendel's principle of independent assortment would hold true for all dihybrid crosses. However each chromosome is composed of a large number of genes and these do not behave independently of one another.

When a cross involves two alleles of two different genes located on the same chromosome, the two genes are transmitted together at meiosis and are said to be linked.

Consider the following example for *Drosophila* involving two linked genes, the gene for wing length (alleles: long = L, vestigial = 1) and the gene for thickness of leg (alleles: thin = T, thick = t). If the two genes were completely linked then a backcross (testcross) would fail to produce any F_2 recombinants. This is shown in Figure 14.8.

However when this cross is carried out, the F_2 is found to contain some recombinants as shown in Table 14.5 which gives a typical set of results. So how can this be explained?

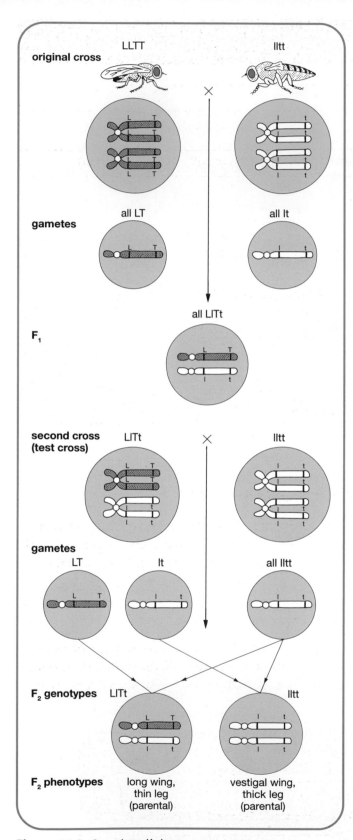

Figure 14.8 Complete linkage

Mechanism by which linked genes are separated

During meiosis when homologous chromosomes form pairs, **crossing over** may occur between adjacent chromatids at points called **chiasmata** (see page 91). If crossing over occurs between two genes then this separates alleles that were previously linked and allows them to recombine in new combinations. Figure 14.9 explains how the recombinants in Table 14.5 arose.

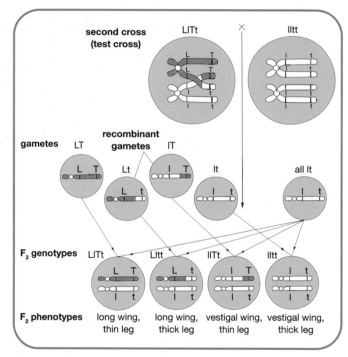

Figure 14.9 Separation of linked genes

Frequency of recombination

Since chiasmata can occur at any point along a chromosome, more crossing over (and therefore more recombination) occurs between two distantly located genes than two that are close together.

Figure 14.10 shows an imaginary example of a homologous pair of chromosomes bearing three linked genes A/a, B/b and C/c. The broken lines represent five possible positions where a crossover could occur between adjacent chromatids. Only crossover 1 would break the linkage between A/a and B/b which are close together. However any one of crossovers 2, 3, 4 and 5 would break the linkage between genes B/b and C/c and produce recombinant gametes.

	long wing, thin leg	vestigial wing, thin leg	long wing, thick leg	vestigial wing, thick leg
original cross	6♂	0	0	3♀
F₁	181	0	0	0
second cross (test cross)	6♂ from F₁	0	0	3♀
F₂	90(parental)	11(recombinant)	13(recombinant)	86(parental)

Table 14.5 Results of dihybrid test cross involving linked genes

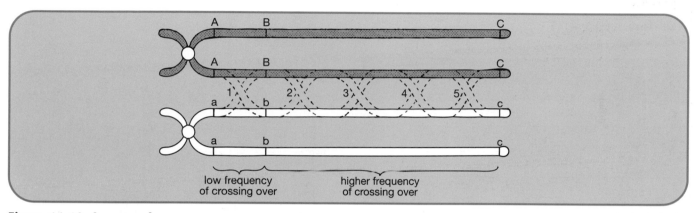

Figure 14.10 Crossover frequency

The distance between a pair of linked genes is therefore indicated by the percentage number of F₂ recombinants that result from a testcross involving them. This percentage is called the **recombination frequency** or **crossover value (COV)** and is calculated as follows:

$$COV = \frac{\text{number of F}_2\text{ recombinants}}{\text{total number of F}_2\text{ offspring}} \times 100$$

Applying this formula to the results in Table 14.5, the COV for the two genes = 24/200 × 100 = 12.

Chromosome maps

For convenience a COV of 1% is taken to represent a distance of one unit on a chromosome. The position of a gene on a chromosome is called its locus (plural = loci). Thus the loci of the genes for long or vestigial wing and thin or thick leg are located 12 units apart on one of *Drosophila*'s chromosomes.

The greater the COV, the longer the distance between two genes; the smaller the COV, the shorter the distance between two genes. By carrying out several testcrosses

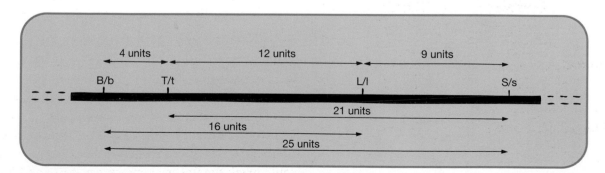

Figure 14.11 Chromosome map

each involving two different linked genes at a time, it is possible to map the **order** and **location** of the genes on a chromosome. Figure 14.11 shows such a map for the fruit fly genes referred to in Tables 14.6 and 14.7.

gene	phenotypic expression
T/t	thick or thin leg
L/l	long or vestigial wing
B/b	presence or absence of bristles
S/s	straight or curved wing

Table 14.6 Linked genes

gene pair in cross	% frequency of recombination (COV)
T/t × L/l	12
T/t × B/b	4
T/t × S/s	21
L/l × B/b	16
L/l × S/s	9
B/b × S/s	25

Table 14.7 COV values

Geneticists have built up comprehensive **chromosome maps** of many organisms including maize and fruit fly (see Figure 14.12).

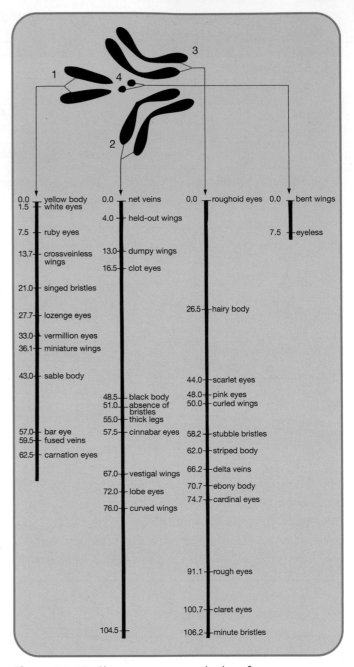

Figure 14.12 Chromosome maps (only a few genes shown)

Testing Your Knowledge

1 a) What are *linked genes*? (1)

b) Referring to the example on page 110, state the genotype of
  i) a true-breeding long-winged, thin-legged fruit fly;
  ii) a true-breeding vestigial-winged, thick-legged fruit fly;
  iii) the F$_1$ generation resulting from a cross between **i)** and **ii)**. (3)

c) State the genotypes of the F$_2$ generation that would be formed by backcrossing this F$_1$ if the two genes were completely linked. (2)

d) In reality, a few members of the F$_2$ have the genotypes Lltt and llTt.
  i) What general name is given to such non-parental forms?
  ii) Copy and complete the following sentence.

In addition to making LT and lt gametes, F$_1$ flies (genotype LlTt) also produce a few _____ and _____ gametes. These recombinants occur as a result of _____ between adjacent chromatids during _____. (5)

2 a) What F$_2$ phenotypic ratio would result from self-pollinating a dihybrid F$_1$ plant (e.g. AaBb) where the two genes involved are on different chromosomes? (1)

b) What F$_2$ phenotypic ratio would result from backcrossing this dihybrid F$_1$ (AaBb)? (1)

c) When F$_1$ dihybrid CcDd was both self-pollinated and backcrossed, neither of these ratios was obtained. What does this indicate about the genes C/c and D/d? (1)

Applying Your Knowledge

1 In tomato plants, round fruit shape is dominant to pear shape and red fruit colour is dominant to yellow.

a) In diagrammatic form, follow, to the F$_2$ generation, a cross between a parent bearing pear-shaped yellow fruit and one which is true-breeding for round red fruit. (Assume that the F$_1$ is self-pollinated and that the two genes are located on different chromosomes.) (4)

b) In your Punnett square, underline FOUR individuals possessing different genotypes and phenotypes. (4)

2 The four fruit flies in Figure 14.13 are all offspring from the same parents.

a) Fly A is homozygous recessive for the alleles of the genes affecting wing type and body colour. Give it a genotype. (1)

b) Assume that in the cross, one parent had the same genotype as A. What gametes must this parent have made? (1)

c) What types of gamete must have been produced by the second parent? (4)

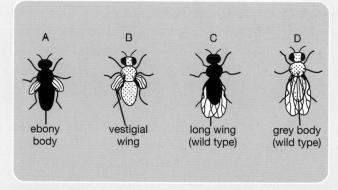

ebony body   vestigial wing   long wing (wild type)   grey body (wild type)

Figure 14.13

d) State the genotypes of the second parent and of each of flies B, C and D. (4)

3 In mice, black fur (B) is dominant to brown (b) and the presence of yellow tips amongst fur (Y) is dominant to absence of yellow tips (y).

Table 14.8 gives the results from a series of breeding experiments. Each F$_1$ female produced two separate families as a result of being mated twice with one of the males.

	black fur, yellow tips	black fur, no yellow tips	brown fur, yellow tips	brown fur, no yellow tips
parents	1			1
F_1	5♂, 5♀			
F_2 family 1	4	0	2	2
2	5	3	1	0
3	3	2	3	1
4	6	1	0	0
5	2	3	2	0
6	7	0	1	0
7	6	2	1	0
8	4	1	3	1
9	5	2	0	0
10	3	1	2	1
F_2 totals				

Table 14.8

a) Express the sex ratio of the F_1 generation in its simplest form. (1)

b) State the genotype of the F_1 mice. (1)

c) Calculate the F_2 totals and then express the F_2 phenotypic ratio in its simplest form. (1)

d) i) If the F_2 ratio had been 13:1:1:5, what would this have indicated about the location of the two genes involved in this cross?

   ii) In light of your answer to part **c)**, what conclusion can you draw about the location of the two genes? (2)

4 Figure 14.14 shows a homologous pair of chromosomes bearing three marker genes R/r, S/s and T/t. Rewrite the following sentences and complete the blanks.

i) Genes R/r, S/s and T/t, which have their loci on the same chromosome, are said to be _____ genes.

ii) Crossing over, the process which can separate the alleles of such genes, occurs at points called _____.

iii) There is a greater chance of crossing over occurring between genes _____ and _____ than between genes _____ and _____.

iv) If a crossover occurred between gene loci S/s and T/t then the recombinant gametes formed would be _____ and _____ and the parental gametes would be _____ and _____. (8)

5 In mice, brown coat colour (B) is dominant to albino (b) and normal movement (M) is dominant to waltzer movement (m). These two genes are located on the same chromosome. A cross between mice which were homozygous for the alleles of these genes was carried out as follows.

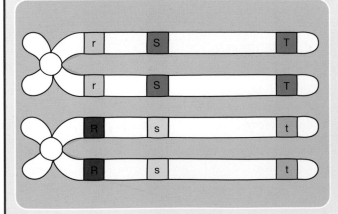

parents   BBMM   ×   bbmm
              (brown coat,   (albino coat,
           normal movement)   waltzer movement)

gametes   BM   ×   bm

F_1 generation   BbMm
                    (brown coat,
                  normal movement)

Figure 14.14

Mice from this F_1 generation were then crossed with mice homozygous for both recessive alleles.

a) Assuming that no crossing over had occurred, state the following:
   i) F_2 genotypes;
   ii) F_2 genotypic ratio;
   iii) F_2 phenotypes;
   iv) F_2 phenotypic ratio. (4)

b) If crossing over had occurred between the two genes then two further F_2 genotypes would have been produced. State what they would have been. (2)

6 A cross was carried out between two maize plants, one true-breeding for brown pericarp and shrunken endosperm, the other true-breeding for white pericarp and full endosperm. F_1 plants (all with white pericarp and full endosperm) were backcrossed against double recessive plants. The F_2 that resulted was:

81 white pericarp, full endosperm
89 brown pericarp, shrunken endosperm
14 white pericarp, shrunken endosperm
16 brown pericarp, full endosperm

a) Using symbols of your choice, present the information in diagrammatic form. (4)

b) State which members of the F_2 are the recombinants. (1)

c) i) Calculate the % number of recombinants present in the F_2.
   ii) Why is this not 50%? (2)

d) How many units apart on their chromosome are the genes that determine pericarp colour and endosperm type? (1)

7 In a certain type of plant, four genes are known to be linked. Table 14.9 gives the COVs obtained from backcrosses involving pairs of these genes.

pair of linked genes	COV
colour of stem (A) × presence of hairs (B)	32
texture of seed coat (C) × colour of stem (A)	9
colour of stem (A) × shape of leaf (D)	6
presence of hairs (B) × texture of seed coat (C)	23
texture of seed coat (C) × shape of leaf (D)	15

Table 14.9

a) Draw a chromosome map to show the position of genes A, B, C and D in relation to one another. (4)

b) Predict the percentage recombination frequency that would be obtained from a backcross involving genes B and D. (1)

8 Outline the procedure that you would adopt to investigate if the gene for grey (G) or yellow (g) body colour is linked to the gene for straight (S) or curved (s) wing in *Drosophila*, the fruit fly. (10)

15 Sex linkage

Sex determination

The nucleus of every normal diploid body cell in a human contains 46 chromosomes. These exist as 22 homologous pairs of **autosomes** which play no part in sex determination and one pair of **sex chromosomes** which determine an individual's sex.

In the female, the sex chromosomes make up a homologous pair, the X chromosomes. Thus a human female has the chromosome complement: 44 + XX.

In the male, the sex chromosomes make up an unmatched pair, an X chromosome and a smaller Y chromosome. Thus a human male has the chromosome complement: 44 + XY.

An egg mother cell is said to be **homogametic** since every egg formed by meiosis contains an X chromosome. A sperm mother cell is **heterogametic** since half of the sperm formed contain an X chromosome and half contain a Y chromosome. When the nucleus of a sperm fuses with the nucleus of an egg at fertilisation, the sex of the zygote is determined by the type of sex chromosome carried by the **sperm**.

Sex ratio

In a large population, approximately 50% of the zygotes formed are XX (female) and 50% are XY (male) giving a **sex ratio** of 1:1 as follows.

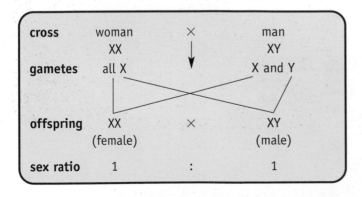

Insects and birds

In the locust, the female has an even number of chromosomes, 18 (16 + XX) and the male has an odd number, 17 (16 + X). Thus half the sperm receive an X chromosome, the other half do not, and sex is determined as before.

The female is not always the homogametic sex. For example in birds and butterflies, the female is heterogametic (XY) and the male is homogametic (XX).

Sex-linked genes

The X and Y chromosomes behave as a homologous pair at meiosis. However the X chromosome differs from the Y chromosome in that the larger X carries many genes not present on the smaller Y. These genes are said to be **sex-linked** (see Figure 15.1).

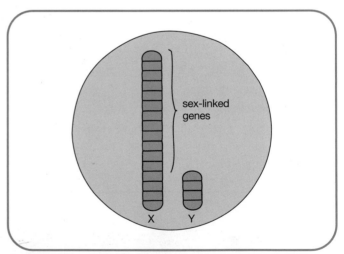

Figure 15.1 Sex-linked genes

When an X chromosome meets a Y chromosome at fertilisation, each sex-linked gene on the X chromosome (whether dominant or recessive) becomes expressed in the phenotype of the organism produced. This is because the Y chromosome does not possess alleles of any of these sex-linked genes and cannot offer dominance to them.

Sex linkage in *Drosophila*

Drosophila possesses the same mechanism of sex determination as humans. Although the Y chromosome is not smaller than the X, it is a different shape and carries very few genes. The X chromosome therefore carries many sex-linked genes. One of these determines eye colour.

The allele for red eye colour (R) is dominant to the allele for white eye (r). Figure 15.2 shows a cross between a white-eyed female and a red-eyed male followed through to the F₂ generation.

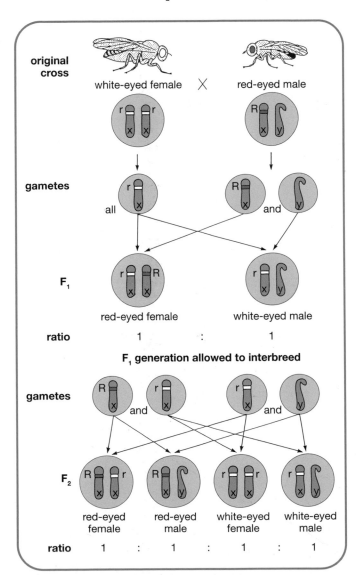

Figure 15.2 Sex linkage in *Drosophila*

A simpler version of this cross is shown in Figure 15.3 where the sex chromosomes are represented by X and Y and the alleles of the sex-linked genes by the superscripts of R and r.

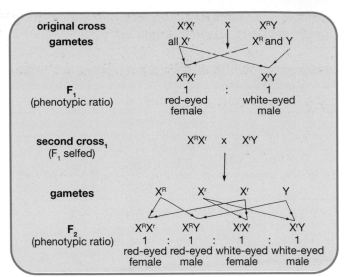

Figure 15.3 Sex-linked inheritance in *Drosophila* using symbols

Since this cross involves a sex-linked gene, the F₂ generation does not show the phenotypic ratio of 3:1 typical of a normal monohybrid cross.

Sex linkage in humans

Red-green colour-blindness

Figure 15.4 shows an example of a test card.

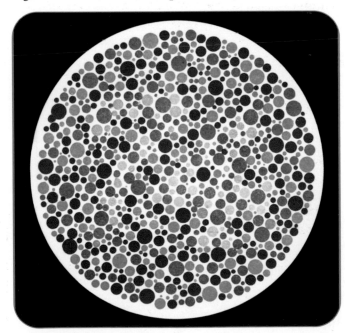

Figure 15.4 Colour-blindness test card

The allele for normal colour vision (C) is dominant to that for red-green colour-blindness (c).

These are the alleles of a sex-linked gene on the X chromosome. The five possible genotypes and their phenotypes for normal colour vision and **red-green colour-blindness** are given in Table 15.1.

genotype	phenotype
X^CX^C	female with normal colour vision
X^CX^c	female (carrier) with normal colour vision
X^cX^c	female with colour-blindness (very rare e.g. 0.5% of European population)
X^CY	male with normal colour vision
X^cY	male with colour-blindness (more common e.g. 8% of European population)

Table 15.1 Red-green colour-blindness in humans

Heterozygous females are called **carriers**. Although unaffected themselves, there is a 1 in 2 (50%) chance that they will pass the allele on to each of their offspring. On average, therefore, 50% of a carrier female's sons are colour-blind. This pattern of inheritance is shown in Figure 15.4.

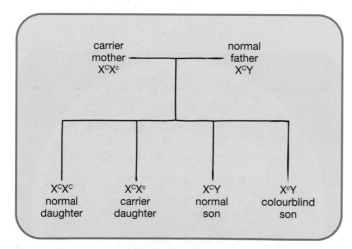

Figure 15.5 Colour-blindness cross using symbols

Red-green colour-blindness is rare in females since two recessive alleles must be inherited; it is more common in males where only one is needed.

Haemophilia

Clotting of blood is the result of a complex series of biochemical reactions involving many essential chemicals. One of these blood-clotting agents is a protein called **factor VIII**. In humans the genetic information for factor VIII is coded for by a gene carried on the X chromosome.

However an inferior version of the factor VIII protein is formed if the gene is changed by a **mutation**. A person who inherits the altered genetic material suffers a condition called **haemophilia**. The sufferer's blood takes a very long time (or even fails) to clot resulting in prolonged bleeding from even the tiniest wound. Internal bleeding may occur and continue unchecked leading to serious consequences.

Since haemophilia is caused by a recessive allele carried on the X but not the Y chromosome, it is a **sex-linked**

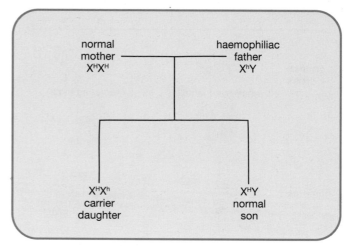

Figure 15.6 Haemophilia cross using superscript symbols

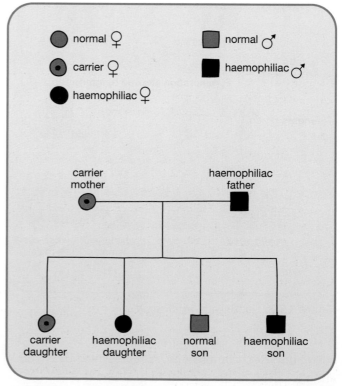

Figure 15.7 Haemophilia cross using alternative symbols

condition. The genotypes of individuals in crosses involving haemophilia are normally represented by the following symbols: X^H (normal blood clotting allele), X^h (haemophilia) and Y (no allele for this gene).

A cross between a carrier female and a normal male would give the same pattern of inheritance as colour-blindness in Figure 15.5. Figure 15.6 shows the possible outcome of a cross between a normal woman and a man suffering haemophilia. Figure 15.7 shows a cross between a carrier woman and a haemophiliac man using a different style of symbol

Testing Your Knowledge

1 a) Which of the human sex chromosomes is the larger one? (1)

   b) What term is used to refer to genes which are present on the X chromosome but absent from the Y chromosome? (1)

   c) If a human male inherits the recessive allele of a sex-linked gene, it is always expressed in his phenotype. Explain why. (1)

2 a) Using symbols, represent the genotype(s) of:
      i)   a true-breeding white-eyed female fruit fly:
      ii) a true-breeding red-eyed male fruit fly;
      iii) the F₁ offspring that would result from a cross between i) and ii). (3)

   b) Draw a Punnett square to show the outcome of crossing F₁ males with F₁ females. (2)

   c) State the phenotypic and sex ratios of the F₂ offspring. (2)

   d) What would have been the phenotypic ratio of the F₂ offspring, had eye colour not been a sex-linked gene? (1)

3 a) Represent symbolically the genotype(s) of:
      i)   a colour-blind human male;
      ii) a human female with normal vision who is a carrier;
      iii) the offspring that could be produced if i) and ii) were their parents. (3)

   b) Name a sex-linked condition which involves defective blood clotting in humans. (1)

   c) Why are such sex-linked conditions expressed much less frequently in the phenotype of females compared with males? (1)

Applying Your Knowledge

1 In fruit flies, red/white eye colour is determined by a sex-linked gene where red (R) is dominant to white (r).

   a) Using the symbols X^R, X^r and Y, draw a diagram to show the F₁ generation that would result from a cross between a white-eyed male and a homozygous red-eyed female. (Clearly state the expected genotypes, phenotypes and sex ratio of the F₁ offspring in your answer.) (4)

   b) Show diagrammatically the F₂ generation that would result if members of the F₁ generation in part a) were allowed to interbreed. (Clearly state the expected genotypes, phenotypes and sex ratio of the F₂ generation in your answer.) (4)

   c) What would have been the phenotypic and sex ratios of the F₂ generation had eye colour not been sex-linked? (1)

2 Using the same format as Figure 15.5, show the possible results of a cross between a colour-blind male and a carrier female. (4)

3 Haemophilia occurs in the family tree shown in Figure 15.8.

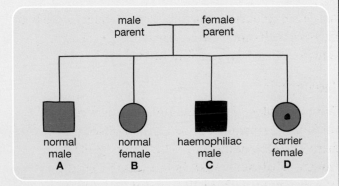

Figure 15.8

a) Using the convention X^H (normal allele), X^h (haemophilia allele) and Y (no allele), give the genotypes of the offspring A, B, C and D. (2)

b) Give the genotype and phenotype of each parent. (2)

c) If C marries a normal female, what proportion of their sons are likely to be haemophiliacs? (1)

4 Decide whether each of the following statements is true or false, and then use T or F to indicate your choice.

a) A man can pass on a sex-linked gene to his sons.

b) In humans, haemophilia is rarer in the female than in the male.

c) Females heterozygous for a sex-linked gene are called carriers.

d) The sex-linked gene for colour-blindness is present on the Y but not the X chromosome.

e) A woman can pass on a sex-linked gene to her daughters. (5)

5 In cats, coat colour is determined by a sex-linked gene where allele B = black and allele G = ginger. In heterozygotes, BG results in tortoiseshell as shown in Table 15.2.

phenotype	female genotype	male genotype
black	$X^B X^B$	$X^B Y$
ginger	$X^G X^G$	$X^G Y$
tortoiseshell	$X^B X^G$	–

Table 15.2

a) Why is no tortoiseshell male given in the table? (1)

b) Show in diagrammatic form TWO crosses that would result in all the female offspring being tortoiseshell. (4)

6 In poultry, the gene for plumage colour is sex-linked. The allele for white feathers (W) is dominant to the allele for red feathers (w). In an attempt to demonstrate that this gene is sex-linked, the following cross was set up.

$$X^W X^W \qquad \times \qquad X^w Y$$

white male   red female

a) Explain why the outcome of this cross would NOT show that the gene is sex-linked. (2)

b) Present in diagrammatic form a cross that would verify that the gene is sex-linked. Explain your answer. (3)

7 Outline the mechanism of sex-linked inheritance in *Drosophila*. Name a sex-linked characteristic and show how it is transmitted from one generation to the next. (10)

16 Mutation

Figure 16.1 Mutant ivy plant

A **mutation** is a change in the structure or amount of an organism's genetic material. It varies in form from a tiny change in the DNA structure of a gene to a large scale alteration in chromosome structure or number. When such a change in genotype produces a change in phenotype, the individual affected is called a **mutant** (see Figure 16.1).

Chromosome mutations
Change in chromosome number
Non-disjunction during meiosis

When a spindle fibre fails during meiosis and the members of one pair of homologous chromosomes fail to become separated, this is called **non-disjunction**. It results in a form of chromosome mutation.

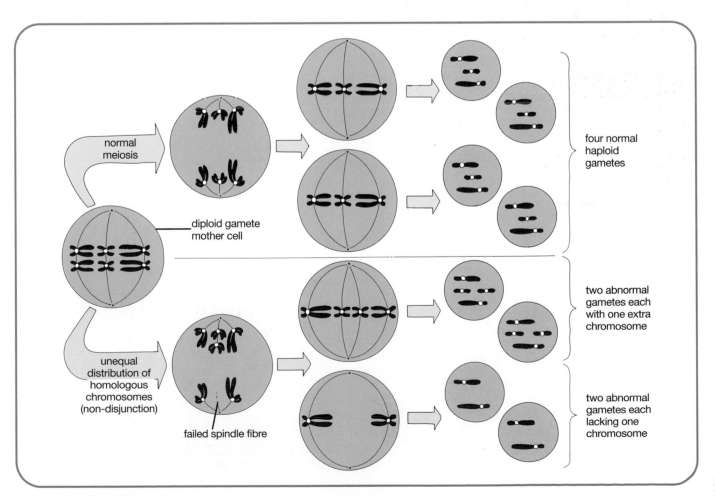

Figure 16.2 Non-disjunction

Normal meiosis and non-disjunction are compared in Figure 16.2 where three pairs of homologous chromosomes are shown. (A human gamete mother cell would contain 23 pairs.) In this example of non-disjunction (where spindle failure occurs during the first meiotic division), two of the gametes receive an extra copy of the affected chromosome and two gametes are formed that lack that chromosome.

Down's syndrome

If non-disjunction of **chromosome pair 21** occurs in a human egg mother cell, then one or more abnormal eggs (n = 24) may be formed. If one of these is fertilised by a normal sperm (n = 23), this results in the formation of an abnormal zygote (2n = 47). The extra copy of chromosome 21 can be seen in the **karyotype** (see Figure 16.3). (A karyotype is a display of a complement of chromosomes showing their number, form and size.)

The affected individual suffers **Down's syndrome** which is characterised by severe learning difficulties and distinctive physical features.

Nearly 80% of the non-disjunctions that lead to Down's syndrome are of maternal origin with frequency being related to maternal age (see Figure 16.4). Since egg mother cells of older women seem to be more prone to non-disjunction at meiosis, pregnant women over the age of 35 are routinely offered a fetal chromosome analysis.

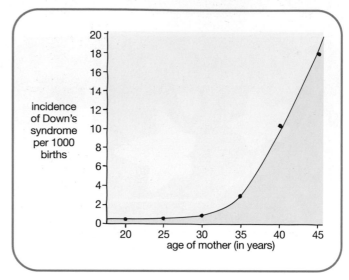

Figure 16.4 Down's syndrome and age of mother

Non-disjunction of sex chromosomes

If human sex chromosomes are affected by non-disjunction during meiosis, then unusual gametes are formed as shown in Figure 16.5.

Turner's syndrome

If a gamete which possesses no sex chromosomes meets and fuses with a normal X gamete, the zygote formed

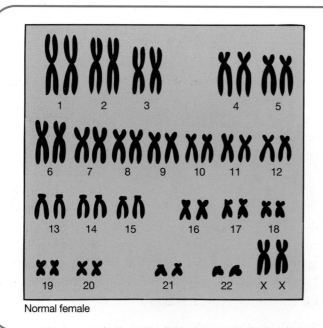

Normal female

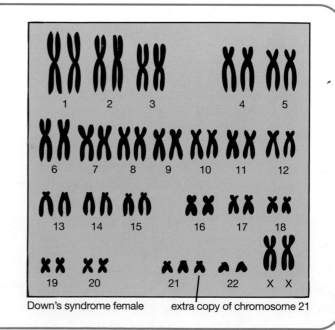

Down's syndrome female   extra copy of chromosome 21

Figure 16.3 Normal and Down's syndrome karyotypes

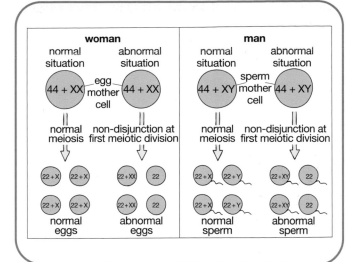

Figure 16.5 Non-disjunction of sex chromosomes

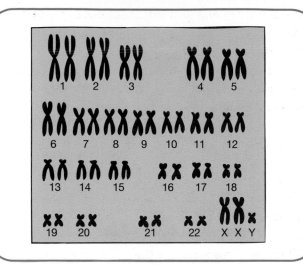

Figure 16.7 Klinefelter's syndrome karyotype

has the chromosome complement 2n = 45 (44 + XO where O represents the lack of a second sex chromosome). Figure 16.6 shows the karyotype.

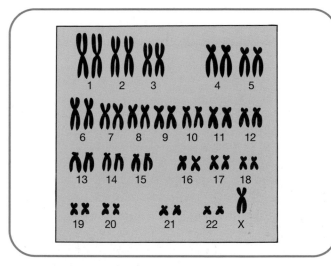

Figure 16.6 Turner's syndrome karyotype

A person with this unusual chromosome complement suffers a condition known as **Turner's syndrome**. Such individuals are always female and short in stature. They are infertile because their ovaries fail to develop normally.

Klinefelter's syndrome

If an XX egg (see Figure 16.5) is fertilised by a normal Y sperm or a normal X egg is fertilised by an XY sperm then the zygote formed has the chromosome complement 2n = 47 (44 + XXY). Figure 16.7 shows the karyotype.

A person with this unusual chromosome complement suffers a condition known as **Klinefelter's syndrome**. Such individuals are always male and possess male sex organs. However they are infertile since their testes only develop to about half the normal size and fail to produce sperm.

Complete non-disjunction and polyploidy

When all the spindle fibres in a gamete mother cell fail during meiosis and the members of all of the homologous pairs fail to become separated, this is called **complete non-disjunction**.

It results in a type of chromosome mutation which leads to the production of abnormal diploid gametes (each containing two complete sets of chromosomes instead of one set). Fertilisation involving these abnormal gametes results in the formation of mutant plants which possess extra complete sets of chromosomes. This type of chromosome mutation is called **polyploidy**.

Economic significance

Polyploid plants are normally larger than their diploid relatives. This often includes increased seed and fruit size which is of economic importance. Many commercially developed crop plants such as wheat, coffee, apples, tomatoes and strawberries (see Figure 16.8) are polyploid and therefore give bigger yields than their non-polyploid relatives.

Polyploid plants with an uneven number of sets of chromosomes are sterile. However this is useful to

123

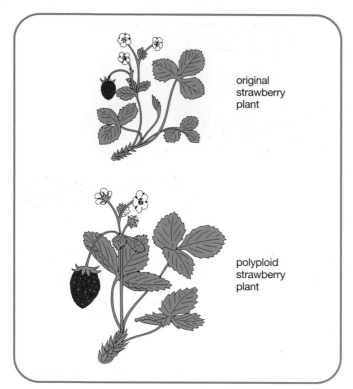

Figure 16.8 Polyploidy in strawberry plants

humans because the affected plants produce **seedless fruit** (e.g. banana).

Where a polyploid plant has arisen from more than one species, it is often found to show an **increase in vigour** and **resistance to disease** as a result of a combination of characteristics from both of its ancestors.

Normal and spelt wheat

Normal wheat (*Triticum aestivum* sub-species *vulgare*) and spelt wheat (*Triticum aestivum* sub-species *spelta*) are close relatives. They differ from one another in one important way. Normal wheat's grains split open easily during threshing and release their starchy contents; spelt wheat's grains fail to do so and are described as showing the 'spelt condition'. Compared with normal wheat, spelt wheat grains are more difficult to work with.

Both normal and spelt wheat are polyploid plants containing multiple sets of chromosomes (six sets of seven). Unlike their diploid relatives, they contain multiple copies of each gene. The ability of wheat grains to split open easily or not is governed by one gene. The three alleles of this gene are: Q (strong suppression of spelt condition), q (weak suppression of spelt condition) and q⁻ (no suppression of spelt

condition). It is thought that allele Q arose from allele q by mutation.

The presence of a single copy of Q or q is insufficient to suppress the spelt condition of the grain. The alleles have an **additive effect**. A minimum of two Q alleles or five q alleles is needed to suppress the condition and produce normal wheat grains that open easily. A genetic constitution containing five copies of the same allele (q) is only made possible by the multiple effect of **polyploidy**.

Testing Your Knowledge

1 With the aid of an example, distinguish clearly between the terms *mutation* and *mutant*. (2)

2 **a)** What name is given to the process by which a spindle fibre fails during meiosis and one or more of the gametes produced receives an extra chromosome? (1)

   b) How many gametes (out of four) will receive an extra chromosome if the spindle failure occurs during the first meiotic division? (1)

3 **a)** If a normal human sperm fertilizes an egg containing an extra copy of chromosome 21, what is the diploid number of the zygote formed? (1)

   b) i) What name is given to the condition suffered by a person who develops from an abnormal zygote of this type?

   ii) What relationship exists between age of mother and incidence of this condition? (2)

4 Rewrite the following sentences choosing the correct alternative from each bracketed choice.

   The sex of a person with Klinefelter's syndrome is (male/female). Such a person began life as a zygote containing the sex chromosomes (XXX/XXY) which could have resulted from an unusual (XX/XY) egg being fertilised by a normal (X/Y) sperm. (3)

5 **a)** What is meant by the term *polyploidy*? (1)

   b) Explain how a polyploid plant containing three separate sets of chromosomes could have arisen. (2)

   c) State TWO ways in which polyploid plants can be of economic importance to farmers. (2)

Change in structure of one chromosome

This type of mutation involves a change in the number or sequence of genes in a chromosome. Such a change is most likely to occur when chromatids break and rejoin during crossing over at meiosis. There are four different ways in which this can happen.

Deletion

The chromosome breaks in two places and the segment in between becomes detached as shown in Figure 16.9. The two ends then join up giving a shorter chromosome which lacks certain genes.

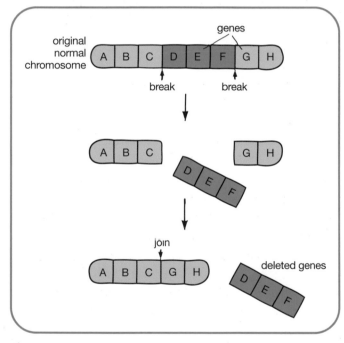

Figure 16.9 Deletion

Deletion normally has a drastic effect on the organism involved. In humans, for example, deletion of part of chromosome 5 leads to the *cri du chat* syndrome. The sufferer has severe learning difficulties and develops a small head with widely spaced eyes. (The condition is so-called because an infant sufferer's crying resembles that of a cat.)

Duplication

A chromosome undergoes this type of change when a segment of its homologous partner becomes attached to one end of the first chromosome or becomes inserted somewhere along its length as shown in Figure 16.10. This results in a set of genes being repeated.

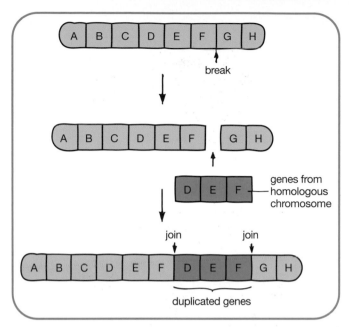

Figure 16.10 Duplication

In some cases, one of the duplicated genes may change (by mutation) and introduce a new characteristic which may or may not be of advantage to the organism.

Translocation

This involves a section of one chromosome breaking off and becoming attached to another chromosome which is not its homologous partner as shown in Figure 16.11.

This condition usually leads to problems during pairing of homologous chromosomes at meiosis. The gametes formed are often non-viable.

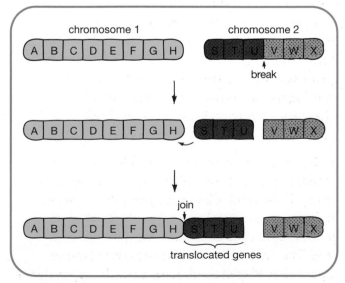

Figure 16.11 Translocation

Inversion

A chromosome undergoing **inversion** breaks in two places as shown in Figure 16.12. The segment between two breaks turns round before joining up again. This brings about a **reversal** of the normal sequence of genes in the affected section of chromosome.

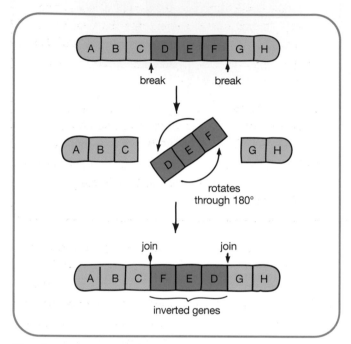

Figure 16.12 Inversion

When a chromosome which has undergone inversion meets its normal non-mutated homologous partner at meiosis, the two have to form a complicated loop in order to pair up. If crossing over then occurs within the loop, chromatids find it difficult or even impossible to become separated. **Non-viable** gametes are often the result.

Tunicate locus in corn

Podcorn (see Figure 16.13) is thought to be very similar to the ancestral form of maize (Indian corn). Each of its kernels (grains) is surrounded by a husk. This characteristic represents the phenotypic expression of the dominant allele Tu (tunicate means wrapped in a tunic) whose locus is situated on chromosome 4. Thus true-breeding podcorn has the genotype TuTu.

Popcorn is a closely related form of maize used in the confectionery industry. Each of its kernels lacks a husk. This represents the phenotypic expression of the recessive allele (tu) at the same locus. Thus true-breeding popcorn has the genotype tutu.

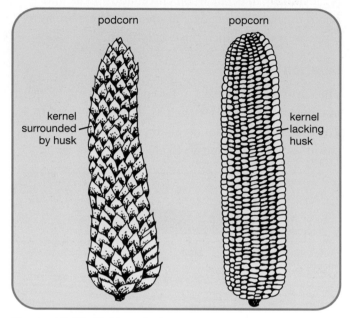

Figure 16.13 Podcorn and popcorn

The tunicate locus is now known to be a compound locus ('super-gene') consisting of three separate components (genes) thought to have arisen by **duplication**. Since each of the dominant components of the super-gene, Tu, contributes in part towards husk formation, it is possible to produce hybrids between pod- and popcorn which have partly formed husks (i.e. semi-tunicate).

1 State FOUR ways in which a chromosome can become changed in such a way that the number or sequence of some of its genes also becomes altered. (4)

2 a) What name is given to a change which involves a chromosome breaking in two places and a segment of genes dropping out? (1)

   b) Is this type of mutation likely to be beneficial or harmful to the organism affected? Explain why. (2)

3 a) Name the type of change which involves a chromosome breaking in two places and the affected length of genes becoming rotated through 180° before becoming reunited with the chromosome. (1)

   b) What effect does this change have on the sequence of the genes in the affected segment? (1)

4 a) What name is given to the type of chromosomal change which involves a segment of genes from one chromosome becoming inserted somewhere along the length of its homologous partner? (1)

   b) Why might this mutation be of benefit to the organism affected? (1)

5 a) Name the type of change which involves a section of one chromosome breaking off and joining onto another non-homologous one. (1)

   b) What effect does this change have on the number of genes present on each of the affected chromosomes? (1)

Gene mutations

Alteration of base type or sequence

This type of mutation (called **gene mutation**) involves a change in one or more of the **nucleotides** in a strand of DNA.

Four examples of gene mutation are shown in Figure 16.14. In each case, one or more codons for one or more particular amino acids have become altered, leading to a change in the protein that is synthesised.

For a protein to work properly, it must have the correct sequence of amino acids. Substitution and inversion are called 'point' mutations. They bring about only a minor change (i.e. one different amino acid) and sometimes the organism is affected only slightly or not at all. However if the substituted amino acid occurs at a critical position in the protein then a major defect may arise (e.g. formation of haemoglobin S and sickle cell anaemia).

Insertion and deletion are called 'frameshift' mutations. Each leads to a major change since it causes a large portion of the gene's DNA to be misread. The protein produced differs from the normal protein by many amino acids and it is usually non-functional. If

organism	characteristic controlled by normal gene	mutant characteristic resulting from gene mutation
fruit fly	long wing grey body red eye	wing yellow body white eye
human	normal blood clotting secretion of normal mucus in lung normal haemoglobin and biconcave red blood cells	haemophilia secretion of abnormally thick mucus which blocks bronchioles (cystic fibrosis) haemoglobin S and sickle-shaped red blood cells
mouse	brown coat	white coat (albino) lacking melanin pigment
ivy	green leaves containing chlorophyll	albino leaves lacking chlorophyll (see Figure 16.1)

Table 16.1 Mutant characteristics

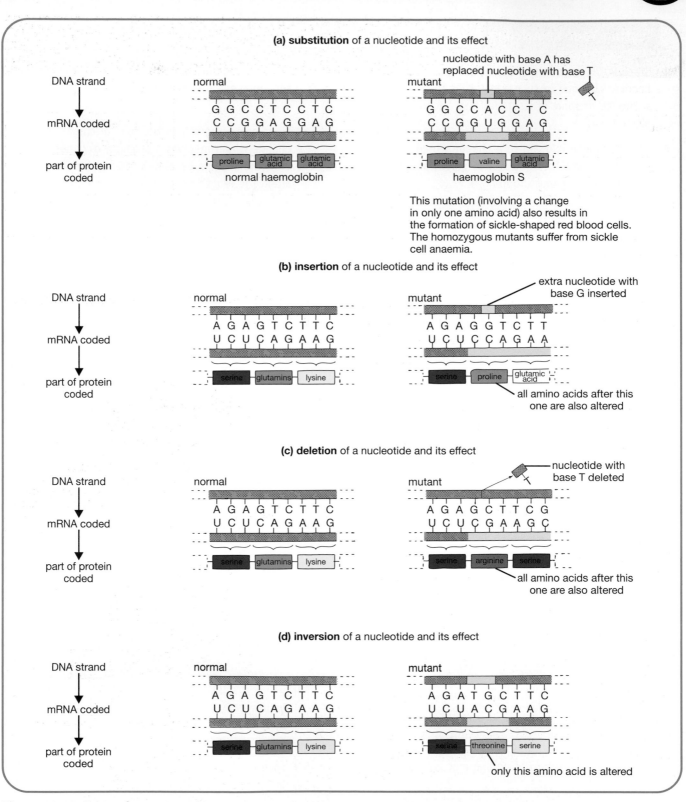

Figure 16.14 Types of gene mutation

organism	mutant characteristic	mutation rate (mutations at gene locus/million gametes)	chance of new mutation occurring
pneumonia bacterium	resistance to penicillin	0.1	1 in 10 000 000
fruit fly	ebony body white eye	20 40	1 in 50 000 1 in 25 000
mouse	albino coat	10	1 in 100 000
human	haemophilia muscular dystrophy	5 80	1 in 200 000 1 in 12 500

Table 16.2 Mutation frequency

such a protein is an enzyme which catalyses an essential step in a metabolic pathway, then the pathway becomes disrupted. An intermediate metabolite may accumulate and cause problems (see phenylketonuria – page 262).

Since most proteins are indispensable to the organism, most gene mutations produce an inferior version of the phenotype (see Table 16.1). If this results in death (an albino plant, for example, cannot photo-synthesise) then the altered gene is said to be **lethal**.

Frequency of mutation

In the absence of outside influences, gene mutations arise **spontaneously** and at **random** but only occur **rarely**. The mutation rate of a gene is expressed as the number of mutations that occur at that gene site (locus) per million gametes. Mutation rate varies from gene to gene and species to species as shown in Table 16.2.

The vast majority of mutant alleles are recessive. Therefore a newly formed mutant allele fails to be expressed phenotypically until two of these recessive alleles meet as a pair in some future generation. However a few mutant alleles are expressed by the first generation to inherit them because they are either dominant (e.g. achondroplasia, see Question 7 on page 131) or sex-linked (e.g. haemophilia).

Mutagenic agents

Mutation rate can be artificially increased by **mutagenic agents**. These include certain chemicals (e.g. mustard gas) and various types of radiation (e.g. gamma rays,

X-rays and UV light). The resultant mutations are said to be **induced**.

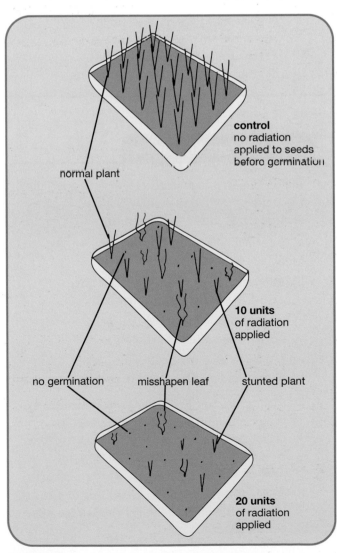

Figure 16.15 Effect of radiation on seeds

The effect of radiation on seeds is shown in Figure 16.15 where an increase in level of radiation is seen to bring about an increase in rate of gene mutation.

Mutation as a source of variation

Mutation is the only source of **new** variation. It is the process by which new alleles of genes are produced. Without mutation all organisms would be homozygous for all genes and no variation would exist.

Most mutations are harmful or even lethal. However on very rare occasions, there occurs by mutation a mutant allele which confers some **advantage** on the organism that receives it. Such mutant alleles (which are better than the originals) provide the alternative choices upon which natural selection can act. They are therefore considered to be the raw material of **evolution** (see chapter 17).

Testing Your Knowledge

1 a) Name FOUR types of gene mutation involving alteration to the base type or sequence in a DNA strand. (4)

   b) i) Classify the four types into 'frameshift' and 'point' mutations.

      ii) Which of these is more likely to lead to the formation of a protein that is greatly changed and non-functional?

      iii) Explain why. (4)

2 Give an example of a lethal mutation and explain why it is lethal. (2)

3 State TWO characteristics of *mutant* alleles with respect to their occurrence and frequency. (2)

4 Name TWO mutagenic agents. (2)

5 Are all mutant alleles inferior versions of the original wild-type allele? Explain your answer. (2)

Applying Your Knowledge

1 In the following four sentences, a small error alters the sense of the message. To which type of gene mutation is each of these equivalent?

   a) Intended: She ordered boiled rice.
      Actual: She ordered boiled ice.

   b) Intended: He walked to the pillar box.
      Actual: He talked to the pillar box.

   c) Intended: She untied the two ropes.
      Actual: She united the two ropes.

   d) Intended: He put a quid in his pocket.
      Actual: He put a squid in his pocket. (4)

2 a) Each of the homologous pairs of chromosomes shown in Figure 16.16 was seen under a microscope during meiosis in two cells. Which type of chromosome mutation had occurred in each case? (2)

   b) Which of these would be more likely to prove lethal to the organism that inherited the affected chromosome? (1)

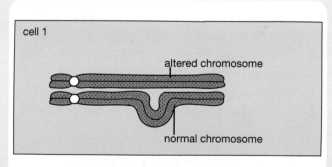

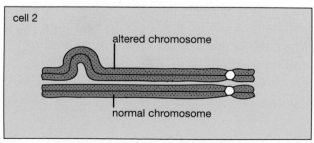

Figure 16.16

3 a) Refer to the graph in Figure 16.4 and estimate the chance of mothers of each of the following ages giving birth to a Down's syndrome baby:
   i) 20;   **ii)** 30;   **iii)** 40 years. (3)

b) Account for this trend. (1)

4 Figure 16.17 shows meiosis in a sperm mother cell. Although the first meiotic division is normal, one of the products is affected by non-disjunction during the second division.

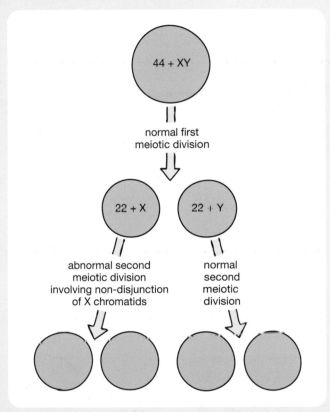

Figure 16.17

a) Using only numbers and letters as required, represent the four sperm that would be formed. (2)

b) Using the same convention, represent the zygote that would be formed if each of these sperm successfully fertilised a normal egg. (2)

c) What percentage of the zygotes in your answer to **b)** would develop into sufferers of Turner's syndrome? (1)

5 Write an essay on chromosome mutations. (10)

6 Although gene mutations are normally rare, the gene controlling colour of grains in maize mutates as often as once in 2000 gametes on average. Express this as a mutation rate. (See Table 16.2 for help.) (1)

7 Achondroplasia (a form of dwarfism in humans) is controlled by a dominant mutant allele (A). All achondroplastic dwarfs are heterozygous (Aa) since the homozygous condition is lethal. Imagine that each of the following sets of parents produce a dwarf child: 1) Aa × Aa; 2) Aa × aa; 3) aa × aa.

a) In which family must a gene mutation definitely have occurred? (1)

b) From such studies it is now known that mutation rate of this gene is 14 per million gametes. Express this figure as a 1 in _____ chance of a new mutation occurring. (1)

8 Describe the main types of gene mutation and their effects on amino acid sequences. (10)

What You Should Know

Chapters 12–16
(See Table 16.3 for Word bank)

alleles	genetic	rarely
backcrossed	independent	recombinant
base	linked	recombination
crossing	maps	same
different	meiosis	separated
evolution	mutagenic	sex-linked
frequency	mutations	sexual
gametes	number	variation

Table 16.3 Word bank for chapters 12–16

1 _____ reproduction is the means by which _____ variation is maintained in a population.

2 _____ is the process by which haploid _____ are formed.

3 During meiosis, new combinations of existing _____ arise by _____ assortment of chromosomes and _____ over between homologous chromosomes.

4 If an F_2 generation with a phenotypic ratio of 9:3:3:1 is obtained as a result of a dihybrid cross where the F_1 is selfed, the two genes involved must be located on _____ chromosomes.

However when the expected ratio of 9:3:3:1 is not obtained, this indicates that the two genes are located on the same chromosome (i.e. are _____).

5 If an F_2 generation with a ratio of 1:1:1:1 is obtained as a result of a dihybrid cross where the F_1 is _____ (testcrossed) to the double recessive, the two genes must be located on different chromosomes.

However when the expected ratio of 1:1:1:1 is not obtained, this also indicates that the two genes are located on the _____ chromosome (i.e. are linked).

6 Linked genes become _____ if crossing over occurs between them. This produces _____ gametes.

7 Since the distance between two linked genes is directly related to the _____ of recombination between them, _____ values can be used to construct gene _____ of chromosomes.

8 Genes present on an X but not on a Y chromosome are _____.

9 _____ are alterations in genotype which involve a change in structure or _____ of chromosomes or _____ type or sequence of a gene's DNA.

10 Mutations occur _____ and at random. Their frequency can be increased artificially by _____ agents.

11 Mutations are the only source of new _____ and provide the raw material for _____.

Natural selection

Historical background

In 1858 Charles Darwin and Alfred Wallace published a joint paper in which they suggested that the main factor producing evolutionary change is **natural selection**.

A year later Darwin amplified this view in his famous book *'The Origin of Species'*. Expressed in modern terms it states that:

- Organisms tend to produce **more offspring** than the environment will support.

- A **struggle for existence** follows and a large number of these offspring die before reaching reproductive age. This happens as a result of factors such as overcrowding, competition, lack of food, inability to escape predators and lack of resistance to disease (see Figure 17.1).

- Members of the same species are not identical but show **variation** in all characteristics (see Figure 17.2). Much of this variation is inherited.

Figure 17.2 Variation within a population of limpets

- Those offspring whose phenotypes are **better suited** to their immediate environment have a better chance of **surviving**, reaching reproductive age and passing on the favourable characteristics to their offspring.

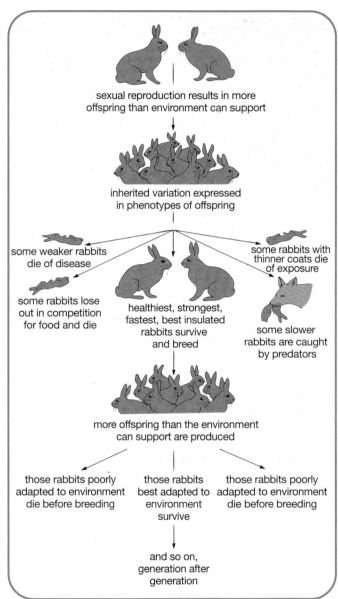

Figure 17.1 Natural selection in rabbits

- Those offspring whose phenotypes are **less well** suited to their immediate environment **die** before reaching reproductive age and fail to pass on the less favourable characteristics.

- This process is repeated generation after generation. The organisms with the phenotypes best suited to the environment are **'selected'** and survive and eventually predominate in the population. (This is often described as the **survival of the fittest**.)

- The organisms least well suited to the environment are 'weeded out' and perish.

Darwin called this 'weeding out' process **natural selection**. Since environmental conditions are constantly changing, natural selection is forever favouring the emergence of new forms.

Natural selection in action

Most mutations produce inferior versions of the original gene. However in each of the following examples of populations there occurs a **mutant allele** of a gene which allows adaptation to the changing environment. This allele gives the mutant form of the organism a **selective advantage**. The change in the environment may be brought about by an abiotic (non-living) factor such as atmospheric pollution or by a biotic (living) factor such as disease.

Industrial melanism in the peppered moth

Two forms of the peppered moth (*Biston betularia*) exist. One form is light brown with dark speckles; the other is completely dark (melanic) in colour. They differ by only one allele of the gene controlling the

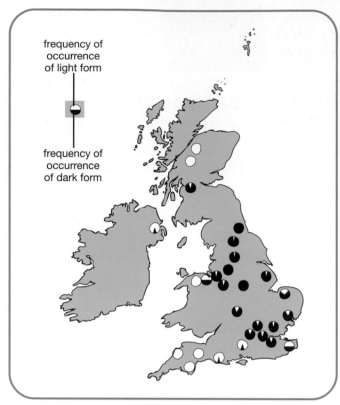

frequency of occurrence of light form

frequency of occurrence of dark form

Figure 17.3 Frequencies of two forms of peppered moth

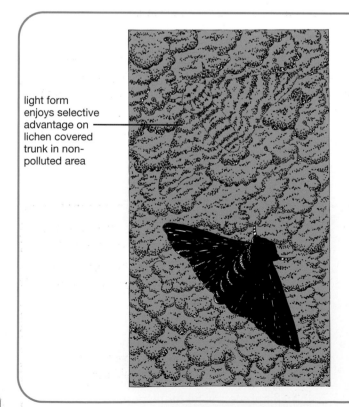

light form enjoys selective advantage on lichen covered trunk in non-polluted area

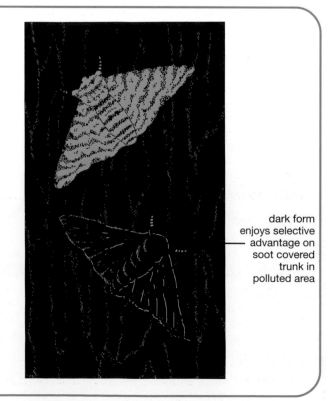

dark form enjoys selective advantage on soot covered trunk in polluted area

Figure 17.4 Natural selection in peppered moth

formation of dark pigment (melanin). Both forms of the moth fly by night and rest on the bark of trees during the day.

Prior to the industrial revolution in the 1800s, the light form was common throughout Britain and the darker form, which occasionally arose by mutation, was very rare indeed.

Surveys in the 1950s (see Figure 17.3) showed that the pale form was most abundant in non-industrial areas whereas the dark form was abundant in areas suffering from heavy industrial air pollution. Experiments and direct observations strongly support the following explanation of these findings.

In non-polluted areas, the tree trunks are covered with pale-coloured **lichens** and the light-coloured moth is well **camouflaged** against this pale background (see Figure 17.4). However the dark form is easily seen and eaten by predators such as thrushes.

In polluted areas, toxic gases kill the lichens and **soot** particles darken the tree trunks. As a result the light-coloured moth is easily seen whereas the dark one is well hidden and is favoured by natural selection.

Nowadays in polluted areas currently undergoing cleaning up campaigns in response to the Clean Air Acts, the pale form is being naturally selected at the expense of the melanic form which is losing its selective advantage. As the environment continues to change and more pale moths survive and breed, the situation which existed before the industrial revolution may return.

Sickle cell trait

Sickle cell anaemia is a genetically transmitted disease of the blood. It is caused by the presence of abnormal **haemoglobin S** which occurs as a result of a mutation (see page 128). In the discussion that follows, H represents the allele for normal haemoglobin and S the allele for haemoglobin S.

The blood of people homozygous (SS) for the mutant allele contains haemoglobin S (which is inefficient at carrying oxygen) and **sickle-shaped red blood cells** (see Figure 17.5). These stick together, interfering with blood circulation. This causes severe anaemia, damage to vital organs and, in the majority of cases, death.

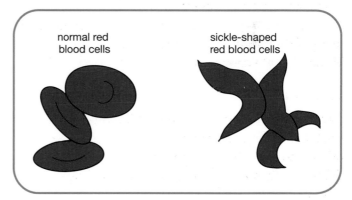

normal red blood cells

sickle-shaped red blood cells

Figure 17.5 Two types of red blood cell

Allele H is **incompletely dominant** to allele S. In heterozygotes (HS), allele S is therefore partially expressed. This milder condition, known as the **sickle cell trait**, is characterised by about a third of the person's haemoglobin being type S. However the red blood cells are normal and the slight anaemia that results does not prevent moderate activity.

Allele S is rare in most populations since it is semi-lethal. However in some parts of Africa up to 40% of the population are genotype HS. Comparison of the two maps in Figure 17.6 shows that a correlation exists between incidence of **malaria** and high frequency of allele S.

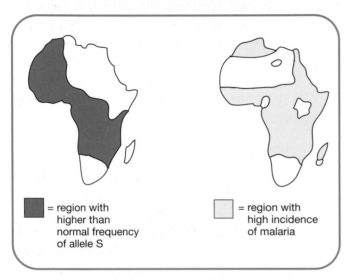

= region with higher than normal frequency of allele S

= region with high incidence of malaria

Figure 17.6 Correlation between allele S and malaria

This is because sickle cell trait sufferers are **resistant** to malaria whereas people with normal haemoglobin are not. Thus in malarial regions, natural selection favours people with genotype HS over those with HH who may die during serious outbreaks of the disease (see Table 17.1). However HS loses its selective advantage in non-malarial areas.

Resistance to antibiotics

Antibiotics have been in use in Britain for over 60 years. As each new antibiotic is introduced, it is found

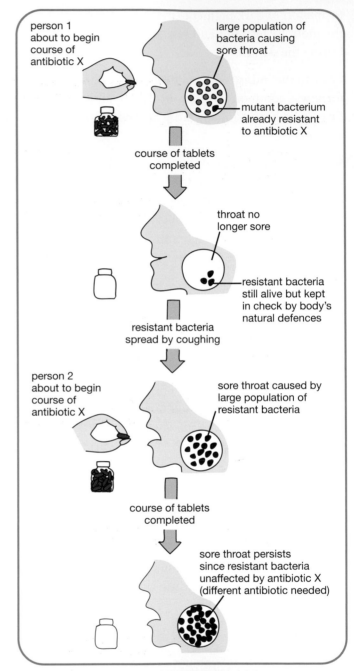

Figure 17.7 Spread of resistant bacteria

	cross		
	HH × HH	**HH × HS**	**HS × HS**
possible result in non-malarial region	HH, HH, HH, HH (100% of offspring survive)	HH, HH, HS, HS (100% of offspring survive)	HH, HS, HS, ~~SS~~ (75% of offspring survive)
possible result in malarial region affected by outbreak of disease	~~HH~~, ~~HH~~, ~~HH~~, ~~HH~~ (0% of offspring survive)	~~HH~~, ~~HH~~, HS, HS (50% of offspring survive)	~~HH~~, HS, HS, ~~SS~~ (50% of offspring survive)

Table 17.1 Genetics of sickle cell anaemia

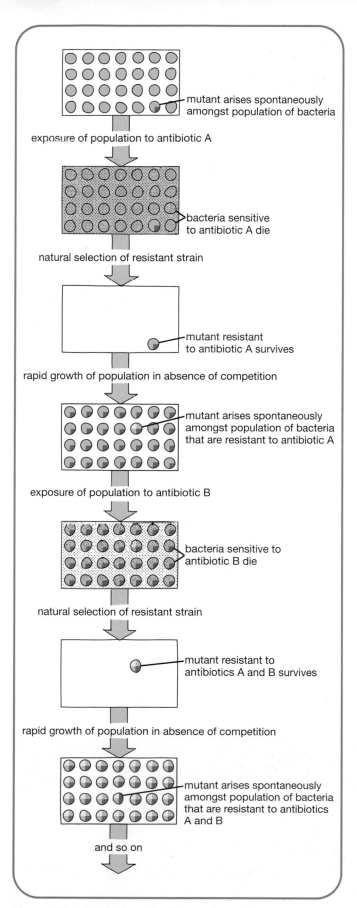

to be most effective against bacteria in its early years. Its effectiveness is found to decrease as the number of bacterial strains that are resistant to it increase and spread.

Micro-organisms such as bacteria occur in huge numbers. Genetic variation (which arises spontaneously by mutation) exists amongst the members of a population. As a result a mutant bacterium may arise that possesses the genetic material giving it resistance to a certain antibiotic without ever having been in contact with the antibiotic. Some disease-causing bacteria, for example, make the enzyme penicillinase which enables them to digest and resist penicillin. Figure 17.7 shows how resistant bacteria can spread from person to person.

High-speed evolution

The sequence of events that can lead to the evolution of mutant bacteria resistant to two different antibiotics is shown in Figure 17.8. As the process continues, some strains showing **multiple resistance** may continue to be favoured by natural selection. If these bacteria are also disease-causing, they will be very difficult to treat.

Conjugation

In addition to the transmission of the gene for antibiotic resistance from one generation of bacteria to the next by asexual reproduction, it is now known that

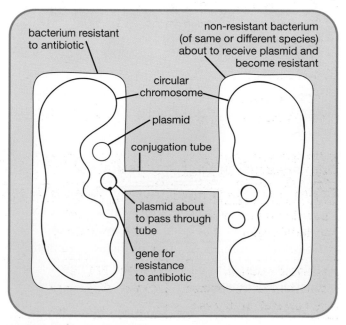

Figure 17.8 Evolution of resistant bacteria

Figure 17.9 Conjugation

drug resistance can be passed from one bacterium to another in a plasmid during the process of **conjugation** (see Figure 17.9).

It is therefore possible for a harmless bacterium that possesses the allele for antibiotic resistance to pass it to a disease-causing bacterium. Already the usefulness of many antibiotics has been greatly reduced by the existence of bacteria with multiple resistance.

'MRSA'

Staphylococcus aureus is a species of bacterium commonly found on the surface of the human skin. It causes one in five of the infections acquired by hospital patients during treatment. In the past it was successfully controlled by a range of antibiotics. However there now exists a strain of *S. aureus* known as MRSA (see Figure 17.10) which is **resistant** to all but one antibiotic. Experts claim that it is only a matter of time until a strain resistant to all known antibiotics appears. For this reason the search for new antibiotic-producing fungi and their products remains as important as ever.

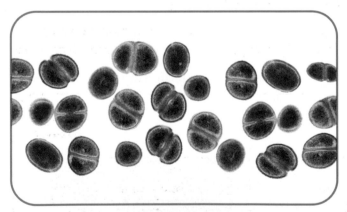

Figure 17.10 MRSA (Methicillin-resistant *Staphylococcus aureus*)

Non-medicinal uses of antibiotics

During intensive meat production, farm animals are often raised in crowded conditions which allow easy spread of infection. The addition of low levels of **antibiotics** to animal feed prevents disease, promotes growth and decreases the time needed to prepare animals for the market.

Unfortunately this practice creates conditions where antibiotic-resistant microbes thrive. If these bacteria

pass to human consumers in contaminated meat products, they may cause food poisoning which will fail to respond to certain antibiotics. Even if the resistant bacteria passed on in the meat are harmless, they can transfer the resistant gene to other harmful bacteria by conjugation. A strain of *Salmonella* is now known to exist which is resistant to **seven** different antibiotics. The use of antibiotics in animal feed is banned in many European countries.

Resistance to insecticides

DDT is a poisonous chemical which has been widely used against many insects. These include mosquitoes, which carry malaria and yellow fever, and insect pests which destroy crops. Within a few years of use, many mutant forms of insects resistant to the insecticide 'appeared'.

These had not arisen in response to the chemical. A tiny number of resistant mutants just happened to be present within the natural insect populations or arose later by a chance mutation.

When the spray was applied, the vast majority of non-resistant insects died and the resistant mutants suddenly enjoyed a **selective advantage** and multiplied. Under such circumstances natural selection enables the mutant to replace its wild type relatives.

Many new pesticides have been developed in recent years but strains of pest resistant to these chemicals are now emerging.

Heavy metal tolerance in grasses

Soil in the waste tips around mines where ores were once extracted contain high concentrations of **heavy metals**. These include copper and lead. Although such

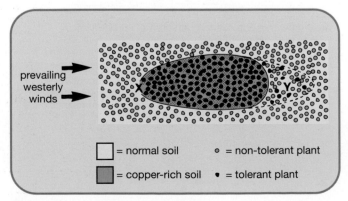

Figure 17.11 Heavy metal tolerance

	percentage survival	
	seeds from tolerant plants	seeds from non-tolerant plants
soil rich in copper	98	0
normal soil	23	89

Table 17.2 Competition experiment results

concentrations are normally toxic to plants and animals, some varieties of grass species are able to tolerate and colonise these polluted sites.

Figure 17.11 refers to two varieties of the grass *Agrostis tenuis* found growing around a Welsh copper mine. Table 17.2 shows results from a competition experiment where seeds from both types of this grass were grown together in the two types of soil.

From the results, it was concluded that the tolerant plants thrive on copper-rich soil where they face no competition from the non-tolerant plants. However the tolerant plants do not do nearly so well on normal soil where they have to compete with the normal non-tolerant plants.

It is now known that **copper tolerance** in *Agrostis tenuis* is an inherited characteristic (involving mutant alleles of several genes). This is a further example of natural selection in action. Only those plants possessing the genetic combination that results in copper tolerance survive on toxic soil and pass this characteristic on to their offspring which are in turn selected.

At point X (the upwind end of the mine), the boundary between the tolerant and non-tolerant populations is sharp because any pollen or seeds carrying non-tolerant genes blown over the boundary fail to produce tolerant plants.

In zone Y (beyond the downwind end of the mine) both varieties of plant occur. Although most of the seeds blown into this area are of the tolerant type, tolerant plants are found to be in the minority because they are unable to compete well with the non-tolerant variety when the soil's copper concentration is low.

Calcifuge and calcicole pairs

A group of plants which has a wide ecological range may evolve a pair of distinct ecotypes each adapted to a

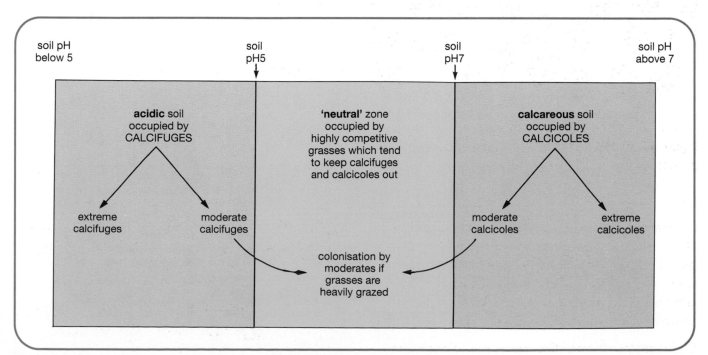

Figure 17.12 Distribution of calcifuges and calcicoles

specific habitat. **Calcifuges**, for example, are plants which grow best on acidic soils whereas their **calcicole** relatives grow best on calcareous (calcium-rich) soils (see Figure 17.12).

Since this difference in soil preference is genetically

determined, it is passed on to subsequent generations. Natural selection continues to favour calcifuges (e.g. *Viola palustris*, the marsh violet) on the acidic soils of marshes and calcicoles (e.g. *Viola hirta*, the hairy violet) on the calcareous soils of calcium-rich pastures.

Testing Your Knowledge

1 a) Distinguish between sickle cell *anaemia* and sickle cell trait. (2)

   b) As many as 40% of the people living in Central Africa are found to be heterozygous for the sickle cell condition. However only 4% of black Americans (descendants of slaves from this region of Africa) suffer the sickle cell trait. Explain why. (2)

2 a) How does a strain of bacteria resistant to an antibiotic arise? (1)

   b) Describe the means by which some types of bacteria are able to resist penicillin. (1)

   c) Give TWO methods by which a bacterium can pass on the genetic material for antibiotic resistance to other bacteria. (2)

   d) Give a possible long-term disadvantage of using antibiotics in feed to promote growth of farm animals. (2)

3 a) What advantage is gained by a type of grass plant that is tolerant to heavy metal in the soil near a disused mine? (1)

   b) Why are only a few tolerant plants found to be growing in the region of soil beyond the downwind end of the mine workings? (1)

Applying Your Knowledge

1 On average, a pair of foxes produce a litter of five cubs once a year. Since the offspring mature within a year and set up territories of their own, millions of foxes could be produced within a few years.

   Using the terms *over-production*, *competition*, *natural selection* and *variation* in your answer, explain why the world is not over-populated by foxes. (4)

2 Figure 17.13 shows how a mutant form of a species that possesses some advantageous characteristic could spread through a population. Imagine that before dying, each mutant form leaves, on average, two offspring as part of the next generation whereas each wild type leaves only one.

   a) Which symbol represents the mutant form? (1)

   b) Draw and complete a box to represent the F_2. State the ratio that applies to the members of its population. (2)

   c) Continue the series of diagrams until you can state the generation in which:

      i)   mutants outnumber wild type for the first time;

      ii) the ratio of wild type to mutants is 1:8. (2)

3 The graph in Figure 17.14 shows the results from a

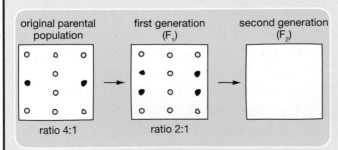

Figure 17.13

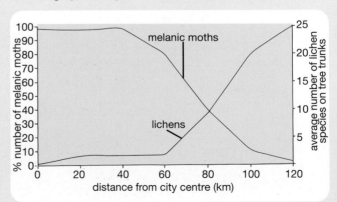

Figure 17.14

survey of lichens and melanic moths done in the vicinity of a large industrial city in the 1950s.

a) State the relationship that exists between the distribution of lichens and melanic moths. (1)

b) Explain in evolutionary terms why the melanic moth graph takes the form shown in Figure 17.12. (2)

c) Predict the form that a graph of the pale-coloured peppered moth numbers would have taken if drawn using the same axes. (1)

d) In the centre of the same city, the percentage number of melanic moths decreased from 95% in 1961 to 89% in 1974. Suggest why. (1)

4 Figure 17.15 shows an experiment carried out to demonstrate the process of natural selection in action in a population of bacteria.

a) Which lettered arrow in Figure 17.15 represents natural selection? (1)

b) What feature gave one of the strains of bacteria a selective advantage over the others? (1)

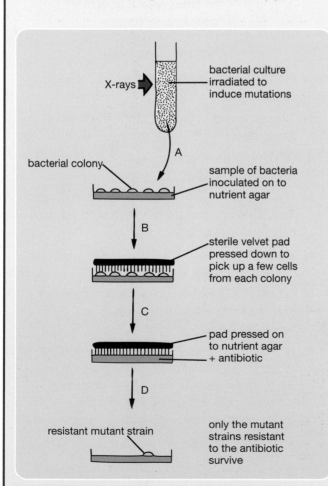

Figure 17.15

c) Did the mutant strain develop its resistance in response to the antibiotic or possess it before coming in contact with the antibiotic? (1)

5 The data in Table 17.3 refer to Scottish cases of a type of chest infection caused by a species of bacterium. Prior to 1980, the bacteria were sensitive to the antibiotic erythromycin.

year	percentage of reported cases successfully treated by erythromycin
1984	85
1986	74
1988	62
1990	48

Table 17.3

a) Account for the trend shown in Table 17.3. (1)

b) Predict what would probably have happened by the year 1998 if doctors had continued to prescribe erythromycin. (1)

c) By the year 1998, the percentage of reported cases successfully treated had risen to 96%.
   i) Suggest how doctors brought about this reversal in the trend.
   ii) Suggest why they were unable to achieve a 100% success rate. (2)

6 Many years ago, farmers in Peru began using vast quantities of pesticide containing DDT on their crops to control insects. They enjoyed record harvests for three successive years and then the problems began. Their crops suffered infestations of aphids and boll weevils which were unaffected by DDT treatment. Within five years the farmers were worse off than they were at the start.

a) Why were the aphids and boll weevils unaffected by DDT treatment after five years? (1)

b) Account for the increase in population numbers of the pests using the terms *natural selection*, *resistant strain*, *susceptible majority*, *pesticide* and *competition* in your answer. (5)

7 In the USA in the 1950s, some brands of toothpaste and chewing gum contained antibiotics. This practice has been discontinued.

a) What is the advantage in the short-term of adding antibiotics to toothpaste? (1)

b) What is the disadvantage in the long-term of continuing with this practice? (1)

8 Figure 17.16 shows the habitat profiles of two species of violet.

key to habitat profile

moisture content of soil	very wet	wet	damp	moist	dry
acidic or calcareous (calcium-rich) soil	very acidic	slightly acidic	neutral	slightly calc.	very calc.

habitat profile of heath dog violet

moisture content of soil	✗	✗	✓	✓	✓
acidic or calcareous (calcium-rich) soil	✓	✓	✓	✗	✗

habitat profile of early dog violet

moisture content of soil	✗	✗	✓	✓	✓
acidic or calcareous (calcium-rich) soil	✗	✗	✓	✓	✓

✓ = plant present   ✗ = plant absent

Figure 17.16

a) Which species is the
   i) calcicole;
   ii) calcifuge? (1)

b) i) Which type of violet could be found growing in a habitat with moist, slightly acidic soil?
   ii) Which of these types of violet would be commonly found on very wet, very acidic soil? (2)

c) i) Describe TWO features of a soil on which both of these types of violet could survive.
   ii) Suggest why, in practice, they are rarely found growing together in the wild on such a soil. (3)

9 Give an account of industrial melanism in the peppered moth. (10)

10 For many years, rats were successfully controlled by a poison called warfarin which interferes with the way that vitamin K is used in the biochemical pathway shown in Figure 17.17.

However in 1958 strains of rat appeared in Scotland that were resistant to warfarin. This resistance is an inherited characteristic as shown by the data in Table 17.4.

a) With reference to the diagram, explain why warfarin is lethal to normal rats. (1)

b) Why are people who suffer thrombosis (internal

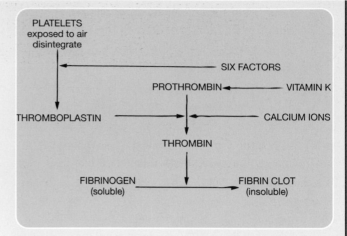

Figure 17.17

rat's genotype	rat's phenotype
W^{s}W^s	sensitive to warfarin
W^{s}W^r	resistant to warfarin (but needs some extra vitamin K in diet)
W^{r}W^r	resistant to warfarin (but needs 20 times normal amount of vitamin K in diet)

Table 17.4

blood clotting) given small amounts of warfarin to take? (1)

c) Suggest how the resistant strain of rat arose. (1)

d) Is allele W^r dominant or recessive to allele W^s? Explain your answer. (2)

e) The resistant strain increased greatly in number over the years. Explain this success in terms of natural selection. (1)

f) If use of warfarin is continued, predict the fate of the normal wild type rat. (1)

g) Construct a hypothesis to account for the fact that most rats resistant to warfarin are heterozygotes. (1)

h) i) Construct a diagram to show the outcome of crossing two heterozygotes.
   ii) Experts claim that the allele W^s cannot disappear completely from the rat population. Are they correct? Explain your answer. (2)

18 Speciation

Gene pool

The total of all the different genes in a population is known as the gene pool. The frequency of occurrence of an allele of a gene in a population (relative to all the other alleles at the same locus) is known as the gene frequency.

Species

A species consists of a group of organisms that share common anatomical and physiological characteristics and have the same chromosome complement and gene pool. This enables them to interbreed and form fertile offspring. They are reproductively isolated from other such groups (different species).

It is thought that there are about 5–10 million species on Earth at present. However species are not constant immutable units. Their number and kinds are always changing. At any given moment some species will be enjoying a stable relationship with the environment, some will be moving towards extinction (see chapter 20) and others will be undergoing speciation.

Speciation

Speciation is the formation of new species (usually many from a few). Evolution is the mechanism by which speciation is brought about and involves changes in genotype (and phenotype) of a population. These changes are adaptive making the organism better at exploiting the environment.

A simplified version of speciation is shown in Figure 18.1.

1   The members of a large population of a species occupy an environment. They share the same gene pool and interbreed freely.

2   The population becomes split into two completely isolated sub-populations by a barrier which prevents interbreeding and gene exchange. This isolating mechanism may be:

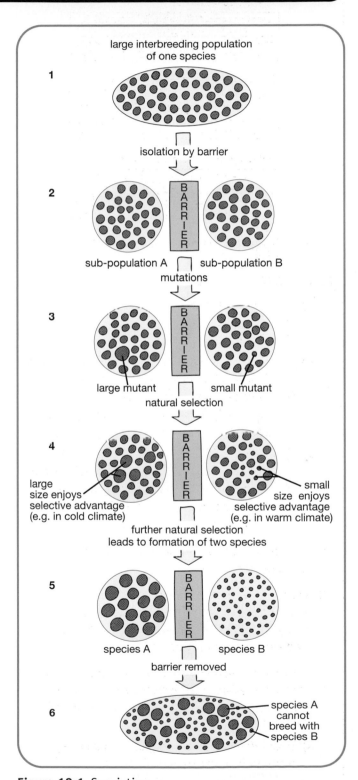

Figure 18.1 Speciation

- **geographical** (see Figure 18.2);

Figure 18.2 Geographical barriers

- **ecological** caused by changes in temperature, humidity, pH etc.;

- **reproductive**

  - different populations become sexually receptive at different times of the year,

  - members of different populations are not attracted to one another or fail to be stimulated by one another's courtship behaviour,

  - pollination mechanisms fail,

  - cross-fertilisation is prevented by physical non-correspondence of sex organs.

3 **Mutations** occur at random. Therefore most of the mutations that occur within one sub-population are different from those that occur within the other sub-population. This results in **new variation** arising within each group which is not shared by both groups.

4 The selection pressures acting on each sub-population are different depending on local conditions such as climate, predators, disease etc. **Natural selection** affects each sub-group in a

different way by favouring those alleles which make the members of that sub-population best at exploiting their environment.

5 Over a very, very long period of time, stages 3 and 4 cause the two gene pools to become so altered that the groups become genetically distinct and **isolated**.

6 If the barrier is removed, they are no longer able to interbreed (since their chromosomes cannot make homologous pairs). **Speciation** has occurred and two separate distinct species have evolved.

Island populations

When a particular type of organism is found only in a certain region (e.g. an island) it is said to be endemic to the region. Endemic species of plants and animals are found to occur on some of the Scottish Islands.

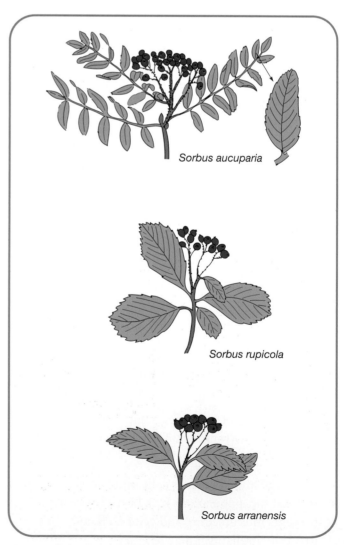

Figure 18.3 Three species of *Sorbus*

Sorbus

Sorbus aucuparia is the scientific name given to the mountain ash (rowan) tree which is commonly found throughout Europe especially on rocky mountainous soil but not on clay or limestone. *Sorbus rupicola* is a shrub (or rarely a small tree) found on crags amongst rocks especially on limestone soil.

Sorbus arranensis (Arran whitebeam) is a small slender tree endemic to the island of Arran where it grows on steep granite stream banks. It is believed to have arisen by hybridisation between *S. aucuparia* and *S. rupicola*. By growing in habitats intermediate to those favoured by its parents and lacking competitors found on the mainland, *Sorbus arranensis* has taken its own course of evolution and is different from both of its parents (see Figure 18.3).

European wren

Figure 18.4 refers to three subspecies of the European wren, *Troglodytes troglodytes*. Geographical isolation from the original mainland population and from one another has caused each island's wren population to take its own course of evolution. Although the populations have not been isolated for long enough to allow distinct species to arise, it is interesting to note that the subspecies on St Kilda (which is further from the mainland than the Hebrides) is already more different.

Similarly four different subspecies of Orkney vole are found on four different islands in the Orkney group. (See also Question 7 on page 148.)

Interruption of gene flow

An environment may be occupied by several populations of an organism which vary in their ability to interbreed as shown in Figure 18.5. At present these four populations belong to the same species because each population can breed with its immediate neighbour(s) allowing genes to flow from A to D via B and C. However if population B or C disappeared then gene flow would be interrupted and A and D would become genetically isolated and form *two distinct species*.

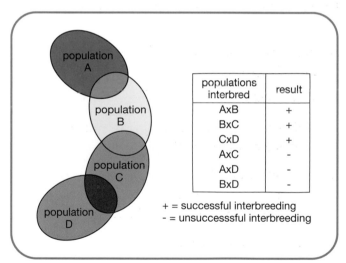

populations interbred	result
AxB	+
BxC	+
CxD	+
AxC	-
AxD	-
BxD	-

+ = successful interbreeding
- = unsuccesssful interbreeding

Figure 18.5 Interruption of gene flow

Continental distribution of organisms

The world's land mass can be divided into several distinct regions according to the types of animals that each possesses. Figure 18.6 shows only a small sample of the fauna (native animals) of each region.

When the full range for each region is considered in detail, it is found that the mammals of regions 1 and 2 are very similar to one another, those of regions 3, 4

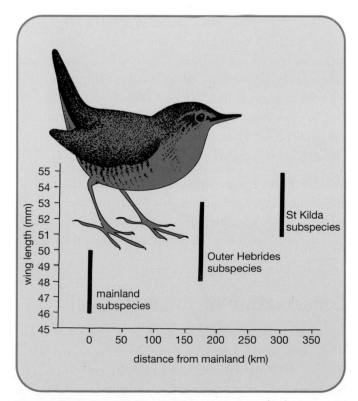

Figure 18.4 Subspecies of wren

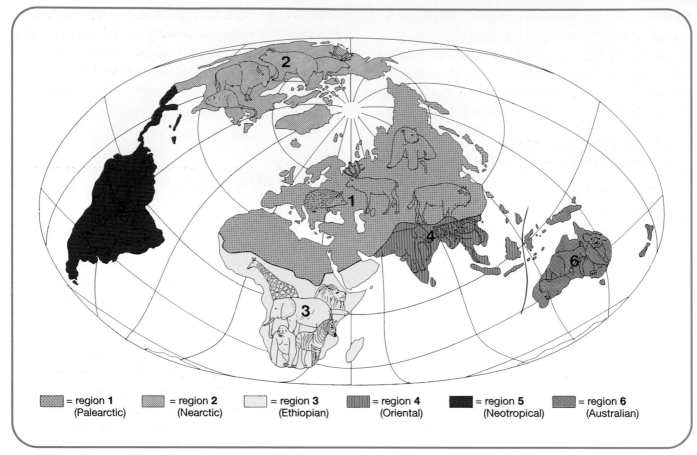

Figure 18.6 Continental distribution of mammals

and 5 fairly different, and those of region 6 very different from all of the others.

Such **continental distribution** of mammals is thought to have occurred as a result of **continental drift**. The world's land masses are thought to have been joined together many millions of years ago (see Figure 18.7) and then to have gradually drifted apart.

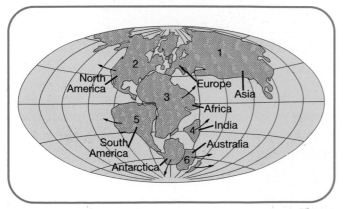

Figure 18.7 Continental drift

It is thought that ancestral stocks of animals originated in the northern hemisphere and that some of these migrated to southern continents still connected by land bridges. The populations became cut off from one another by geographical barriers such as **water** (e.g. Indian Ocean and other seas isolating region 6 from 4), **mountains** (e.g. Himalayas isolating region 4 from 1) and **deserts** (e.g. Sahara isolating region 3 from 1).

Regions 1 and 2 with their very similar animals are thought to have become separated only relatively recently. Region 6 with its very different animals has been isolated from the other land masses for the longest time.

Conservation of the marsupials

Whereas the most advanced group of mammals, the **placentals**, retain their young in the uterus until a late stage of development, the more primitive marsupials bear their young at a very early stage and rear them in a pouch.

Since there are no native placentals in Australia, it is thought that only marsupials reached region 6 (Australia) via region 5 (South America) and Antarctica before separation, of the land masses. When the more successful group, the placentals, arrived later in region 5 only a few marsupials survived the competition.

However in Australia, in the absence of placentals, the marsupials were able to diversify and fill every available **ecological niche** (see page 150). In this case, the isolating barrier (water) led therefore to the conservation of the rich assortment of Australian marsupials that survive to this day.

Testing Your Knowledge

1 a) Give THREE characteristics shared by all the members of a species. (3)

   b) Explain what is meant by the statement 'Species are not constant immutable units.' (2)

2 Distinguish between the terms *speciation* and *evolution*. (2)

3 a) Give TWO examples of each of the following isolating mechanisms which split a population into separate groups during speciation:
   i)   reproductive;
   ii) geographical;
   iii) ecological. (3)

   b) Identify TWO further processes that must take place before speciation is complete. (2)

4 a) What name is given to the process by which the world's land masses are thought to have gradually moved apart over a long period of time? (1)

   b) Refer to Figure 18.6 and then state the type of barrier that isolates: **i)** region 3 from region 1;

   ii) region 4 from region 1; **iii)** region 6 from region 4. (3)

   c) **i)** Which region in Figure 18.6 has been isolated from the others for the longest time?
      ii) What effect has this long period of isolation had on the types of species that have evolved there compared with other regions? (3)

   d **i)** Which TWO regions in Figure 18.6 have become separated only relatively recently?
      ii) On 18th January 1987, an American man successfully jogged across the five miles of frozen Bering Sea between Alaska and Siberia. Does this information support your answer to part **i)**? Explain why?
      iii) What effect has the recent separation of the two regions you gave as your answer to **i)** had on the types of species that have evolved there? (3)

Applying Your Knowledge

1 When a horse is crossed with a donkey, the result is a sterile animal called a mule. Do a mule's parents belong to the same species? Explain your answer. (2)

2 Arrange the following stages into the correct order in which they would occur during speciation:

   A isolation of gene pools
   B formation of new species
   C mutation
   D occupation of territory by one species with one gene pool
   E natural selection (1)

3 Figure 18.8 shows 12 breeds of dog.

Figure 18.8

147

a) How many species are shown in the diagram? Explain your answer. (2)

b) What type of isolating mechanism prevents breeds 2 and 11 from reproducing? (1)

c) Imagine that the 12 breeds comprised the entire dog population of the world and that breeds 3, 4, 5, 6, 7, 8, 9 and 10 became extinct. How many species would result if the remaining dogs were left to breed naturally? Explain your answer. (2)

4 *Cerastium arcticum* is the scientific name for mouse-ear chickweed which is a common plant on the Scottish mainland. *Cerastium arcticum edmondstonii* is a subspecies of mouse-ear chickweed endemic to the Shetland Isles and found nowhere else in the world.

a) Suggest how this subspecies originated. (2)

b) Under what conditions may it evolve into a separate species of *Cerastium?* (1)

5 Give a possible explanation for each of the following observations.

a) When Australia was first discovered by European explorers in the seventeenth century, it possessed a rich variety of marsupial mammals but the only placentals were man (Aborigine) and the semi-domesticated dog (dingo). (2)

b) When New Zealand was discovered at about the same time, it possessed no marsupials. The only mammals were man (Maori) and two species of bat. (2)

6 Write an essay on the process of speciation including reference to the part played by isolating mechanisms. (10)

7 The data in Table 18.1 and Figure 18.9 refer to three subspecies of the fieldmouse (*Apodemus sylvaticus*).

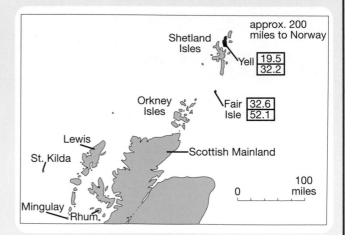

Figure 18.9

Some scientists suggest that the fieldmouse became established on the Shetland Isles and Fair Isle millions of years ago while these islands were attached to the mainland. Following separation, each island population then took its own course of evolution.

However this theory is now disputed by the work of Professor R.J. Berry. From detailed studies of the skeletons of different subspecies of *Apodemus*, this scientist has devised a way of measuring how close or distant the relationship is between two subspecies. This is expressed as 'genetic distance'

scientific name			location	average body length (mm)	average tail length (mm)	ventral colour	pectoral spot
genus	species	subspecies					
Apodemus	sylvaticus	sylvaticus	Scottish Mainland	92	83	dull white with slate-grey throat and belly	buff coloured area between front legs
Apodemus	sylvaticus	granti	Yell (Shetlands)	101	88	dull bluish-white	small spot between front legs
Apodemus	sylvaticus	fridariensis	Fair Isle	113	99	dull bluish-white	normally absent

Table 18.1

(the higher the value, the more distant the relationship).

The genetic distances of *Apodemus sylvaticus granti* and *Apodemus sylvaticus fridariensis* are given as boxed figures on the map. The upper figure indicates the genetic distance between the island population and the Norwegian mainland species and the lower figure that between the island population and the Scottish mainland species.

a) State ONE quantitative and ONE qualitative difference between the fieldmouse native to Fair Isle and the one from the Scottish mainland. (2)

b) Suggest why scientists do not classify the three subspecies given in the table as separate species. (1)

c) i) Berry suggests that the first population of *Apodemus* became established on Yell after being brought from Norway on Viking ships. What evidence from the data supports this theory?

   ii) Fossil evidence shows that *Apodemus* became extinct in Britain during the last ice age (about a million years ago). When the ice melted, the Shetland Isles became separated from the mainland and have remained so ever since. Does this information lend support to or cast doubt on Berry's theory? Give a reason for your answer. (3)

d) St Kilda and several Hebridean Islands (e.g. Lewis, Mingulay and Rhum) each possesses its own distinct subspecies of *Apodemus*. Suggest how this could have come about. (2)

19 Adaptive radiation

Ecological niche

An organism's **ecological niche** is the role that it plays within a community. Although often applied solely to the organism's mode of feeding, this term refers to its whole way of life including important factors such as habitat, competitors, enemies and the use that it makes of the resources present in its environment.

Darwin's finches

In 1835 Darwin visited the Galapagos Islands which lie about 600 miles west of the South American mainland. He found them to be inhabited by many different species of finch which varied greatly in **beak size** and **shape** according to diet (see Figure 19.1). On the other hand he found the mainland to have only one species of finch. It is thought therefore that the islands were colonised by a flock of the mainland species carried there by freak weather conditions.

On the mainland, the one species of finch was unable to exploit ecological niches already occupied by other birds (e.g. woodpeckers, warblers etc.) However the finches that arrived on the Galapagos found themselves in an environment lacking competitors and offering a wide range of unoccupied ecological niches. Reduced selection pressure enabled these finches to increase in number and occupy the islands.

Sub-populations became isolated by barriers (see page 143) and speciation occurred. Each group underwent its own course of evolution by becoming **diversified** and **adapted** to suit an available ecological niche (see Figure 19.2).

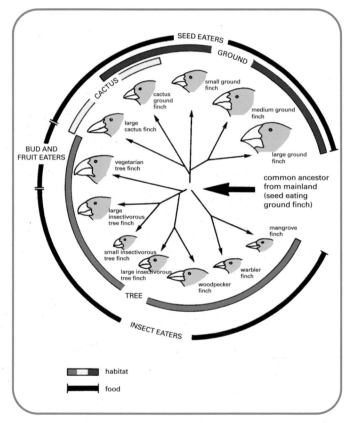

Figure 19.2 Adaptive radiation in finches

Figure 19.1 Darwin's finches

Adaptive radiation

This evolution of a group of related organisms along several different lines by adapting over a long period of time to a wide variety of environments is called adaptive radiation. It promotes the optimum use of many different types of environment by the various species that evolve and reduces competition between them.

Marsupials

In the absence of competition from placentals, the early marsupial colonists of Australia were able to undergo adaptive radiation. Over a long period of time the variety of types present today evolved (see Figure 19.3). Similar ecological niches in other ecosystems are occupied by placentals as shown in Table 19.1.

British buttercups

Table 19.2 refers to three species of buttercup. Although two or more of these species may be found growing on the same ploughed field or ridged grassland, each is restricted to its own particular habitat as shown in Figure 19.4.

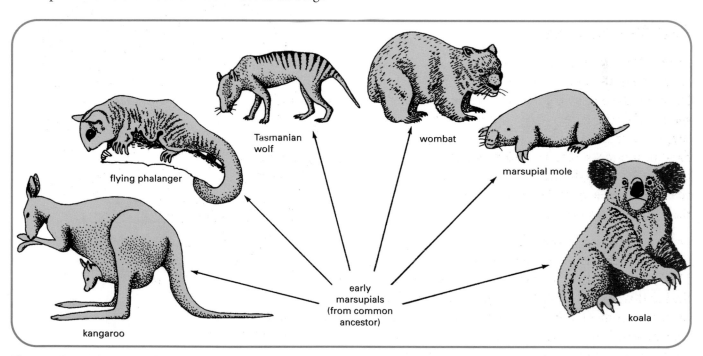

Figure 19.3 Adaptive radiation in marsupials

marsupial	ecological niche		example of placental that occupies similar ecological niche in another ecosystem
	habitat	mode of feeding	
kangaroo	grassland	herbivorous (grass)	deer
koala	eucalyptus and gum trees	herbivorous (leaves)	sloth
flying phalanger	trees	omnivorous (fruit and insects)	flying squirrel
marsupial mole	underground burrow	insectivorous (insects and worms)	European mole
Tasmanian wolf	rocky terrain	carnivorous (flesh of birds and mammals)	wolf
wombat	underground burrow	herbivorous (grass)	capybara

Table 19.1 Marsupial and placental occupants of ecological niches

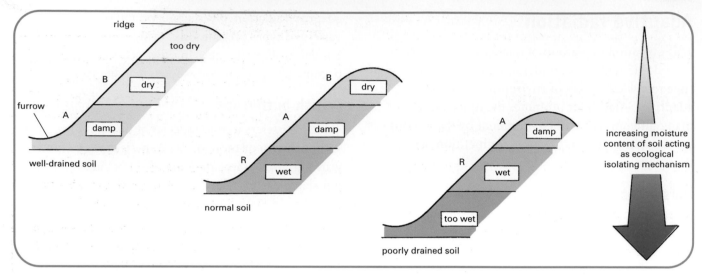

Figure 19.4 Distribution of three buttercup species

common name	bulbous buttercup	meadow buttercup	creeping buttercup
scientific name	*Ranunculus bulbosus*	*Ranunculus acris*	*Ranunculus repens*
diploid chromosome number (2n)	16	14	32
symbol in Figure 19.4	B	A	R
habitat	dry well-drained soil	damp soil	heavy wet soil

Table 19.2 Three species of British buttercup

The fact that each of these closely related species of buttercup has become adapted to suit one particular ecological niche provides evidence that they have undergone adaptive radiation from a common ancestor. Each species has evolved along a separate line and no longer competes with related species.

Homology and divergent evolution

Structures are said to be **homologous** if they have the same evolutionary origin (common ancestor) and are structurally alike. They need not perform the same function.

A group of related organisms that have undergone adaptive radiation are found to possess homologous structures. Mammals (and other terrestrial vertebrates), for example, all possess the **pentadactyl** (five digit) limb. In the course of evolution it has become adapted to suit different functions (see Figure 19.5). This process is called **divergent evolution** (divergence).

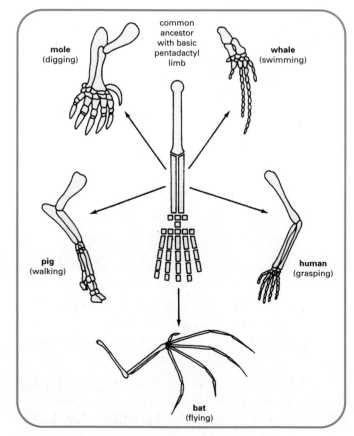

Figure 19.5 Homologous pentadactyl limbs

Like fossils, the existence of homologous structures provides powerful evidence to support the theory of evolution.

Convergent evolution

Some animals possess similar structural features and may even resemble one another fairly closely yet are found to have arisen from **different** evolutionary origins.

The three burrowing mammals referred to in Table 19.3 all have powerful digging claws and are similar in appearance. However they have evolved along separate evolutionary lines. By becoming adapted over a long period of time to occupy similar ecological niches in different parts of the world, they have coincidentally evolved very similar features. Examples such as these are described as products of **convergent evolution** (**convergence**).

common name	reproductive type	native continent
European mole	placental	Europe
strand mole	placental	Africa
marsupial mole	marsupial	Australia

Table 19.3 Convergent evolution of moles

Testing Your Knowledge

1   **a)** What is meant by the term *ecological niche*? (2)
    b) i) Describe the ecological niche of the mainland ancestor of Darwin's finches.
       ii) Suggest why this common ancestor was unable to exploit a wide variety of ecological niches in the forests of mainland South America.
       iii) Why were the Galapagos finches able to exploit several ecological niches? (4)

2   **a)** What name is given to the process by which a group of organisms evolve along several different lines by becoming adapted to different environments? (1)
    b) i) Name TWO marsupials and describe TWO aspects of the ecological niche of each one.
       ii) For each of these marsupials, name a placental that occupies an equivalent niche in another ecosystem.
       iii) Why do these placentals not occupy the Australian ecological niches? (7)

3   **a)** Figure 19.4 shows the dry, damp and wet habitat of buttercup plants. Following adaptive radiation, how many species of buttercup are found to occupy each habitat? (1)
    b) Why is this of advantage to the three species of buttercup? (1)

Applying Your Knowledge

1   Figure 19.6 refers to seven species of honeycreeper birds that live on the islands of Hawaii.

    a) Explain how the information in Figure 19.6 illustrates adaptive radiation. (2)

    b) Suggest why the seven different species failed to evolve on the mainland. (1)

    c) What type of barrier isolated the different groups during speciation? (1)

    d) Give ONE feature typical of a beak adapted to a

diet of: **i)** nectar; **ii)** insect larvae extracted after cracking open hard plant material. (2)

    e) From the information given, state how many of the species of honeycreeper are:
    i) herbivorous;
    ii) carnivorous;
    iii) omnivorous. (1)

    f) What does each species of honeycreeper gain by being a specialised feeder? (1)

→

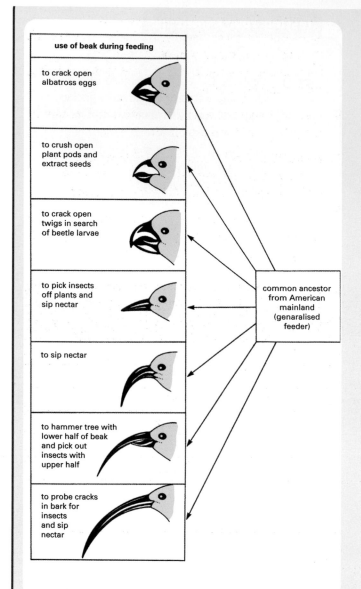

use of beak during feeding

to crack open albatross eggs

to crush open plant pods and extract seeds

to crack open twigs in search of beetle larvae

to pick insects off plants and sip nectar

to sip nectar

to hammer tree with lower half of beak and pick out insects with upper half

to probe cracks in bark for insects and sip nectar

common ancestor from American mainland (genaralised feeder)

Figure 19.6

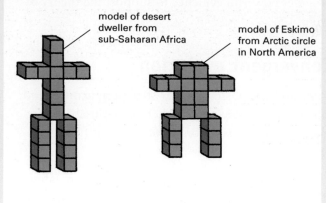

model of desert dweller from sub-Saharan Africa

model of Eskimo from Arctic circle in North America

Figure 19.7

i) Explain the meaning of the term adaptive radiation with reference to the above examples.

ii) Why is it unlikely that humans will evolve into separate species? (3)

3 Unlike a tiger's fang-like canine teeth, those of a walrus take the form of tusks (see Figure 19.8) used for digging up shellfish. Like human canine teeth, those of the tiger and walrus are composed of enamel, dentine and cement.

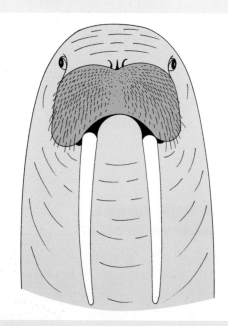

Figure 19.8

2 a) i) Calculate the total volume of each of the human models shown in Figure 19.7.

ii) Calculate the total surface area of each of the models.

iii) Relate the difference between the two to the environment in which each lives. (4)

b) It is thought that the first humans evolved in a climate much less extreme than either of the above two and that they were intermediate in structure to the above two varieties of humankind.

a) Copy and complete Table 19.4. (2)

animal	description of canine tooth	function
	tusk	
		stabbing prey
human		

Table 19.4

b) Do these teeth provide an example of convergent or divergent evolution? Explain your answer. (2)

4 The marsupial wolf of Australia (now nearing extinction) is almost identical in body shape and life style to the placental wolf of the northern hemisphere. However they are not regarded as being close relatives. Explain why. (2)

5 Give an account of adaptive radiation and its role in the process of speciation. (10)

6 Figure 19.9 shows a map of the Galapagos Islands. Each figure in brackets refers to the number of different species of finch found on the island.

Table 19.5 gives further information about six of the islands. Table 19.6 and Figure 19.10 refer to ground-living finches on two of the islands.

name of island	relative size of island	location of island relative to centre of group	percentage number of finch species only found on this island
Culpepper	small	distant	75
Pinta	small	distant	33
Española	small	distant	67
Isabella	large	central	22.2
Santa Cruz	medium	central	0
Santa Fe	small	central	14.3

Table 19.5

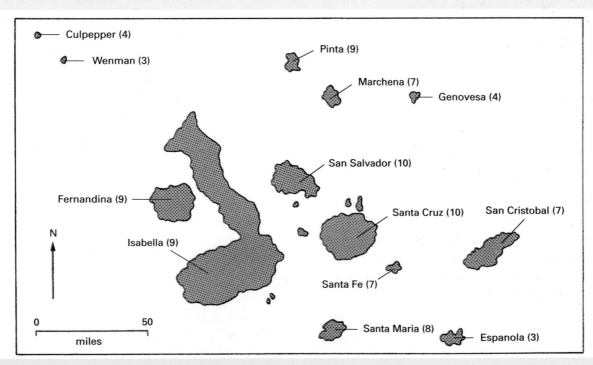

Figure 19.9

species of ground finch		island	
body size	food	Culpepper	Pinta
small	cactus	present (11.3 × 9.0)	present (9.7 × 8.5)
large	cactus	absent	present (14.6 × 9.7)
medium	seeds	present (15.0 × 16.5)	absent
large	seeds	absent	present (16.0 × 20.0)

Table 19.6 (number in brackets = length × depth of beak in mm)

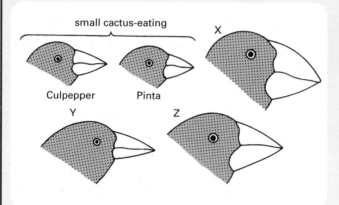

Figure 19.10

a) Calculate the number of finch species which are found only on Culpepper and on no other island. (1)

b) One of the islands has two species of finch which are found on it and nowhere else. Identify the island. (1)

c) Which island is populated by finch species which are all found on other islands? (1)

d) i) In general, what relationship exists between percentage number of finch species found on only one island and that island's geographical location relative to the centre of the group?

   ii) It has been suggested that this relationship could be dependent upon island size. Why is this unlikely?

   iii) Suggest a more likely explanation to account for the relationship. (4)

e) Identify birds X, Y and Z using the information in Table 19.6 (1)

f) i) If a large number of finch type Z were transported from Culpepper to Pinta, which native type on Pinta would face most competition?

   ii) Predict the possible outcome over a long period of time. (2)

20 Extinction and conservation

Extinction of species

Ever since life evolved on Earth, new species better suited to the environment have appeared and older less successful forms have died out. **Fossil evidence** shows that life on Earth has been punctuated by several waves of **mass extinction**.

The graph in Figure 20.1 shows the pattern of mass extinction that is thought to have occurred over a period of more than 250 million years. At point 1 on the graph, a major group of animals called the trilobites vanished completely, the corals came close to extinction and the fish were severely affected. Such waves of mass extinction are closely related to changes in **global climate** that have occurred. During an ice age, for example, sea levels drop and vast areas of frozen land appear. As a result, many animals native to warm, sheltered, marine waters and estuaries perish.

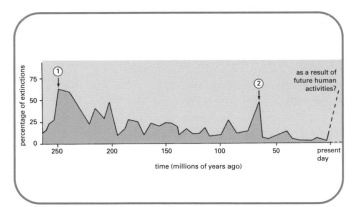

Figure 20.1 Waves of mass extinction

When the environment becomes disrupted, species that were previously successful may suddenly find themselves, at a major disadvantage and become extinct. Other species which had been enjoying minimum success in a hostile environment may find themselves well suited to the new environment and evolve rapidly.

Mammals, for example, existed as a relatively insignificant group during the age of the dinosaurs and only became the Earth's dominant life form when some major change in the environment, such as global cooling, caused the dinosaurs to die out (point 2 on the graph). Unlike 'cold-blooded' (ectothermic) dinosaurs, mammals are 'warm-blooded' (endothermic) animals able to withstand extremes of temperature in their surroundings.

Their successful expansion was further promoted by the development of **milk** for their young and the improved level of **parental care** that is associated with suckling.

Effect of human activities

In an attempt to satisfy the demands of the ever increasing population, humans chop down, plough up, dam and often pollute natural habitats (see Figure 20.2). These activities are causing the current wave of extinction to run at a level many times greater than the natural rate. Since the early seventeenth century, hundreds of species of birds and mammals have been wiped out by **over-hunting** and **habitat destruction**. Figure 20.3 shows a few of the many species now threatened with extinction.

"Hello! We can't be far from civilisation."

Figure 20.2 "Hello! We can't be far from civilisation."

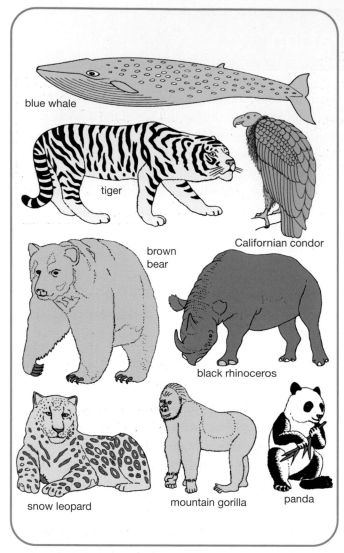

Figure 20.3 Endangered species

In addition to these 'high profile' animals, a much larger number of plants and invertebrate species (e.g. insects) are in critical danger as humans continue to destroy ecosystems such as tropical rain forest. It is estimated that *Homo sapiens* had already wiped out one million species by the end of the twentieth century.

If this process of destruction continues, the future will be bleak. Species necessary for mankind's survival will be lost forever and many ecological niches will become populated by **opportunist species** such as cockroaches, rats and weed plants.

Genetic diversity (biodiversity)

The Earth still supports an extraordinary variety of interdependent life forms upon which natural selection has been acting for millions of years. Each member of a species possesses hundreds or even thousands of genes. Since two or more alleles exist for most genes, the number of genetic permutations possible amongst the members of a species is enormous.

This potential for **genetic diversity (biodiversity)** amongst its living inhabitants, is one of our planet's most valuable resources.

Need for conservation of species

Humans are able to destroy in a decade what has taken millions of years to evolve. At last we have begun to appreciate that nature is not inexhaustible. It is essential that other species are **conserved** for their genetic potential to provide mankind with future food, fuel, medicines and raw materials. In addition we must have a supply of **alternative alleles** to introduce into domesticated varieties in order to make them more vigorous and more resistant to disease in years to come.

Protecting genetic diversity

On-site

On-site protection means identifying an ecosystem containing a valuable gene pool and trying to maintain it in its natural state. This could mean making it into a

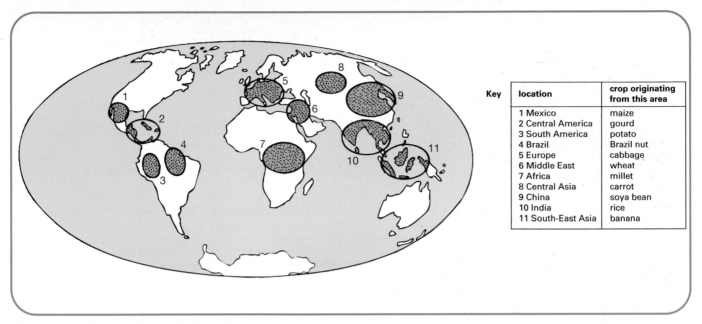

Figure 20.4 Centres of genetic diversity

Key	location	crop originating from this area
	1 Mexico	maize
	2 Central America	gourd
	3 South America	potato
	4 Brazil	Brazil nut
	5 Europe	cabbage
	6 Middle East	wheat
	7 Africa	millet
	8 Central Asia	carrot
	9 China	soya bean
	10 India	rice
	11 South-East Asia	banana

wild life reserve or a national park in an attempt to conserve its species.

Figure 20.4 shows regions of the world called **centres of diversity** from where modern day food plants originated. Since these regions still possess wild relatives of our food plants, they are prime targets for conservation.

Off-site

Off-site protection means identifying a potentially valuable species and maintaining it in an artificial environment.

This may take the form of a **rare breed farm** where varieties of farm animal (or their ancestors) which have otherwise become extinct are bred to provide a reservoir of alternative genes (see Figure 20.5).

Conservation of species by **captive breeding** is also carried out in zoos which have helped to save several animal species threatened with extinction. Such carefully controlled breeding programmes often involve **zoo exchanges** and **artificial insemination** which help to prevent continuous inbreeding.

Seeds and sex cells of rare, threatened and valuable species can be stored in **cell (gene) banks** (see Figure 20.6). In Norway a gene bank has been built 120m deep inside a mountain on an Arctic island. The site is permanently frozen and high enough up to remain unaffected by any future rise in sea level. This seed

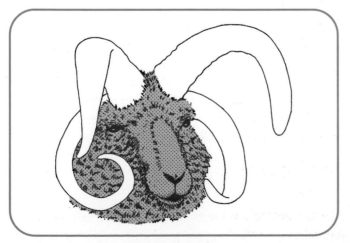

Figure 20.5 Four-horned Manx Loghtan ram

Figure 20.6 Gene bank

wild relative of tomato	native habitat	useful features controlled by genes
Lycopersicon cheesmanii	Galapagos Islands	ability to absorb water from sea water; possession of jointless fruit stalks allowing easy mechanical harvesting
L. chilense	Chile and Peru	drought resistance
L. esculentum cerasiforme	many tropical and sub-tropical countries	ability to tolerate high temperature and humidity
L. hirsutum	Ecuador and Peru	resistance to various pests and viruses
L. pennellii	Peru	resistance to drought; high sugar and vitamin A and C content
L. peruvianum	Peru	resistance to various pests; high vitamin C content
L. pimpinellifolium	Peru	resistance to many diseases; improved fruit quality; high vitamin C content

Table 20.1 Wild relatives of tomato plant

bank's aim is to store seeds of every food crop from every country in the world.

The seeds in a gene bank are kept in conditions of low temperature and humidity to maintain their dormancy. It is by this means that many thousands of species including wild varieties of crop plants are being conserved for future use.

Table 20.1 lists some of the wild relatives of the domesticated tomato plant (*Lycopersicon esculentum*) which are held in a cell bank. These possess many useful characteristics under genetic control which are already being introduced into the domestic tomato to produce improved varieties.

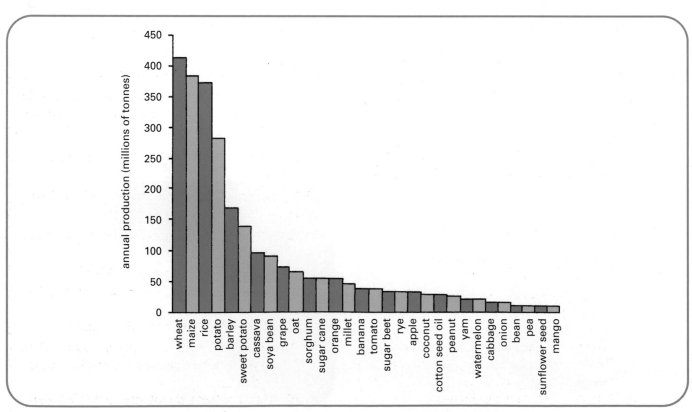

Figure 20.7 Reliance on 30 plant species

The future

Although the Earth possesses at least 75 000 species of edible plants, humans depend on only 30 of these to produce 95% of the world's food supply (see Figure 20.7). Experts forecast that global temperatures will rise by 3°C in the next 50 years as a result of the 'Greenhouse Effect'. If this occurs, many of today's crops may fail in the warmer climate. It is for this reason that new cell banks are being developed. However it is not possible to store every variety of every species. Priority is being given to wild **drought-resistant** relatives of cereals and leguminous plants which form **root nodules** containing nitrogen-fixing bacteria. It is hoped that it will be possible in the future to use the material stored in cell banks to develop new varieties able to survive the Greenhouse Effect.

Testing Your Knowledge

1 Give TWO reasons why the conservation of other species is directly related to the future survival of mankind of Earth. (2)

2 What is the difference between *on-site* and *off-site* protection of a species? (2)

3 a) What is a cell (gene) bank? (1)
   b) Why are cell banks important? (2)
   c) Why are drought-resistant cereals and leguminous crops given priority in cell banks? (2)

Applying Your Knowledge

1 Figure 20.8 shows five of Britain's extinct mammals.

Figure 20.8

a) Match these animals with the following descriptions.
   i) Ancestor of modern cattle. Became extinct in Britain in the 11th century due to habitat destruction and over-hunting.
   ii) Herbivorous mammal hunted to extinction in Britain in the 10th century. Continues to survive in some parts of Europe.
   iii) Ancestor of modern pig. Hunted to extinction in Britain in the 17th century.
   iv) Large omnivorous mammal. Became extinct in Britain during the 10th century due to habitat destruction and over-hunting.
   v) Ancestor of modern dog. Hunted to extinction in Britain in 18th century. (1)

b) Which of these mammals was successfully reintroduced to Scotland in 1952 and now survives as a herd in the Cairngorms? (1)

c) Which of these mammals is extinct worldwide? (1)

d) Which of these mammals might possess alleles for disease resistance which could be introduced into a domesticated variety of farm animal? (1)

2 Table 20.2 summarises the evolution of plants on Earth based on fossil records.

a) Which geological era first provides definite fossil evidence of plant life on Earth? (1)

b) i) To which group did the earliest known plants belong?
   ii) Is this group of plants now extinct? (2)

c) i) During which geological period did vascular plants evolve?
   ii) What is meant by a *vascular* plant? (2)

d) i) Identify TWO groups of vascular plants which have since become extinct according to the table.
   ii) Which of these contributed to Britain's coal deposits? (3)

Geological era	geological period	millions of years from present	major developments as indicated by fossil record
Cenozoic	Quaternary	present to 1	herbaceous angiosperms continue to increase; forests in decline; some tree species threatened with extinction
	Tertiary	1 to 65	angiosperms dominate; herbaceous plants increase; forests extend worldwide and then go into decline
Mesozoic	Cretaceous	65 to 140	advanced conifers decline; angiosperms dominate, some as trees
	Jurassic	140 to 205	advanced conifers dominate; primitive conifers disappear; earliest angiosperms (flowering plants) appear
	Triassic	205 to 245	seed ferns disappear; advanced conifers increase
Paleozoic	Permian	245 to 270	giant clubmosses decline; advanced conifers appear
	Carboniferous	270 to 350	coal-forming seed ferns and clubmosses dominate on land; primitive conifers evolve
	Devonian	350 to 400	early vascular plants (mosses and ferns) appear on land
	Silurian	400 to 440	algae dominate; first land plants appear
	Ordovician	440 to 500	algae dominate
	Cambrian	500 to 600	multicellular algae evolve
Proterozoic		600 to 1500	bacteria and simple algae present
Archeozoic		1500 to 4000	no fossils known; bacteria and unicellular algae may have been present

Table 20.2

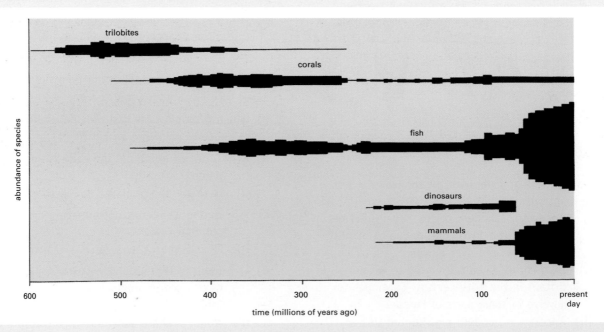

Figure 20.9

e) i) Which type of plant is now in decline and may be threatened with extinction in the future?

ii) Give ONE reason why this is the case. (2)

3 Figure 20.9 shows a kite diagram which charts the abundance of species of five animal groups present on Earth over a timescale of 600 million years.

a) i) Which animal group is thought to have been present on Earth approximately 600 million years ago?

ii) Which group first appeared about 490 million years ago? (2)

b) A wave of extinction occurred about 250 million years ago.

i) Which animal group was reduced almost to extinction?

ii) Which group was reduced but not seriously threatened?

iii) Which group was completely wiped out? (3)

c) Approximately when did i) the dinosaurs; ii) the mammals appear on Earth? (2)

d) i) When did the dinosaurs become extinct?

ii) What effect did this extinction have on the number of species of mammal present thereafter on Earth?

iii) Give a possible explanation for the change in number of mammalian species over the last 60 million years. (3)

4 The sex of certain reptiles is known to be determined by the temperature at which their eggs are incubated. In the alligator, 30°C produces females and 33°C males. It is possible that this was also true of the dinosaurs. It is known that the end of the age of the dinosaurs coincided with a lowering of the Earth's temperature.

Based on the above information, construct a hypothesis to account for the extinction of the dinosaurs. (2)

5 Figure 20.10 shows a 'hamburger'. Explain in your own words the message the artist is attempting to convey. (2)

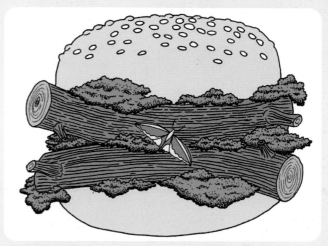

Figure 20.10

6 It is estimated that at the end of the 19th century, mankind was responsible for the extinction of species at the rate of one in every four years. This rate had risen to 1000 per year by the end of the 20th century.

a) By how many times was the annual rate of extinction greater in 1999 compared with 1899? (1)

b) By the year 1999, the world's tiger population stood at about 5000 individuals. Over a period of 100 years, it has been reduced by 95% of its total.

i) Calculate the number of tigers present in the world in 1899.

ii) Give TWO reasons for the current demise of the tiger. (2)

7 Write an essay on the extinction of species. (10)

8 Figure 20.11 shows five statements (1–5) made by members of an animal rights group expressing their concern about animals being 'held captive' in zoos.

→

163

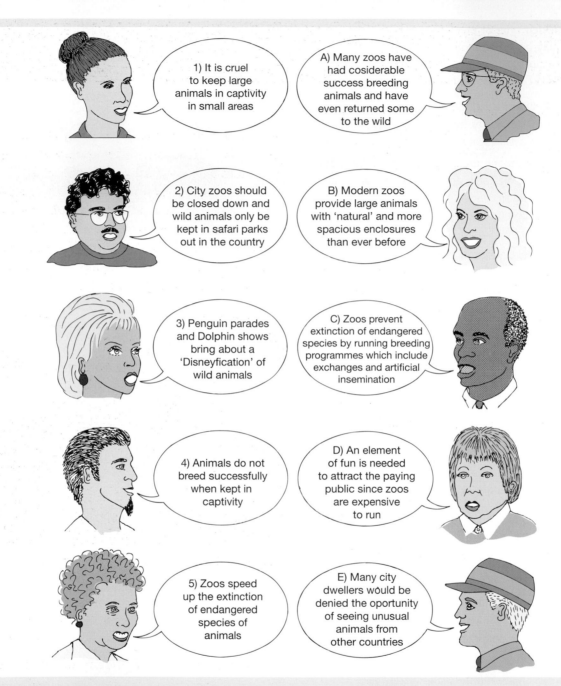

Figure 20.11

Figure 20.11 also shows the replies (A–E) given by five supporters of zoos.

a) Match speech bubbles 1-5 with replies A–E. (1)

b) Say whether you are for or against city zoos and give TWO reasons to support your view. (2)

9 Give an account of the methods of conserving species. (10)

21 Artificial selection

Natural selection (chapter 17) is the process by which those individuals best adapted to the environment survive and pass their genes on to succeeding generations. The 'fittest' survive and the weaker members of the population perish.

Artificial selection is the deliberate selection by humans of organisms with characteristics useful to mankind. Such selection imposed by humans on other species has resulted in the evolution of a wide range of crops and domesticated animals. These varieties would probably not have arisen naturally since they would not enjoy a selective advantage in the wild.

Selective breeding

Many useful varieties of plants and animals have been derived from species by **selective breeding**. Variation exists amongst the members of every species. For thousands of years, people have selected, repeatedly, from each generation of plants and animals, those individuals possessing **desirable characteristics**. These individuals have been used as the parents of the next generation, and so on down through the ages. The plants and animals lacking the beneficial characteristics have been prevented from breeding.

Plants

Selection over many, many generations has improved the quality of the chosen types of plant with respect to their usefulness to humans as a source of food. Cultivated varieties of wheat, barley and potatoes, for example, have been bred that produce **higher yields** and are **more resistant** to disease than their predecessors. In the cabbage plant (*Brassica oleracea*) several varieties have been developed from a common ancestor (see Figures 21.1 and 21.2).

Figure 21.2 Cultivated varieties of *Brassica*

Animals

Similarly, a wide variety of domesticated animals have resulted from selective breeding (see Figure 21.3).

Certain characteristics of domesticated animals used to provide humans with food have been greatly enhanced through selective breeding. Friesian cattle, for example, have been selected for **milk yield** and **butterfat content** of milk (see Figure 21.4). Aberdeen Angus cattle have been selected for **increased meat production** over the years.

The breeder does not create the new variation. The variation already existed amongst the members of the species. Breeders simply accumulate certain desirable parts of the genotype within pedigree strains.

By employing artificial selection, humans often direct another species along an evolutionary path that it

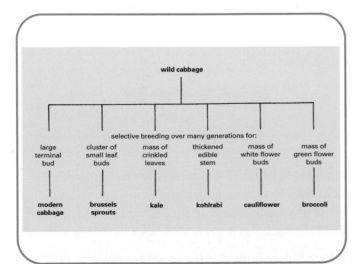

Figure 21.1 Selective breeding of *Brassica oleracea*

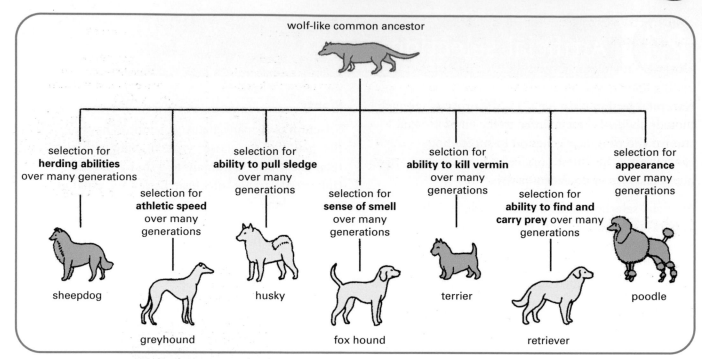

Figure 21.3 Artificial selection of pedigree dog strains

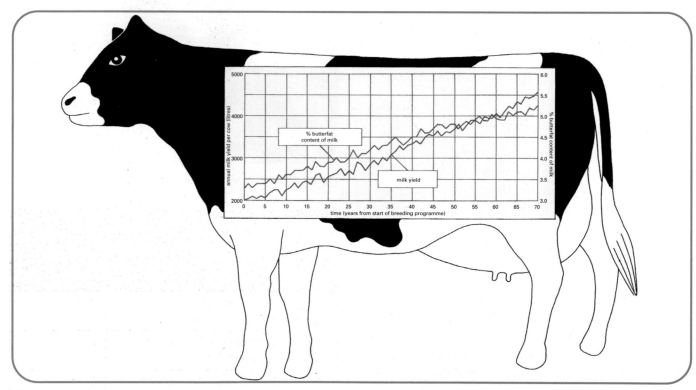

Figure 21.4 Selective breeding of cattle

would not otherwise have taken. Slow moving cattle with heavy udders, for example, would be easy prey to predators.

Inbreeding

Once a desirable strain has been established, farmers

often try to produce a uniform crop or herd with the desirable characteristics by employing **inbreeding**. This is the process by which an individual is bred with its close relatives and is prevented from mating at random.

Inbreeding ensures that the members of each generation of the selectively bred strain (of plant or animal) receive the genes for the desirable characteristics.

However, inbreeding increases the chance of individuals arising that are **double recessive** for some harmful allele (e.g. poor resistance to disease). Such individuals are biologically and economically inferior and are said to show **inbreeding depression**. It is for this reason that inbreeding is not carried out indefinitely. Instead, new alleles are introduced by **hybridisation**.

Hybridisation

A **hybrid** is an individual resulting from a cross between two genetically dissimilar parents of the same species. Breeders often deliberately employ **hybridisation** to cross members of one variety of a species which have certain desirable features with those of another variety possessing different desirable features in an attempt to produce at least some hybrids which have both. A race horse breeder, for example, might try crossing a strong stallion with a fast mare.

Hybrid vigour

Hybridisation often produces some hybrids that are stronger, more fertile or better in some other way than either parent. This is called **hybrid vigour**. It is especially pronounced when it involves a cross between two strains that have become homozygous as a result of many generations of inbreeding.

Hybridisation re-establishes **heterozygosity** and poorer recessive alleles may be masked by better dominant ones as shown in the following imaginary example.

Bedding plants

Professional plant breeders maintain two different true-breeding parental lines of annual bedding plants such as marigold, pansy and snapdragon. One line might be true-breeding for flower colour (CC) and uniform growth habit (GG), the other for large flower size (SS) and number (NN).

When these are crossed (see Figure 21.5), all the F_1 hybrids possess all the desirable features and **outclass** both parents. Since the hybrids are heterozygous and therefore not true-breeding, the hybridisation process using the original parental lines must be repeated every year.

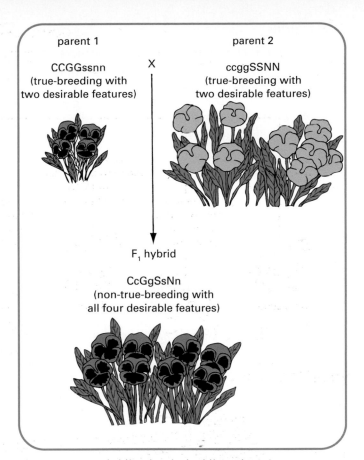

Figure 21.5 Hybridisation in bedding plants

	trait	% improvement shown by hybrid
dairy cattle	birth weight	3–6
	number of live calves per calving	2
	feed conversion efficiency	3–8
	milk yield	2–10
beef cattle	birth weight	2–10
	calves weaned per cow mated	9–15
	feed conversion efficiency	1–6

Table 21.1 Hybrid vigour in cattle

Farm animals

Hybridisation is commonly practised by farmers on their animal stocks. Table 21.1 lists some examples of traits showing hybrid vigour in cattle.

The following cross shows the outcome of hybridisation between two different varieties of sheep.

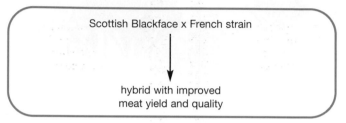

Scottish Blackface x French strain

hybrid with improved
meat yield and quality

Testing Your Knowledge

1 Compare the processes of *natural selection* and *artificial selection*. (4)

2 Describe how farmers carry out selective breeding of a strain of animal. (2)

3 **a)** Give TWO characteristics of the potato plant that have been enhanced by selective breeding. (2)

   b) Name FOUR varieties of crop plant that have been selectively bred from wild cabbage, their common ancestor. (2)

4 **a)** Identify TWO characteristics of cattle that have been enhanced by selective breeding. (2)

   b) i) Name TWO varieties of dog that have been developed from their wolf-like common ancestor by selective breeding.

   ii) Describe the purpose for which each has been bred. (3)

5 **a)** What is the difference between the processes of *inbreeding* and *hybridization*? (2)

   b) i) What is meant by the term *hybrid vigour*?
   ii) Give an example of hybrid vigour in bedding plants. (2)

Genetic engineering

Altering the genome

The single haploid set of chromosomes typical of a species is called its **genome**. Recombinant DNA technology (**genetic engineering**) involves the transfer of one or more genes from the genome of one organism (e.g. human) to the genome of another organism (e.g. bacterium).

The naturally-occurring mechanism of genetic recombination described on page 104 takes place within only one species. Genetic engineering, on the other hand, provides mankind with unlimited opportunities to create new combinations of genes from more than one species which would not occur naturally.

Location of genes

In genetic engineering, it is essential to be able to locate the required gene(s). This can be done by the following methods.

Recognition of characteristic banding patterns on chromosomes

In some animals, unusual **giant chromosomes** are found to exist. The chromosomes in the salivary gland cells of fruit fly larvae, for example, are ten times longer and a hundred times thicker than normal.

Some parts of these chromosomes are found to take up more stain than others producing distinct **bands** which vary in thickness (see Figure 21.6). The particular pattern of banding is a **constant characteristic** of each type of chromosome. The following example shows how it is possible to relate the location of an individual gene to a particular band on a chromosome.

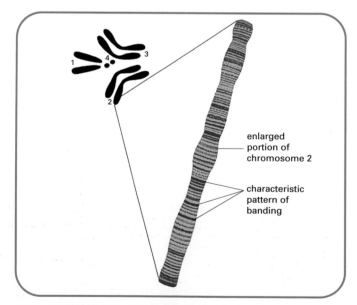

enlarged
portion of
chromosome 2

characteristic
pattern of
banding

Figure 21.6 Pattern of banding on chromosome

The gene for red/white eye colour in fruit fly is sex-linked. To find the location of this gene on the X chromosome, red-eyed males (X^RY) are first irradiated to induce mutations. They are then crossed with white-eyed females (X^{r}X^r). Almost all of the females produced are found to be red-eyed as expected. However a few are white-eyed because they have received an X chromosome which has suffered a **deletion** at the locus of the R allele (see Figure 21.7).

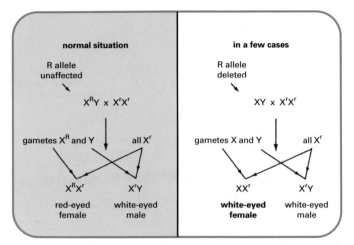

Figure 21.7 Crosses to locate a deleted gene

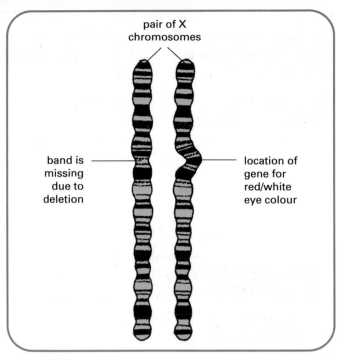

Figure 21.8 Location of deleted gene on X chromosome

If the X chromosomes from the salivary glands of larvae produced by these white-eyed females are examined, some of them are found to be unusual (see Figure 21.8). Instead of matching one another band for band, one X chromosome is found to have a band missing. It is therefore concluded that this is the location of the gene for red/white eye colour.

Gene probes

Each gene codes for a specific protein (or polypeptide). When genetic engineers wish to transfer a gene that codes for some useful protein from one organism to another, they often use a gene probe to locate the required gene.

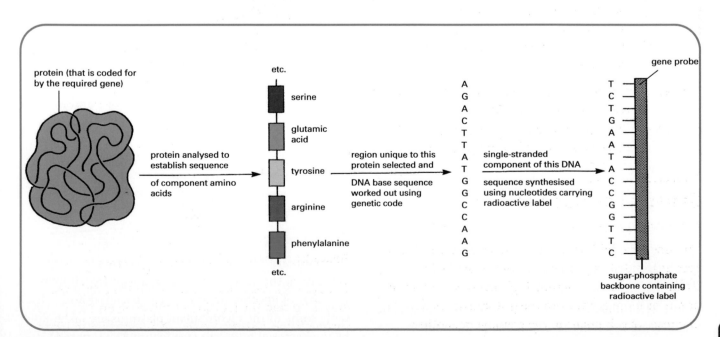

Figure 21.9 Construction of a gene probe

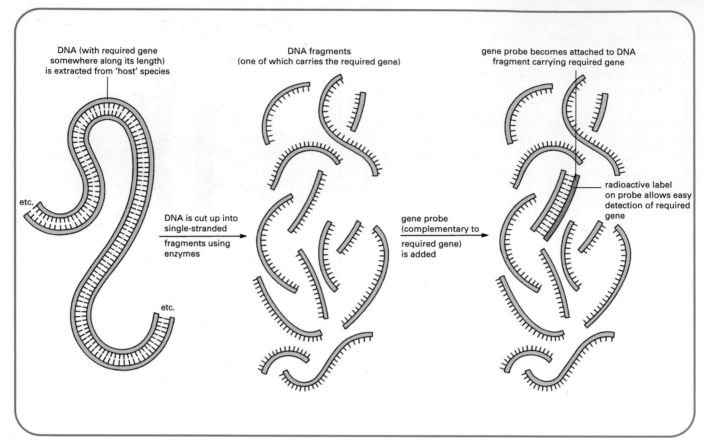

Figure 21.10 Use of a gene probe

A **gene probe** is a short length of single-stranded DNA (or RNA) containing an easily identified chemical **label** (such as radioactive phosphate). One method of constructing a gene probe is shown in Figure 21.9. The probe is complementary to a single-stranded DNA segment from the gene that the genetic engineers wish to locate. Figure 21.10 shows how a gene probe is put to use.

Genetic engineering

Genetic engineers are able to select a particular gene for a desirable characteristic (e.g. gene for human insulin), splice it into the DNA of a **vector** (e.g. **plasmid** from a bacterial cell) and insert the vector into the host cell (e.g. bacterium such as *Escherichia coli*). When the reprogrammed host cell is propagated, it expresses the 'foreign' gene and produces the product (e.g. insulin).

Figure 21.11 illustrates the general principles of genetic engineering involving several 'tools of the trade' as follows.

Endonuclease

An **endonuclease** (full name – restriction endonuclease) is an enzyme extracted from bacteria which is used to cut up the DNA (containing the required gene) into fragments and cleave open the bacterial plasmids. Each endonuclease recognises a **particular sequence** of bases on DNA and makes its cuts at these sites (see Figure 21.11, parts 1 and 2).

Ligase

Ligase is an enzyme which seals 'sticky ends' together. It is used to seal a DNA fragment into a bacterial plasmid forming a recombinant plasmid (see Figure 21.11, part 3).

Vector (carrier)

Recombinant plasmids are used as **vectors** to carry DNA from the genome of one organism (e.g. human) into that of another organism (e.g. bacterial host cell). Only some of the recombinant plasmids are taken up by the bacterial host cells (see Figure 21.11, part 4).

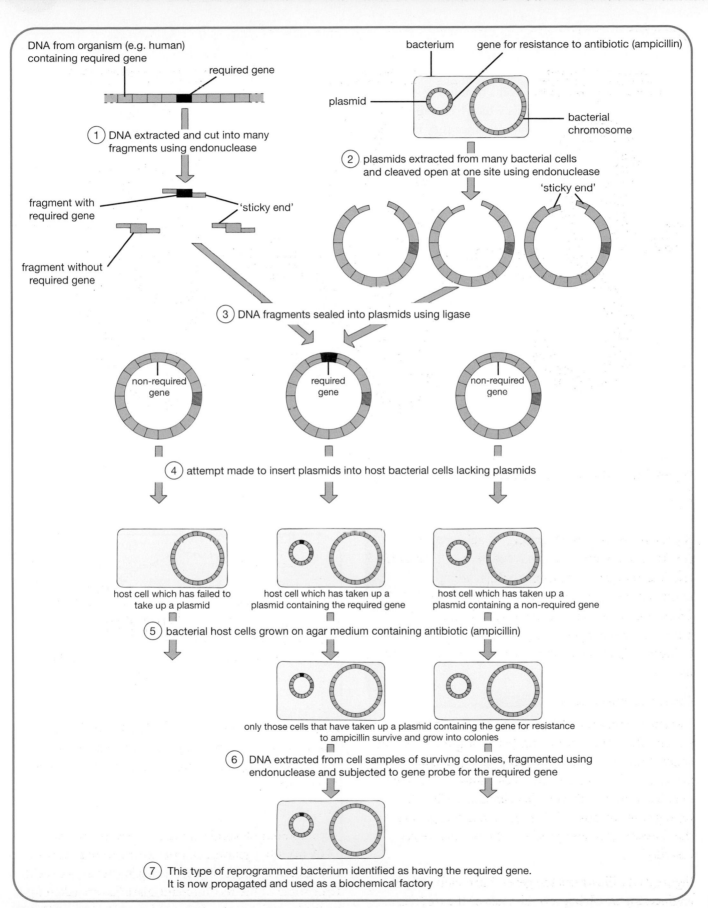

Figure 21.11 Genetic engineering

Antibiotic

Since each recombinant plasmid also contains the gene for resistance to ampicillin, this **antibiotic** can be used to select for the bacterial host cells containing 'foreign' DNA. Those bacterial host cells which have failed to take up a recombinant plasmid die since they lack the resistance gene (see Figure 21.11, part 5).

Gene probe

The surviving colonies are screened for the required gene using a gene probe (see Figure 21.11, part 6).

Once such a colony has been identified, it is propagated in optimum growing conditions to produce many identical cells each with one or more recombinant plasmids containing the required gene which will code for the required product.

Procedures similar to those outlined above have been successfully employed to produce insulin (see Figure 21.12), interferon and **human growth hormone**. Since each of these products is identical to the human type, it does not cause side effects when put to use in the human body.

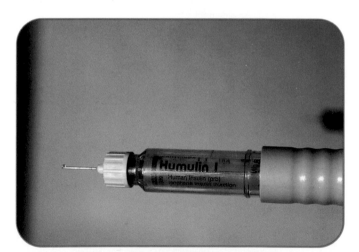

Figure 21.12 Insulin pen

Development of transgenic varieties

Much of the earlier work in genetic engineering concentrated on producing reprogrammed micro-organisms. However recombinant DNA technology is now advancing rapidly into the realm of multicellular organisms.

Crop plants

Agrobacterium tumefaciens is a primitive bacterium present in soil which invades wounded plant tissue (see Figure 21.3). The bacterium injects a plasmid into a plant cell and some of the plasmid's genetic material becomes incorporated into the plant's DNA. *Agrobacterium* is therefore a '**natural genetic engineer**'. It provides scientists with opportunities to introduce into crop plants desirable genes from other organisms which have been inserted first into one of *Agrobacterium's* plasmids.

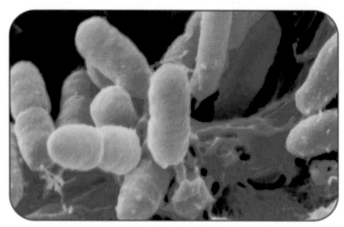

Figure 21.13 *Agrobacterium tumefaciens*

"I'm not having it if it's been genetically altered."

Figure 21.15 Genetic engineering controversy

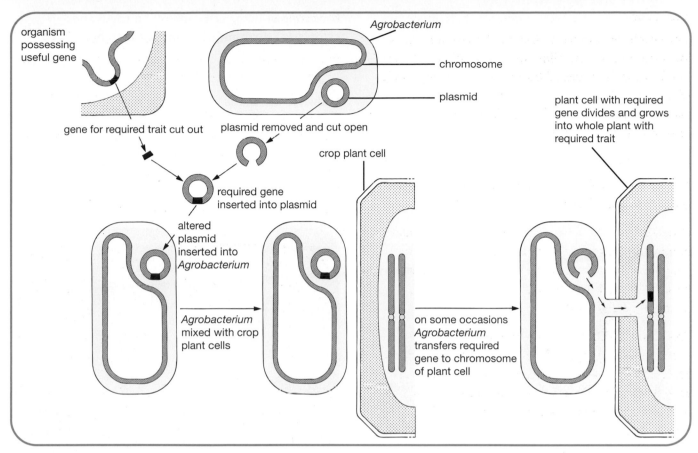

Figure 21.14 Use of *Agrobacterium* in genetic engineering

Figure 21.14 shows a simplified version of this process. It has already been successfully performed on several types of broad-leaved crop plants. These altered plants are called **transgenic varieties** (see Table 21.2 and Figure 21.15).

Farm animals

Genetic engineering involving animals is still at an early stage of its development. However scientists have already produced some strains of **genetically modified** farm animals. Examples include sheep which produce milk containing human proteins (e.g. blood-clotting factor needed by haemophiliacs) and hens that lay eggs containing antibodies needed to make cancer-fighting drugs.

In the future, DNA technology will enable scientists to develop an infinite variety of organisms that could never have been produced by selective breeding. However some of these applications are controversial and require extensive public debate. What is possible scientifically is not necessarily acceptable ethically to

transgenic crop plant	role of inserted gene	beneficial effect
apple	part blockage of production of chemicals which promote ripening	fruit's shelf life is extended
pea	production of insecticide protein	leaves resist attack by caterpillars
soya	production of protein which gives resistance to weedkiller	crop survives but weeds die when weedkiller is applied
strawberry	production of antifreeze chemical	fruit is protected against frost damage

Table 21.2 Transgenic broad-leaved crops

society. It is of critical importance that scientists proceed in this area with the utmost caution.

Production of new crop plants by somatic fusion

Two different species cannot interbreed successfully. At best a cross between them produces a sterile hybrid whose chromosomes are unable to form pairs during meiosis. However new techniques are enabling scientists to overcome this problem of **sexual incompatibility**.

Cells of two different plant species are selected and their cell walls digested away using the enzyme cellulase. This leaves structures called **protoplasts** each

consisting of living cell contents (nucleus and cytoplasm) surrounded by the cell membrane.

In some cases, isolated protoplasts from two different species will fuse to form a **hybrid protoplast** (also known as a **somatic cell hybrid**). This process (see Figure 21.16) is called **somatic fusion** because it involves the fusion of two body (somatic) cells not two sex cells (gametes). It is brought about by subjecting the naked protoplasts to chemical treatment or to an electric current.

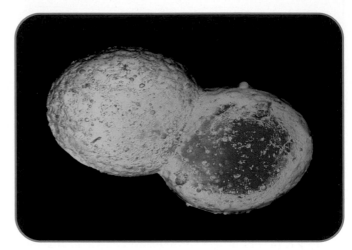

Figure 21.16 Somatic fusion

The somatic cell hybrid formed is then induced to form a cell wall and divide into an undifferentiated cell mass (callus). In the presence of hormones, calluses develop into hybrid plants containing a mixture of the parents' genetic traits.

Resistance to potato leaf roll

Potato leaf roll is a disease caused by a virus which is spread by aphids. It can severely reduce the yield of a crop of the potato plant (*Solanum tuberosum*). *Solanum brevidens* is a wild non-tuber-bearing species of potato plant which is resistant to potato leaf roll virus.

Scientists have managed to overcome the lack of sexual compatibility between these two species of potato plant by uniting them by somatic fusion (see Figure 21.17). This has resulted in the production of a new variety of potato plant which is both tuber-bearing and resistant to potato leaf roll virus.

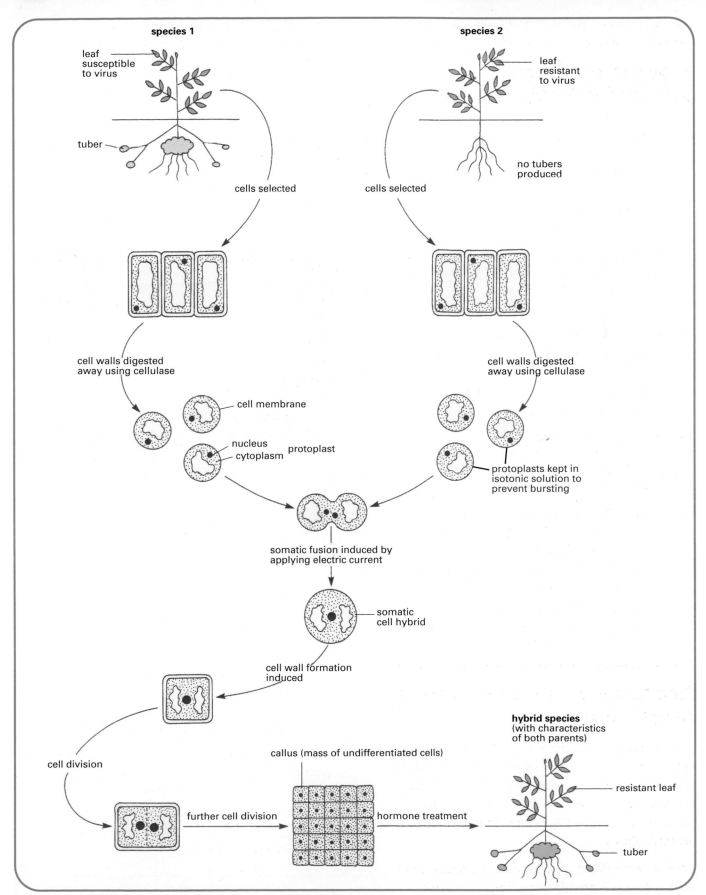

Figure 21.17 Somatic fusion in potato plant species

Scientists are also trying to unite wheat with pea plants by somatic fusion. If successful the somatic cell hybrid might combine the food-producing features of wheat with the ability of pea plants to form root nodules containing nitrogen-fixing bacteria. This would enable the plant to grow on soil poor in nitrate. Expensive fertilisers would not need to be added to the soil.

Already hybridisation between potato and tomato plants has been successfully performed by somatic fusion. However none of the hybrids produced so far is able to make both potato tubers and tomato fruit on the one plant.

Testing Your Knowledge

1 a) i) What name is given to the process by which protoplasts from two different plant species are induced to fuse together into one cell?
   ii) Why do scientists not simply unite the two plant types by sexual reproduction? (2)
   b) i) How do scientists obtain suitable protoplasts to use for somatic fusion?
   ii) By what means are the protoplasts induced to fuse together? (2)

2 a) Name the treatment given to the callus derived from a somatic cell hybrid to make it grow into a hybrid plant. (1)
   b) The new hybrid species formed by somatic fusion often expresses a mixture of the two parental species' traits. Why is this only sometimes of benefit to humans? (2)

Applying Your Knowledge

1 A breed of pig has been produced which reaches bacon weight in 150 days instead of the normal 185 days. In addition, this is achieved on 20% less food. Can this improvement process be continued indefinitely? Explain your answer. (2)

2 State ONE advantage and ONE disadvantage of producing a crop that is uniform and true-breeding. (2)

3 Table 21.3 refers to a breeding experiment involving cattle. What can be concluded from the data? (2)

	genotype	weight gained by phenotype per day (kg)
parent 1	AAbb	0.82
parent 2	aaBB	0.82
F₁ generation	AaBb	1.00

Table 21.3

4 The data in Tables 21.4 and 21.5 refer to breeding experiments using maize (*Zea mays*), see Figure 21.18.

	parents		successive generations (resulting from repeated self-fertilisation)							
	P₁	P₂	F₁	F₂	F₃	F₄	F₅	F₆	F₇	F₈
number of generations selfed	17	16	0	1	2	3	4	5	6	7
average ear length (mm)	84	107	162	141	147	121	94	99	110	107
average yield (metric tonnes/hectare)	2.0	2.1	4.3	4.0	3.4	3.1	2.8	2.6	2.5	2.3

Table 21.4

year	millions of hectares used for maize planting	average yield (metric tonnes/hectare)
1932	46	2.4
1946	36	3.4
1957	30	4.1

Table 21.5

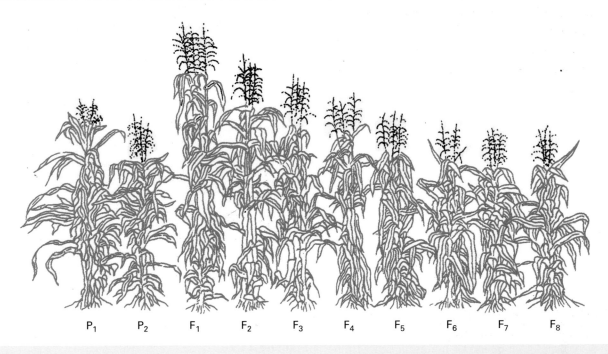

Figure 21.18

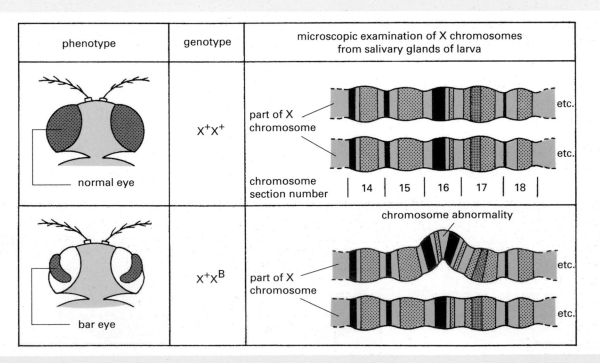

Figure 21.19

a) Identify the generation formed by hybridisation. (1).

b) State THREE pieces of information which provide evidence of hybrid vigour. (3)

c) Which of the following was a true-breeding inbred line: P_1, F_1, F_3, F_5? (1)

d) Using an example from the data to illustrate your answer, explain what is meant by the term *inbreeding depression*. (1)

e) In 1933, American farmers began planting maize grains produced by hybridisation. This practice has continued over the years and has brought about the changes shown in Table 21.5.
   i) State TWO ways in which farmers have benefited by using hybrid maize.
   ii) Explain why farmers must buy expensive hybrid grain every year from supply houses to sow their maize crop instead of simply using grain kept back from the previous year's crop. (3)

5 Figure 21.19 refers to *Drosophila*, the fruit fly, where + = allele for normal eye and B = allele for bar eye.

a) Is the allele for bar eye dominant or recessive to normal eye? Explain your answer. (1)

b) What type of chromosome mutation caused the chromosomal abnormality? (1)

c) On which numbered section of the X chromosome is the gene for bar eye located? (1)

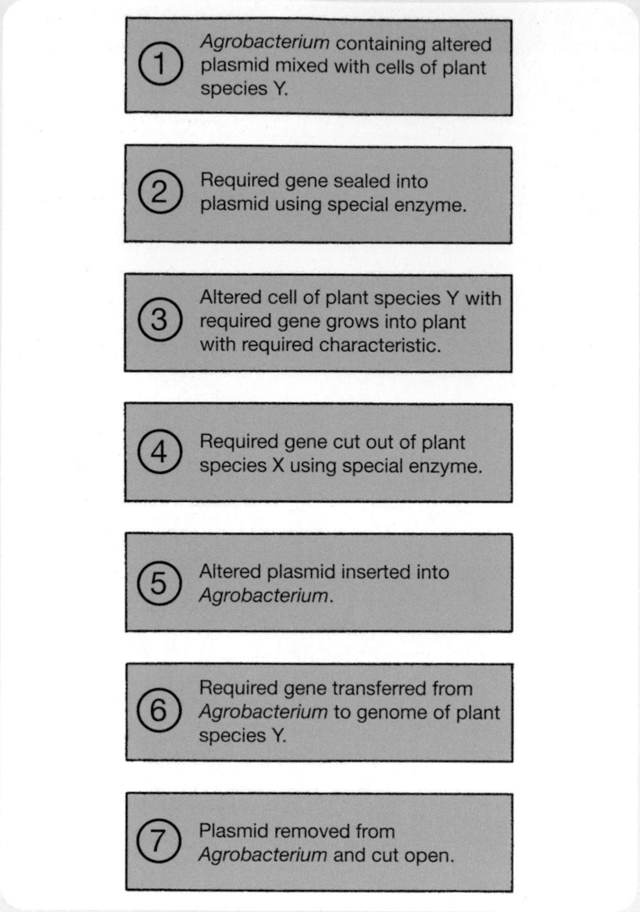

① Agrobacterium containing altered plasmid mixed with cells of plant species Y.

② Required gene sealed into plasmid using special enzyme.

③ Altered cell of plant species Y with required gene grows into plant with required characteristic.

④ Required gene cut out of plant species X using special enzyme.

⑤ Altered plasmid inserted into Agrobacterium.

⑥ Required gene transferred from Agrobacterium to genome of plant species Y.

⑦ Plasmid removed from Agrobacterium and cut open.

Figure 21.20

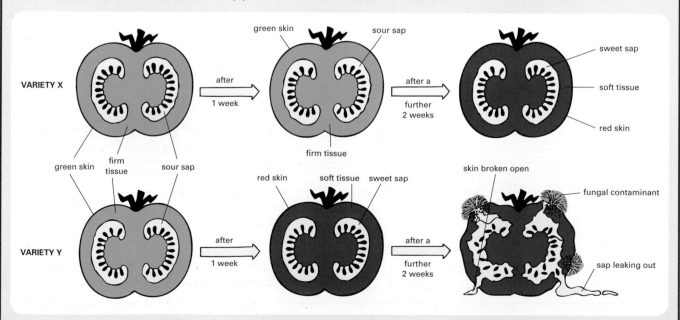

VARIETY X — green skin, firm tissue, sour sap — after 1 week → green skin, sour sap, firm tissue — after a further 2 weeks → sweet sap, soft tissue, red skin

VARIETY Y — green skin, firm tissue, sour sap — after 1 week → red skin, soft tissue, sweet sap — after a further 2 weeks → skin broken open, fungal contaminant, sap leaking out

Figure 21.21

6 Figure 21.20 shows seven boxed statements which refer to the production of a transgenic plant species by genetic engineering.

a) Arrange the stages into the correct order starting with box 4. (1)

b) i) Which type of enzyme is being referred to in boxes 4 and 7?
   ii) Why must the same enzyme be used both times? (2)

c) Name the type of enzyme being referred to in box 2. (1)

d) Thale cress (scientific name, *Arabdopsis*) is a garden weed closely related to rape and cabbage plants. Scientists are analysing all of its genes in order to identify the ones responsible for resistance to fungal attack and oil quality of seeds. Explain why. (2)

7 When a tomato fruit is fully grown but still green and unripe, it begins to produce ethylene. Ethylene is a growth substance which promotes the ripening of fruit.

Genetic engineers have developed a variety of tomato plant which makes very little ethylene by inserting a gene which almost completely blocks ethylene production. Table 21.6 gives the results from one of their experiments. Figure 21.21 shows a typical fruit from each variety of tomato plant.

a) Draw a line graph of the results. (2)

b) Match the terms *control* and *genetically modified variety* with the two lettered varieties in the diagram. (1)

c) State how many units of ethylene were produced on day 6 by i) the genetically modified variety; ii) the control. (1)

d) Calculate the percentage decrease in ethylene production at day 6 caused by the blocked gene in the genetically modified variety. (1)

e) Suggest why it is important that ethylene production is not completely blocked in the modified variety. (1)

f) What is the benefit gained from this application of genetic engineering? (1)

8 Production of the human growth hormone somatotrophin is controlled by a gene. Lack of somatotrophin leads to dwarfism in humans. This problem has now been solved by the production of

time from green fruit reaching full size (days)	ethylene production (units)	
	genetically modified variety	control
0	1	1
1	4	5
2	7	25
3	5	70
4	6	71
5	8	68
6	7	70
7	6	52
8	7	41
9	7	32
10	8	34
11	6	31
12	7	32
13	7	29
14	8	32
15	7	31
16	6	28
17	8	25
18	7	26
19	7	23
20	6	21
21	6	19

Table 21.6

somatotrophin by genetic engineering. Give a step by step account of how this could have been achieved. (10)

9 In nature, hybridisation between sugar beet and cabbage plants never occurs because they are unrelated species.

a) Name a method by which scientists could attempt to combine the genetic material of the two in the laboratory to produce a useful hybrid. (1)

b) The cells to be used in an experiment of this type must have their walls removed. Why? (1)

c) i) Naked protoplasts are kept in a solution of salt or sugar isotonic to their cell contents. Explain why.

   ii) Describe what would happen if naked protoplasts were immersed in distilled water. (2)

10 Give an account of selective breeding and hybridisation of food crops and domesticated animals. (10)

What You Should Know

Chapters 17-21

(See Table 21.7 for Word bank)

adaptive	diversity	isolated
advantage	engineering	niche
antibiotics	environment	probes
artificial	evolution	rare
banding	extinct	selection
banks	genome	somatic
barriers	hybridisation	species
continuous	inbreeding	suited
divergence	incompatibility	vigour

Table 21.7 Word bank for chapters 17–21

1 Natural _____ favours those members of a population best _____ to an environment.

2 Rare mutant forms sometimes enjoy a selective _____ if some biotic or abiotic factor brings about a change in the _____ making it favour their survival at the expense of their competitors.

3 The rapid appearance of bacteria resistant to _____ is an example of high-speed _____.

4 The members of a _____ form a natural interbreeding group which is reproductively isolated from other species.

5 The process of speciation depends on _____ to gene exchange dividing a population into two or more _____ groups, each of which takes its own course of evolution.

6 _____ radiation is the _____ over a very long period of time of a group of related organisms along several different lines by each becoming adapted to suit a particular ecological _____.

7 Evolution is a _____ process. As new species appear, other less successful ones become _____.

8 Important wild varieties of crop plant and endangered species are often conserved in cell _____ and _____ breed farms.

9 Breeders use _____ selection to selectively breed organisms useful to mankind. Loss of genetic _____ is associated with _____ of domesticated plants and animals.

10 The use of _____ in plant and animal breeding often produces offspring which show hybrid _____.

11 Genes can be located using gene _____ or by recognizing _____ patterns on chromosomes.

12 Using genetic _____, scientists are able to take genetic material from one species and seal it into the _____ of another species producing an organism which would never have arisen otherwise.

13 Sexual _____ between two species of plant can be overcome by using _____ fusion.

22 Maintaining water balance – animals

Adaptation

An **adaptation** is a characteristic possessed by an organism which makes it well suited to its environment. The members of a species possess certain **evolutionary adaptations** typical of that species. These adaptations are the phenotypic expressions of changes that have occurred in the species' genotype over a very long period of time. They have been favoured by natural selection because possession of them gives the species some advantage and increases its chance of survival.

Some adaptations are **structural** (i.e. involve specialised structures possessed by the organism), some are **physiological** (i.e. depend on the way the organism's body works) and some are **behavioural** (i.e. depend on the way the organism responds to stimuli).

Water balance in aquatic animals

Animals such as the jellyfish whose cell contents are **isotonic** (equal in water concentration) to the surrounding environment do not have the problem of maintaining water balance since their cells neither gain nor lose water by osmosis.

A water balance problem does exist however in aquatic animals whose cell contents are **hypertonic** (lower in water concentration) compared with the surrounding environment since they constantly gain water by osmosis.

A water balance problem also exists in aquatic animals whose cell contents are **hypotonic** (higher in water concentration) compared with the surroundings since they constantly lose water to their environment by osmosis.

In each case, these animals possess certain physiological adaptations which enable them to maintain their water (and ion) concentration at the correct level. This control of water (and ion) balance is called **osmoregulation**.

Osmoregulation in freshwater bony fish

The scaly skin of a fish is impermeable to water. However in a freshwater fish (see Figure 22.1), the delicate membranes of the mouth and gills are selectively permeable and the fish has to deal with the problem of a continuous influx (inflow) of water by osmosis. To prevent the body tissues swelling up and becoming damaged, this excess water must be removed.

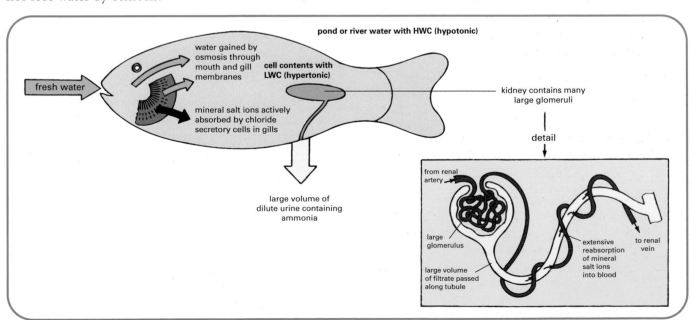

Figure 22.1 Osmoregulation in freshwater bony fish

Over millions of years of evolution, freshwater bony fish have become adapted in several ways to carry out this function.

- Their kidneys possess **many large glomeruli** (see Figure 22.1) which allow a **high filtration rate** of the blood. This results in the production of very dilute glomerular filtrate.

- The kidney tubules are very efficient at **reabsorbing mineral salts** (ions) from this glomerular filtrate back into the bloodstream. The fish therefore excretes a **large volume of very dilute urine** which contains only a trace of salts and some nitrogenous waste. (Ammonia, the nitrogenous waste, is toxic in high concentrations. However in this type of fish, it is diluted to a harmless level in the urine.)

- The membranes of a freshwater bony fish's gills contain **chloride secretory cells** which **actively absorb salts** (ions) from the water as it passes through the gills. The salt concentration of the body is therefore maintained by these cells replacing the small amount lost in the large volume of urine. This process involves **active transport** of ions against a concentration gradient and therefore requires **energy.**

Osmoregulation in saltwater bony fish

Although its scaly skin is impermeable to water, this type of fish (see Figure 22.2) suffers constant dehydration of its hypotonic tissues. This is because water is continuously being lost to the surrounding hypertonic sea water by osmosis through its selectively permeable gill and gut membranes.

This problem is overcome in several ways.

- **Sea water is drunk** to replace losses.

- The kidney contains only a **few small glomeruli** (see Figure 22.2) or none at all. This results in a **low filtration rate** of blood and only a **little water** being lost in urine.

- Nitrogenous waste is converted to a **non-toxic** form which requires a minimum volume of water for its removal.

- The chloride secretory cells in the gill membranes work in the reverse direction to those of the freshwater fish. In a marine bony fish they **actively excrete** the excess salt (gained from drinking sea water) back out into the sea. Again this **active transport** against a concentration gradient requires **energy.**

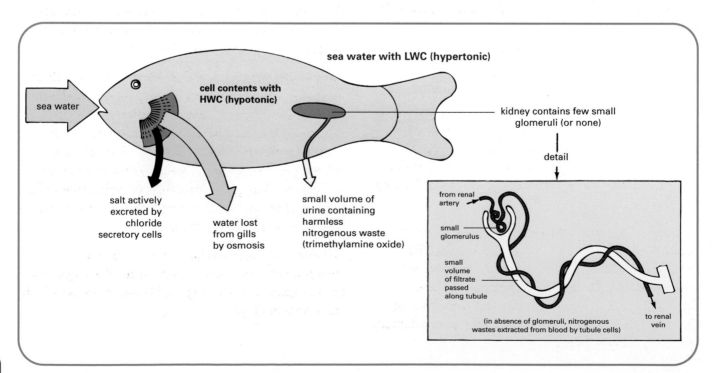

Figure 22.2 Osmoregulation in saltwater bony fish

Adaptations of migratory fish

A **migratory** fish (e.g. Atlantic salmon and eel) spends part of its life in fresh water and part in salt water and is able to osmoregulate successfully in both environments.

Salmon

The eggs of salmon are laid amongst the gravel in a river bed. After hatching, the young fish spends the first three years of its life feeding and growing until it becomes a **smolt** (the migratory stage). In response to some stimulus (probably a full moon), the smolts migrate downstream together. This is called the **smolt run** (see Figure 22.3).

Figure 22.3 Smolt run

After lingering for a short time in the river estuary to become **acclimatised** to salt water, the smolts leave the river together and head in vast numbers for their feeding grounds off the west coast of Greenland (see Figure 22.4). In these food-rich waters the young salmon increase their body weight by up to 50 times in one year by feeding on shoals of capelin (small herring-like fish).

After one to six years, the salmon respond to some (unknown) stimulus and migrate back across the Atlantic. The returning fish gradually stop feeding, develop reproductive ability and, using sensitive olfactory (smell) receptors, locate and swim back up the river (see Figure 22.5) in which they were originally spawned, ready to breed.

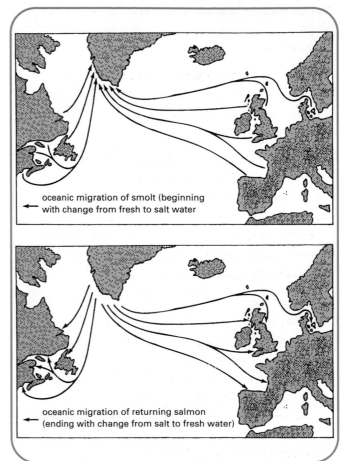

Figure 22.4 Oceanic migrations of Atlantic salmon

On entering the river they must become acclimatised for a second time to a dramatic **change in the water concentration** of the external environment.

The precise means by which migratory fish alter their method of osmoregulation is poorly understood but depends at least partly upon the action of several hormones. When a migratory fish moves from salt water to fresh water (and vice versa), its hormonal balance becomes altered. This is accompanied by an appropriate **change in glomerular filtration rate** and a **reversal of the direction** in which the chloride secretory cells actively transport salt ions.

The ability to adapt to changes in the external environment's water concentration enables migratory fish to exploit a wider range of habitats than fish which are unable to do so.

Figure 22.5 Salmon returning home to breed

Water conservation in a desert mammal

Since the desert is a region of very low rainfall where drought conditions persist for most of the year (accompanied by high daytime and low night-time temperatures), desert mammals such as the kangaroo rat have only a very **limited supply of water** available to them. To survive they have to be able to practise rigorous **water conservation**.

In its natural habitat, the kangaroo rat does not drink water at all. It is able to gain all of its water from its food ('dry' seeds) and remain in water balance as shown in Figure 22.6. This is made possible by the possession of certain physiological and behavioural adaptations.

Physiological adaptations

- A kangaroo rat's mouth and nasal passages tend to be **dry**, thereby reducing water loss during expiration.

- Its bloodstream contains a **high level** of antidiuretic hormone (see page 307).

- Its kidney tubules possess **very long** loops of Henle. These adaptations promote water reabsorption so effectively that it can produce urine 17 times as concentrated as its blood. (The most hypertonic urine that humans can produce is only four times as concentrated as blood.)

- A kangaroo rat's large intestine is extremely efficient at **reabsorbing water** from waste material and producing faeces with a very low water content.

- It does **not** sweat.

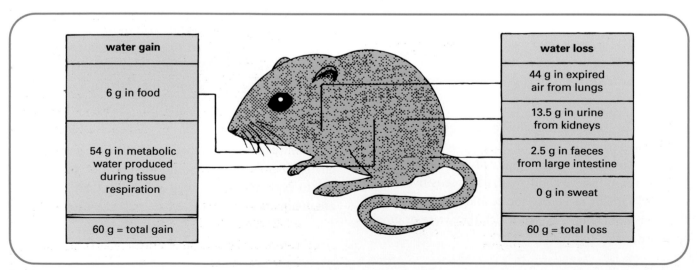

water gain	water loss
6 g in food	44 g in expired air from lungs
	13.5 g in urine from kidneys
54 g in metabolic water produced during tissue respiration	2.5 g in faeces from large intestine
	0 g in sweat
60 g = total gain	60 g = total loss

Figure 22.6 Water balance in kangaroo rat (fed 100 g seeds over one month)

Behavioural adaptations

A **behavioural adaptation** is an instinctive behaviour pattern which helps the animal to survive in its natural habitat.

Water conservation is further promoted by the fact that the kangaroo rat remains in its **underground burrow** in a fairly **inactive state** during the extreme heat of the day. Unlike the very hot dry air above ground, the air inside the burrow is **cooler and more humid**. Thus the air being inhaled by the rat is almost as damp as the air being exhaled and minimum net water loss occurs.

The animal is **active at night** and goes out foraging for food when the conditions are cool. It therefore has no need to produce sweat to give a cooling effect and water is conserved.

Testing Your Knowledge

1 Rewrite the following sentences choosing only the correct answer from each choice in brackets.

   The tissues of a freshwater fish are (hypotonic/hypertonic) to the surrounding aquatic environment and so the fish constantly (loses/gains) water by osmosis. The problem is solved by the fish producing a (small/large) volume of (dilute/concentrated) urine. (4)

2 Briefly describe the means by which osmoregulation is brought about in a saltwater bony fish. (4)

3 a) Why is the Atlantic salmon described as a *migratory* fish? (1)

   b) i) What is the *smolt run*?

      ii) Why do smolts linger in the river estuary for a short time before heading out to sea? (2)

   c) Compare a 2-year old salmon in a Scottish river with a 5-year old salmon feeding off the coast of Greenland with respect to:
      i) rate of glomerular filtration;
      ii) direction in which chloride secretory cells transport salt ions. (2).

4 a) Name TWO ways in which a desert rat i) gains; ii) loses water. (4)

   b) Give TWO physiological and TWO behavioural adaptations shown by a desert mammal which help it to conserve water. (4)

Applying Your Knowledge

1 a) With reference to the bar chart in Figure 22.7, briefly outline the water balance problem encountered by each type of fish in its natural habitat. (2)

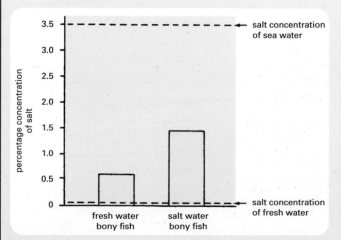

Figure 22.7

   b) Copy and complete Table 22.1 which refers to adaptations possessed by each of these types of fish to solve its osmoregulatory problems. (6)

   c) State TWO differences between diffusion and active transport. (2)

2 In fresh water, a salmon's rate of urine production is 2 cm³/kg of body mass/hour. In salt water, the rate drops to 0.5 cm³/kg of body mass/hour. A smolt was found to weigh 0.1 kg on leaving its native river and 3 kg on returning as an adult several years later.

   a) By how many times is a salmon's rate of urine production greater in fresh water than in saltwater? (1)

   b) Express as a whole number ratio the mass of the adult salmon to that of the smolt. (1)

   c) Calculate the volume of urine produced per day by the smolt in:

	fresh water bony fish	salt water bony fish
relative size of glomeruli		
relative number of glomeruli		
relative filtration rate of blood		
relative volume of urine produced		
relative concentration of urine		
direction in which salts are actively transported by chloride secretory cells		

Table 22.1

   i)   fresh water;
   ii) salt water once it had begun its migration. (2)

  d) i)   Calculate the increase in volume of urine produced daily by the adult salmon on migrating from salt water to its native river.
     ii)   Express this increase as a percentage. (2)

3 Figure 22.8 shows the migratory patterns of eels. Spawning takes place in the sea and after about three years, the young eels (elvers) travel to the rivers of Europe where they feed and grow to adulthood.

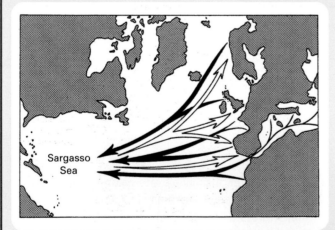

Figure 22.8

When sexually mature, the adults migrate downstream to the sea. They then set off on a journey of thousands of miles to a region of the Atlantic Ocean called the Sargasso Sea where they spawn and begin the life cycle again.

a) Which type of arrow (shaded or unshaded) represents:

   i)   the migration of adult eels to their breeding grounds;
   ii) the movement of young eels to the regions where they do most of their feeding and growing? (1)

b) Both of these movements involve a change in environment from one osmotic extreme to another. In each case, describe how the eel adjusts its osmoregulatory system with reference to:
   i)   filtration rate of kidney;
   ii) volume of urine excreted;
   iii) direction of activity of chloride secretory cells. (3)

c) In what way is eel migration different from that of salmon? (1)

d) What benefit is gained by fish able to adapt to changes in the external environment and migrate compared with those unable to do so? (1)

4 Several specimens of two species of crab (A and B) were submerged in sea water (3.5% salt solution) and given time to acclimatise. A sample of each crab's blood was then taken and analysed to measure its salt content.

The experiment was repeated using 3% salt solution and successively more dilute salt solutions. Average values of blood salt content for both species of crab in each external salt solution were calculated and the data graphed as shown in Figure 22.9.

a) Which species of crab has blood which is always isotonic to the surrounding water? (1)

b) i)   Which species of crab has blood which remains hypertonic to the surrounding water

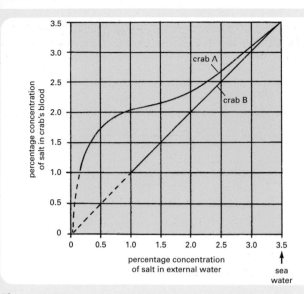

Figure 22.9

as the salt content of this external medium decreases below that of sea water?

ii) At which % concentration of salt in the external water is this crab's blood most hypertonic to its surroundings? (2)

c) i) Which crab is able to exert some degree of osmoregulation?

ii) Why is oxygen required for this process? (2)

d) Copy and complete Table 22.2 by identifying which species of crab is suited to life in each of the habitats and briefly explaining why. (?)

habitat	species of crab	reason
river estuary		
deep sea water		

Table 22.2

5 Read the passage and answer the questions that follow it.

Camels lose water steadily through their skin and their breath, as well as in urine and faeces. Although camels can go for days without drinking water, they normally survive during this time by feeding on plants. The special compartments attached to a camel's stomach are filled with foul-smelling liquid (containing fermenting food) which is isotonic to mammalian blood.

If a man loses more than 12% of his body water, he is found to be in a critical state because his blood has become thick and sticky and his heart is having difficulty pumping it. A camel can lose up to 25% of

its body mass without straining its heart because the water is lost from tissues other than the blood.

A camel's hump consists of fat and acts as an energy reserve. Fat does yield metabolic water during tissue respiration but so much extra oxygen is needed to burn fat that any gains in water are lost as water vapour during the increased rate of breathing that occurs. The camel's coat provides insulation against the heat of the desert day. The animal's body temperature begins at 34°C at dawn and does not reach 40°C until around midday. Thus the rate of sweating is kept to a minimum during the first part of the day. The body temperature drops back down to 34°C during the night.

a) Compared with a kangaroo rat, which behavioural adaptation is a camel unable to employ to conserve water? (1)

b) Compared with a kangaroo rat, which additional method of water loss occurs from a camel's body? (1)

c) i) Is hypothesis 1 in Figure 22.10 supported by the information in the passage? Justify your answer.

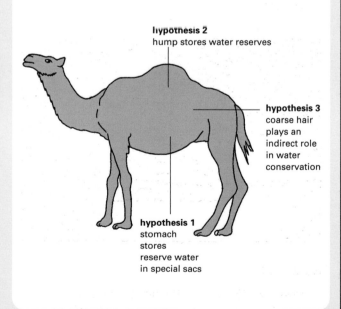

Figure 22.10

ii) Would a human being, almost dead from dehydration, be able to save his/her life by drinking the liquid from a camel's stomach compartments? Explain your answer. (2)

d) i) Is hypothesis 2 supported by the passage? Explain your answer.

   ii) Why does a camel have a hump? (2)

e) Is hypothesis 3 supported by the passage? Explain your answer. (2)

f) Explain why a camel which has lost 20% of its body mass by dehydration is still capable of the same physical exertion as a well-watered camel. (1)

6 Give an account of water conservation in a kangaroo rat. (10)

23 Maintaining a water balance – plants

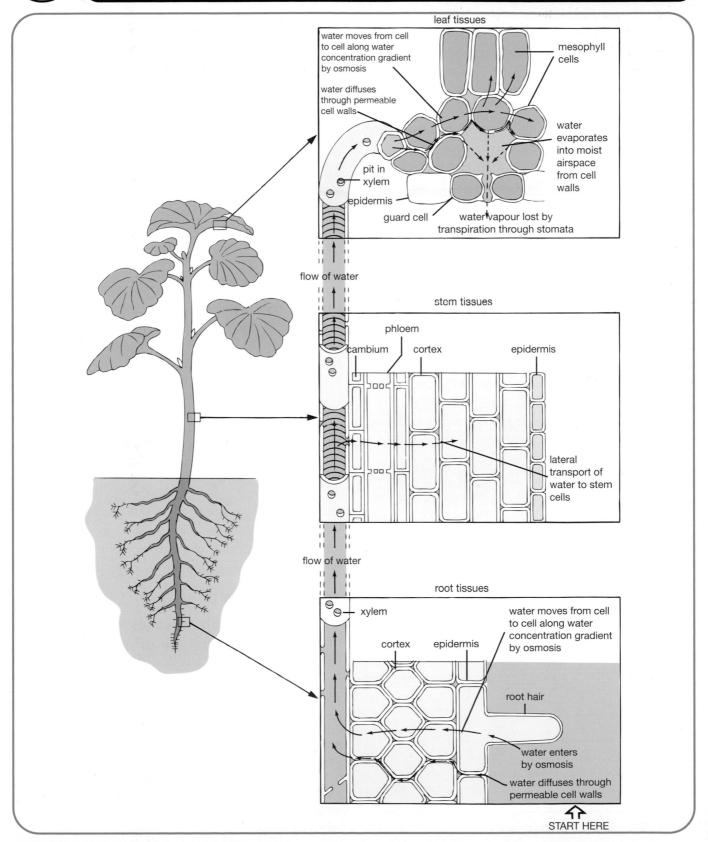

leaf tissues

water moves from cell to cell along water concentration gradient by osmosis

mesophyll cells

water diffuses through permeable cell walls

water evaporates into moist airspace from cell walls

pit in xylem

epidermis

guard cell

water vapour lost by transpiration through stomata

flow of water

stem tissues

phloem

cambium   cortex   epidermis

lateral transport of water to stem cells

flow of water

root tissues

xylem

water moves from cell to cell along water concentration gradient by osmosis

cortex   epidermis

root hair

water enters by osmosis

water diffuses through permeable cell walls

START HERE

Figure 23.1 Water uptake in transpiration stream

Transpiration stream

The continuous passage of water (and nutrient ions) flowing up through a plant from the roots to the leaves is called the **transpiration stream**.

Entry of water

Osmosis

A **root hair** (see Figure 23.1) is an extension of an epidermal cell which presents a **large absorbing surface** to the surrounding soil solution. Since the cell sap inside a root hair is a region of lower water concentration than the soil solution outside, water passes into a root hair by **osmosis** through the selectively permeable plasma membrane.

Water then passes from the root hair (which now has a higher water concentration) into a neighbouring cortex cell (which by comparison has a lower water concentration) and so on across the root to the xylem vessels along a **water concentration gradient**.

Diffusion

Many water-filled spaces occur between the cellulose fibres which make up cell walls. Much water therefore diffuses across the root via the permeable cell walls and intercellular spaces without entering the living cells.

Root pressure

If a plant's stem is severed, the cut stump continues to exude water. The force with which this water is pushed up the stem by the roots is called **root pressure**. It can be measured using a **manometer** as shown in Figure

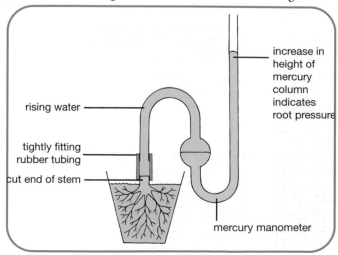

Figure 23.2 Measuring root pressure

increase in height of mercury column indicates root pressure

rising water

tightly fitting rubber tubing

cut end of stem

mercury manometer

23.2. This movement of water is due to the flow of water by osmosis across the root. It is thought to play only a minor role in the ascent of water via the transpiration stream.

Capillarity

Adhesion is the force of attraction between **unlike** particles. In the experiment shown in Figure 23.3, **capillarity** (force of **capillary** attraction) occurs because liquid water molecules adhere to the solid glass particles. Water rises to the highest level in the narrowest tube because it presents the largest relative surface area of glass in contact with adhering water molecules.

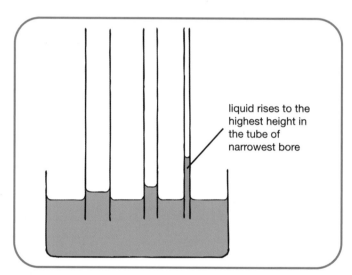

liquid rises to the highest height in the tube of narrowest bore

Figure 23.3 Capillarity

Since xylem vessels are also tubes of very fine bore, water similarly adheres to the vessel walls. Capillarity is therefore thought to play a minor role in the upward movement of water in a plant.

Upward transport in xylem

Water passes up through the plant in the xylem tissue which consists of a continuous system of hollow vessels whose structure exactly suits their functions of water transport and support. Pits in the walls of xylem vessels allow **lateral transport** of water to other tissues all the way up the plant.

Exit of water

Transpiration is the process by which water is lost by evaporation from the aerial parts of a plant. Most transpiration occurs through tiny holes in the leaves called **stomata**.

The contents of leaf cells closest to a moist air space from which water is being lost (via a stoma) have a lower water concentration than cells nearer the xylem vessels. Thus a **water concentration gradient** also exists in a leaf and water passes from xylem to leaf cells by osmosis as shown in Figure 23.1. Some of this water is used to maintain cell turgor and as a raw material in photosynthesis.

The walls of the cells in contact with a vein also draw water from the xylem vessels. In fact most of the water that crosses a leaf does so via the cell walls (as in the root) without entering the cells.

Transpiration pull

Water from the walls of the leaf cells lining the moist air spaces evaporates and is lost by transpiration via the stomata. In order to replace these losses, cells draw water from the xylem vessels and set up a **transpiration pull**. This is the major force which brings about the ascent of water in the transpiration stream.

Cohesion is the force of attraction between like particles. Transpiration pull is explained in terms of the **cohesion–tension theory** which proposes that such raising of water against gravity is possible because water molecules **cohere** strongly together. This allows thin tense columns of water to be pulled up the xylem vessels to the top of the tallest tree without the water columns breaking.

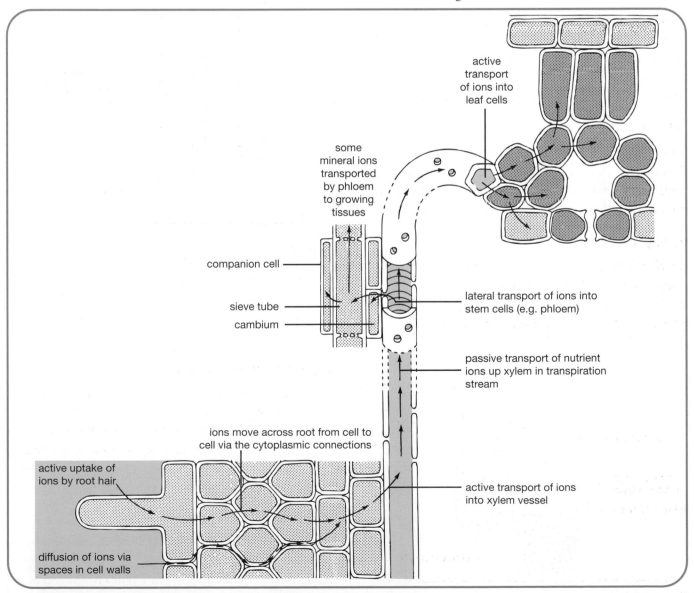

Figure 23.4 Nutrient ion uptake

The diameter of a tree is found to decrease during periods of maximum transpiration. This is thought to occur because the columns of water molecules (drawn thin by the tension exerted by the transpiration pull) still continue to adhere to the xylem vessel walls and pull them inwards.

Mineral (nutrient) uptake

Some nutrient ions enter the plant by active uptake (see Figure 23.4) and then move from cell to cell via tiny intercellular cytoplasmic connections (see Figure Ap 2.2 on page 352). Other nutrient ions enter by diffusion through the water-filled spaces in the cell walls.

In both cases those mineral ions not retained for use by root cells are actively pumped into the xylem vessels and then depend on the transpiration stream for passive transport up the plant. Some active transport of ions occurs laterally into cambium, phloem and metabolising tissues along the way.

On arriving in the leaf via the transpiration stream, the remaining ions are actively transported into leaf cells. Iron and magnesium, for example, are needed for the formation of chlorophyll.

Importance of transpiration stream

The transpiration stream is the source from which cells in all parts of the plant gain water and nutrient ions. Green plants need water for photosynthesis and the maintenance of turgor (especially important for support in non-woody land plants). Nutrient ions are needed for healthy growth and for various metabolic processes (see page 289).

Evaporation of water from the leaves during transpiration has a cooling effect in hot weather.

Stomata

Distribution of stomata

In monocotyledonous plants (e.g. grass) stomata are found fairly equally distributed on both leaf surfaces. In dicotyledonous plants (e.g. geranium) they are almost entirely restricted to the lower epidermis.

Stomatal mechanism

Guard cells (see Figure 23.5) differ from normal epidermal cells in three ways. Guard cells are sausage-shaped and possess chloroplasts. In addition the inner regions of their cell walls (i.e. those facing the stomatal pore) are thicker and less elastic than the outer regions of their cell walls (see Figure 23.6).

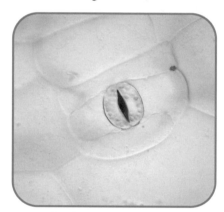

Figure 23.5 Guard cells

Opening and closing of stomata occur as a result of changes in turgor of guard cells.

When water enters a pair of flaccid guard cells, turgor increases. Due to their larger surface area and greater elasticity, the thin outer parts of the two guard cells' walls become stretched more than the thick inner parts. As the two guard cells bulge out, the thick inner walls become pulled apart opening the stoma.

Testing Your Knowledge

1 **a)** Starting in the soil solution, describe a possible route through a plant that could be taken by:
   i) a molecule of water which is eventually lost from the leaf surface;
   ii) an ion of magnesium which is eventually used in the formation of chlorophyll. (8)

   b) What name is given to the passage of water and nutrient ions up through a plant? (1)

2 **a)** Identify THREE factors that are thought to cause the ascent of water in a leafy land plant. (3)

   b) Which of these factors is the major force? (1)

3 **a)** What is meant by the term *transpiration*? (2)

   b) Is transpiration an active or a passive process? Explain your answer. (2)

   c) Give TWO ways in which transpiration is beneficial to a plant. (2)

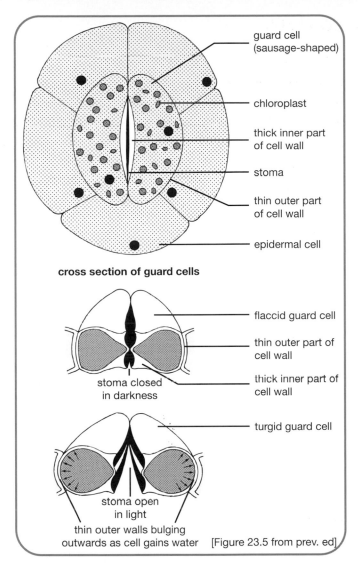

cross section of guard cells

Figure 23.6 Stomata and stomatal mechanism

When water leaves a pair of guard cells, they lose turgor and return to a flaccid condition. This results in the closing of the stoma.

Investigating opening and closing of stomata using *Commelina communis*

The leaves used in this experiment are taken from a well-watered plant that has been in light for several hours so that its stomata are fully open at the start.

A piece of lower epidermal peel is mounted on a microscope slide in a drop of each of the following liquids: distilled water, 0.2, 0.4 and 0.6 molar sucrose solution. The widths of 10 different stomata are measured for each liquid using an eyepiece **graticule** (see Figure 23.7) and averages are calculated. Figure 23.8 shows the type of graph obtained when the results are plotted.

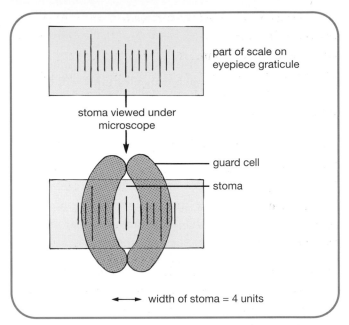

Figure 23.7 Measuring stomatal width

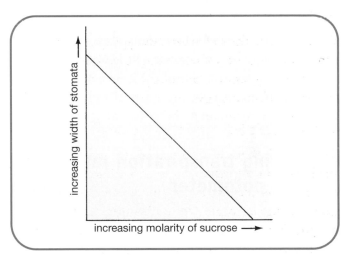

Figure 23.8 Effect of sucrose concentration on stomatal width

The results show that the stomatal width decreases as molarity of the bathing solution increases. At high concentrations of sucrose solution, the stomata are found to be closed.

It is therefore concluded that the more hypertonic the bathing solution, the more water is lost from the guard cells by osmosis. The more flaccid the guard cells become, the narrower the stomatal opening that remains between them.

Disadvantage of transpiration

Guard cells **gain turgor** in daylight and the stomata open. This physiological adaptation is ideal since the

193

plant needs to take in carbon dioxide for photosynthesis.

However open stomata also allow the loss of water by evaporation. Although such transpiration keeps the plant cool in hot weather, excessive transpiration during a dry sunny spell creates a potential **water balance problem**: the quantity of water lost at the leaf surfaces may exceed that being absorbed by the roots.

Under such circumstances leaf cells including guard cells lose turgor. Stomata are therefore often found to close for a time during the day in very hot weather. Since this cuts down water loss, the cells gradually regain turgor and the stomata may reopen later in the day.

Guard cells also **lose turgor** in darkness. Stomata therefore close at night and water is conserved. By controlling its water loss according to external conditions, a plant is capable of a limited degree of **osmoregulation**.

During a very dry spell when stomata are shut, cells still continue to lose turgor because a little water is lost directly through the cuticle of the leaves. At first **wilting** occurs but if the drought continues, the plant dies since it is unable to maintain its water balance.

Comparing transpiration rates using a potometer

A **bubble potometer** (see Figure 23.9) is an instrument used to measure **rate of water uptake** by a leafy shoot.

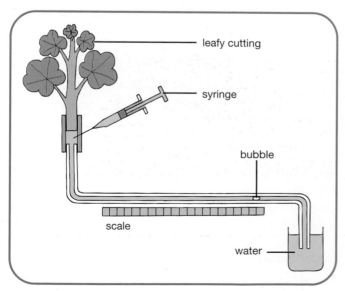

Figure 23.9 Bubble potometer

This rate of water uptake is only approximately equal to transpiration rate since some water may be retained by the leafy shoot for other processes (e.g. photosynthesis).

To set up a bubble potometer, the stem of a leafy shoot is cut **under water** and attached to a length of tightly fitting rubber tubing. With the cut end of the stem and the rubber tubing still immersed, one end of the capillary tubing is inserted into the rubber tubing and gently pushed until water flows along the length of tubing.

The open end of the capillary tube is transferred to the beaker of water and the end lifted to allow an **air bubble** to enter. Once the bubble has appeared on the horizontal arm of the potometer, its rate of movement along the scale is measured (e.g. in mm/min).

The syringe is used to inject water and return the bubble to the start of the scale allowing the experiment to be repeated for this 'normal' environmental condition.

The plant is then subjected in turn to each environmental condition and allowed time to **equilibrate** before its rate of water uptake is measured. Repeat runs are done in each case and the average rate of movement of the bubble calculated.

Table 23.1 summarises the reasons for adopting certain techniques and precautions during the investigation. Table 23.2 gives a specimen set of results for three different environmental conditions.

Discussion of results

From this experiment it is concluded that wind increases rate of transpiration. This occurs because the air outside the stomata is continuously being replaced with drier air which accepts more water vapour from the plant.

Increased **humidity** of the air surrounding the plant results in decreased transpiration rate. This occurs because the concentration gradient of water vapour between the inside and the outside of the leaf is decreased. Rate of diffusion of water molecules therefore slows down (see Figure 23.10).

design feature or precaution	reason
stem cut under water and end of stem, rubber tubing and capillary tubing all connected under water	to prevent air entering xylem and forming air locks
tightly fitting rubber tubing used	to prevent leakage of water and ensure that system is air-tight
time allowed for plant to equilibrate between different environmental conditions	to ensure that rate of movement of bubble is governed by factor being investigated and not the previous one
repeat measurements of rate of movement of bubble taken for each condition and average calculated	to obtain a more reliable result for each condition
all factors kept equal except for one change in environmental conditions	to ensure that only one variable factor is altered at a time

Table 23.1 Experimental design features and precautions

environmental condition	additional apparatus needed to create condition	average rate of movement of bubble (mm/min)
normal day	none	5
windy day	electric fan	20
humid day	transparent plastic bag	1

Table 23.2 Bubble potometer results

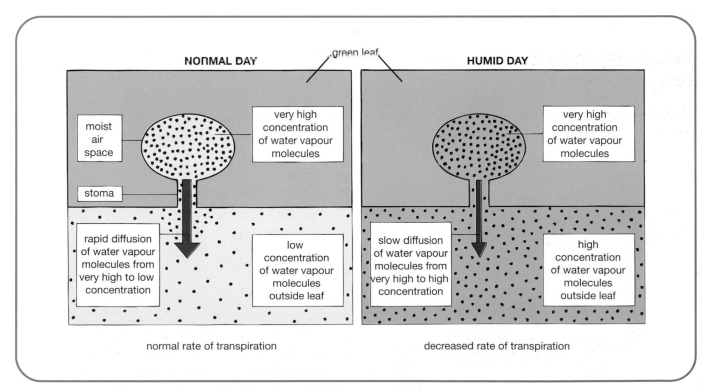

Figure 23.10 Molecular model of transpiration

Bubble atmometer

A **bubble atmometer** (see Figure 23.11) is an instrument used to measure **rate of evaporation** from a non-living surface.

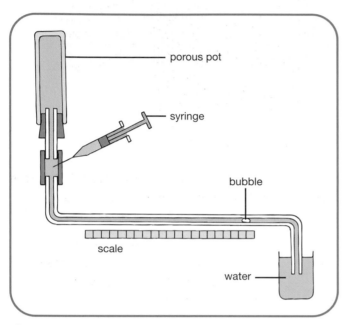

Figure 23.11 Bubble atmometer

When a bubble potometer and a bubble atmometer are subjected to the same conditions, the changes in rate of water loss from a living and a non-living system can be compared. In light, for example, both lose water rapidly but in darkness only the atmometer continues to do so since the plant has closed its stomata.

Thus the presence of an atmometer shows when the potometer is acting as a free evaporator and when it is affected by **physiological** factors such as stomatal closure.

Further factors affecting transpiration rate

In addition to light, availability of soil water, humidity of the surrounding air and wind speed, transpiration rate is also affected by the following factors.

Temperature

Transpiration increases with increase in temperature due to the faster evaporation rate of water molecules.

Air pollution

Transpiration rate is decreased if stomata are blocked with dirt.

Testing Your Knowledge

1 **a)** What is a *stoma*? (1)

   b) Give THREE structural differences between a guard cell and a neighbouring epidermal cell. (3)

2 **a) i)** Describe the state of the stomata when a plant is in bright light.

   ii) Why is this physiological adaptation normally of survival value to a plant? (2)

   b) i) Describe the mechanism by which a plant is able to osmoregulate in response to temporary shortage of soil water during a hot dry day.

   ii) Under what circumstances could such maintenance of water balance break down? (4)

3 **a)** To what use is each of the following put: **i)** a potometer; **ii)** an atmometer? (2)

   b) State THREE precautions that must be adopted when setting up the bubble potometer shown in Figure 23.9. (3)

   c) Name TWO factors other than light intensity which affect rate of transpiration and for each explain why. (4)

Xerophytes

'Normal' plants which live in habitats where water is abundant and excessive transpiration does not occur are called **mesophytes**.

Xerophytes are plants which live in habitats where a mesophyte would not survive because its transpiration rate would be excessively high. Such habitats are characterised by either hot, dry conditions and lack of soil water (e.g. desert) or exposed, windy conditions (e.g. moorland).

Xerophytes are able to maintain a water balance in such extreme habitats because they have evolved certain adaptations as follows.

Structural adaptations which reduce transpiration rate

The leaf shown in figure 23.12:

- has a **thick cuticle**. (This acts as a waterproof barrier and reduces water loss.)

- has a **reduced number of stomata**. (This cuts down the stomatal pore area through which water vapour is lost.)

- is **rolled** and **hairy**. (These features reduce transpiration rate by trapping moist, relatively immobile air between the stomata and the outer atmosphere.)

The leaf shown in figure 23.13 has **stomata sunken in pits**. The air present in the pits above the stomata is humid and therefore decreases transpiration rate.

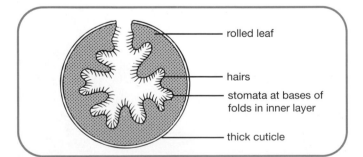

Figure 23.12 Transverse section of marram grass leaf

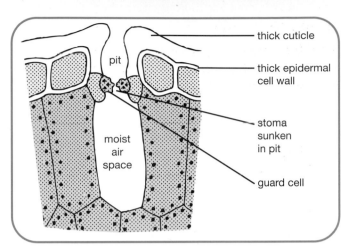

Figure 23.13 Detail of *Hakea* leaf

Some leaves (e.g. pine needles) are **small** and **circular** in cross-section thereby reducing the surface area of transpiring leaf exposed to the atmosphere.

In many cacti the leaves are reduced to protective **spines** (see Figure 23.14) and water loss is limited to the stem which carries out photosynthesis and possesses relatively few stomata.

Figure 23.14 Protective spines

Structural adaptations for resisting drought

Many cacti have **long roots** allowing absorption of subterranean water. Others possess extensive systems of **superficial roots** which grow parallel to the soil surface enabling them to absorb maximum water on those rare occasions when rain does fall. Cacti store this water in **succulent tissues** (see Figure 23.15) and may even have a **folded stem** which allows expansion and contraction (subject to water availability) without it becoming damaged.

197

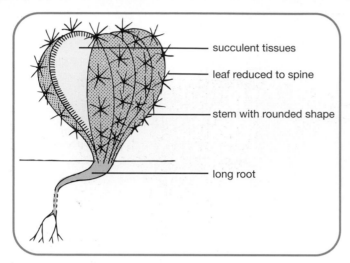

Figure 23.15 Cactus plant (with part of stem cut out)

- succulent tissues
- leaf reduced to spine
- stem with rounded shape
- long root

Physiological adaptation

Most cacti reduce water loss by showing **reversed stomatal rhythm** (i.e. closed during the day and open at night). During the night carbon dioxide is taken in and stored for use in photosynthesis during the day when the stomata are closed.

Hydrophytes

Hydrophytes are plants which live completely submerged (e.g. water-milfoil) or partially submerged (e.g. water-lily) in water. They have evolved many adaptations which help them to survive in their aquatic environment.

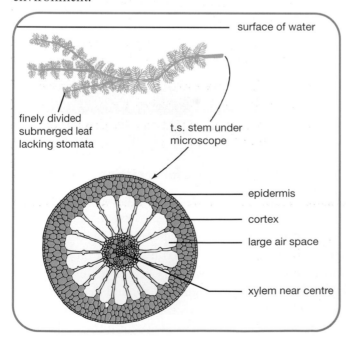

- surface of water
- finely divided submerged leaf lacking stomata
- t.s. stem under microscope
- epidermis
- cortex
- large air space
- xylem near centre

Figure 23.16 Water-milfoil

Air spaces

The organs of a hydrophyte possess an extensive system of intercommunicating **air-filled cavities** (see Figure 23.16). This **aeration tissue** gives a submerged plant **buoyancy** keeping its leaves near the surface for light. Much of the oxygen formed during photosynthesis is stored in these air spaces ready for use in respiration when required.

Reduction of xylem

Since water provides a submerged plant with support and is readily available for absorption all over its surface, a hydrophyte is found to possess little strengthening or water-conducting tissue. Any xylem present is normally found at the centre of the stem. This allows the stem **maximum flexibility** in response to water movements.

Specialised leaves

A hydrophyte's submerged leaves are **narrow** in shape or **finely divided** (see Figure 23.16). This adaptation helps to prevent them from being torn by water currents.

Stomata must be in contact with the air to bring about gaseous exchange. Submerged leaves therefore lack stomata and floating leaves (see Figure 23.17) have all of their stomata on their upper surfaces (see Figure 23.18). Such floating leaves often possess **long leaf stalks** (petioles). These allow the leaves to move up and down with changing water levels and prevent the stomata from being flooded when the water level rises.

Figure 23.17 Water-lily

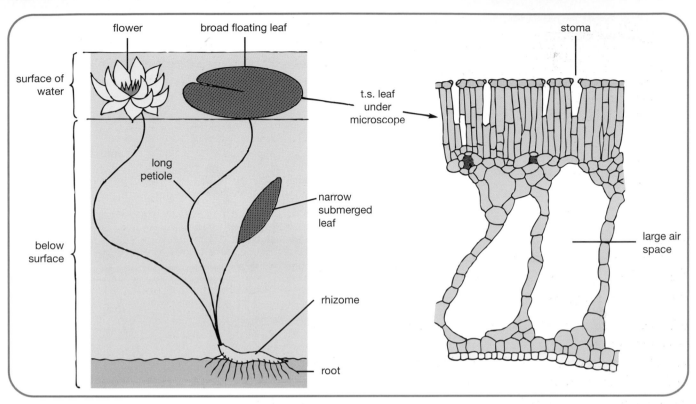

Figure 23.18 Structure of water-lily

Testing Your Knowledge

1 a) Name TWO particular environments which are populated almost exclusively by xerophytes. (1)

b) Why are mesophytes unable to survive in these habitats? (1)

2 a) Give TWO examples of structural adaptations which reduce transpiration rate of xerophytes. (2)

b) For each of these, describe how it brings about its effect. (2)

3 a) Give TWO examples of structural adaptations which help a xerophyte to resist drought. (2)

b) For each of these, describe how it plays its role. (2)

4 a) What is a *hydrophyte*? (1)

b) Give TWO reasons why hydrophytes normally have very little xylem tissue compared with mesophytes. (2)

c) If a hydrophyte does have xylem, where is it located in its stem? Explain why. (2)

5 a) Why is it of advantage to water-milfoil to have leaves that are finely divided? (1)

b) Where are stomata found on a floating leaf of water-lily? Explain why. (2)

Applying Your Knowledge

1 a) Explain why the stem of the leafy shoot shown in the experiment in Figure 23.19 must be cut under water. (1)

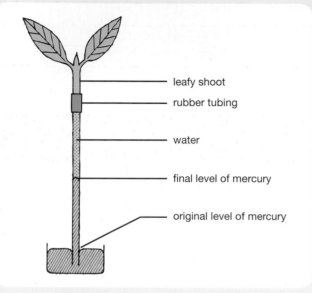

Figure 23.19

leafy shoot

rubber tubing

water

final level of mercury

original level of mercury

b) State a further precaution which must be adopted when setting up this experiment. (1)

c) To what important theory does the result obtained in this experiment lend strong support? Explain your answer. (2)

2 a) Describe what happens to the tree's diameter during the 24-hour period shown in the graph in Figure 23.20. (2)

b) Briefly explain why the process of transpiration is held responsible for this effect. (2)

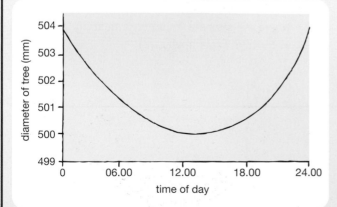

Figure 23.20

3 The readings on the balances in Figure 23.21 show the final masses of two sets of apparatus (X and Y) which have been allowed to lose water for two days in a laboratory.

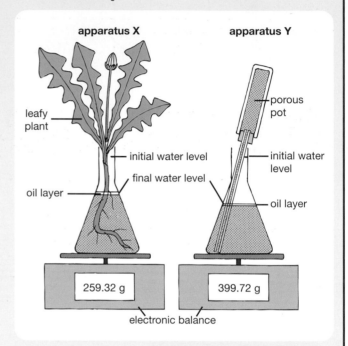

apparatus X   apparatus Y

leafy plant

porous pot

initial water level

initial water level

final water level

oil layer

oil layer

259.32 g   399.72 g

electronic balance

Figure 23.21

The partly completed Table 23.3 records the initial mass of apparatus X and the volume of water which had to be added to restore the water in each flask to its initial level.

a) Copy and complete Table 23.3. (2)

b) Calculate the average mass of water lost per hour by the plant. (1)

c) Explain why the volume of water needed to restore the initial level in X differed from the volume lost by the plant. (1)

d) Predict the effect on water loss of subjecting X and Y to the following conditions for two days:
  i) covering each with a transparent plastic bag;
  ii) keeping each in the same laboratory in total darkness.
  Explain your answer in each case. (4)

e) i) Apparatus X is called a weight potometer. Name apparatus Y.
  ii) Why is apparatus Y included in this experiment? (2)

	apparatus X	apparatus Y
initial mass (g)	283.80	
mass after 2 days (g)		
mass of water lost (g)		56.00
volume of water lost (cm³)	24.48	56.00
volume of water required to restore initial level (cm³)	25.00	56.00

Table 23.3

	total distance travelled by bubble (mm)			
time (in min)	leaf A	leaf B	leaf C	leaf D
start	0	0	0	0
2	1	30	6	20
4	2	80	12	50
6	3	128	18	92
8	3	168	24	130
10	3	200	30	158

Table 23.4

4 Four large leaves of similar size were removed from a horse chestnut tree and treated as follows:

leaf A	vaseline applied to both surfaces
leaf B	no vaseline applied
leaf C	vaseline applied to lower surface only
leaf D	vaseline applied to upper surface only

Each leaf was then attached to a bubble potometer of uniform design. The four potometers were kept in identical environmental conditions and the distance travelled by the bubble recorded for each over a period of ten minutes as shown in Table 23.4.

a) Present the data as line graphs. (4)

b) Which leaf showed the greatest rate of transpiration? Explain why. (1)

c) Which leaf would be unable to lose water by transpiration? (1)

d) What information in the table justifies the claim that the rate of movement of the bubble is not exactly equal to the rate of transpiration? (1)

e) What does the movement of the bubble in a potometer actually measure? (1)

f) Which surface of this type of leaf possesses more stomata? Explain how you arrived at your answer. (2)

5 Figure 23.22 illustrates the leaf of Scots pine. This tree often grows on exposed windy hillsides.

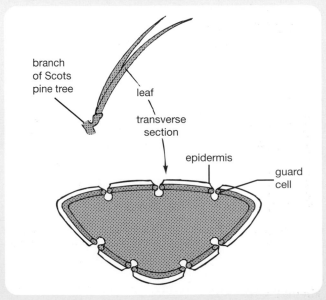

branch of Scots pine tree

leaf

transverse section

epidermis

guard cell

Figure 23.22

a) Name THREE structural adaptations shown by the leaf that suit the plant to life in its natural environment. (3)

b) Explain why Scots pine does not need to shed its leaves annually in autumn. (2)

6 The data in Table 23.5 refer to a series of experiments in which a large leafy shoot attached to a bubble potometer was subjected to various conditions of several abiotic factors. Table 23.6 records the results from a further series of experiments in which the effect of a wider range of one of the abiotic factors was studied.

a) Consider each of the following pairs of experiments in turn and for each pair draw a conclusion about the effect of a named abiotic factor on the time taken by the bubble to travel 100 mm: i) 1 and 2; ii) 3 and 7; iii) 6 and 10. (3)

b) Why is it impossible to draw a valid conclusion about the effect of an abiotic factor on the time taken by the bubble from a comparison of experiments 3 and 8? (1)

c) Which experiment in Table 23.5 should be compared with experiment 5 to ascertain the effect of darkness on time taken by the bubble to travel 100 mm? (1)

d) Which TWO experiments should be compared in order to find out the effect of wind speed on time taken by the bubble when the plant is in light at 25°C and in air of 95% humidity? (1)

e) Which TWO experiments should be compared in order to find out the effect of temperature on transpiration rate by the plant when exposed to windspeed of 15 m/s, in light and air of 75% humidity? (1)

f) i) Name the one variable factor studied in the series of experiments given in Table 23.6.
   ii) State the relationship that exists between this variable factor and transpiration rate.
   iii) Account for this relationship in terms of water vapour molecules. (5)

7 Figure 23.23 shows a type of plant adapted to life in semi-desert environments which receive hardly any rain.

		experiment number									
		1	2	3	4	5	6	7	8	9	10
abiotic factor	wind speed (m/s)	0	0	0	0	15	15	15	15	15	15
	temperature (°C)	5	25	5	25	5	25	5	25	5	25
	air humidity (%)	75	75	95	95	75	75	95	95	75	75
	light (L)/dark (D)	L	L	L	L	L	L	L	L	D	D
time taken by bubble to travel 100 mm		3 min 3 s	1 min 35 s	4 min 28 s	3 min 2 s	1 min 56 s	0 min 30 s	3 min 22 s	1 min 57 s	24 min 10 s	22 min 4 s

Table 23.5

		experiment number				
		11	12	13	14	15
abiotic factor	wind speed (m/s)	5	5	5	5	5
	temperature (°C)	20	20	20	20	20
	air humidity (%)	75	80	85	90	95
	light (L)/dark (D)	L	L	L	L	L
time taken by bubble to travel 100 mm		1 min 34 s	1 min 56 s	2 min 20 s	2 min 40 s	3 min 2 s

Table 23.6

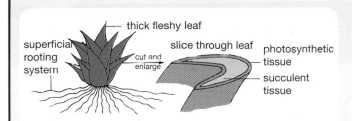

Figure 23.23

a) What general name is given to plants that can survive in desert conditions? (1)

b) i) Identify TWO structural features of this plant that enable it to make full use of rain on the rare occasions that it does fall.
   ii) Describe the role played by each of these features. (4)

c) Predict TWO ways in which this plant's stomata would differ from those of a mesophyte. (2)

8 Figure 23.24 shows the stems of three different plants cut in transverse section. Although they are not drawn to scale, the diameter of each is given in the diagram.

a) Copy and complete Table 23.7. (6)

b) Describe TWO ways in which the air spaces in stem B help the plant to survive in its natural environment. (2)

9 Write an essay on the transpiration stream and its importance in the life of a leafy plant. (10)

10 Give an account of the adaptations shown by xerophytes and hydrophytes that enable them to survive in their natural habitats. (10)

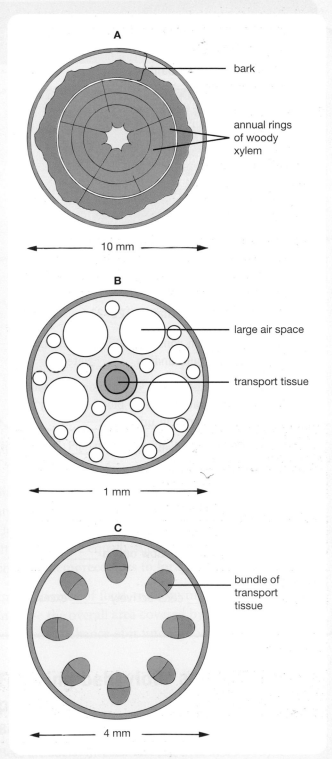

Figure 23.24

plant type	letter in diagram	two structural features of stem that led to choice of letter
hydrophyte		
young mesophyte		
old mesophyte		

Table 23.7

Figure 24.3 Waggle dance in bees

richness of the food supply present at the indicated source. This ensures that the attention of newly-recruited foragers is kept focussed on those locations offering the richest food supply.

Ants

When ants go foraging, each covers an area of ground and often 'meanders' back and forth over it. This search pattern increases the animal's chance of finding food as close to the colony as possible.

Once food has been located, the successful forager leaves **scent marks** on the ground as it heads directly back to the colony. Other ants quickly follow the trail to the food and reinforce the scent marks as long as the food supply lasts. Since the scent marks quickly fade, no energy is wasted following an old trail that no longer leads to food (see Figure 24.4).

Higher animals

The foraging information contained in the bee dance and the ant trail are social communication devices based on instinctive behaviour. Higher animals also forage for food. However their search patterns are more complex since their behaviour is not purely instinctive but also involves aspects of **learning**.

Among predatory animals which have to hunt and catch prey, a vast variety of methods of obtaining food exists. These often involve sophisticated physiological adaptations.

A bat (see Figure 24.5) emits **high-pitched sounds** which bounce off objects in the immediate

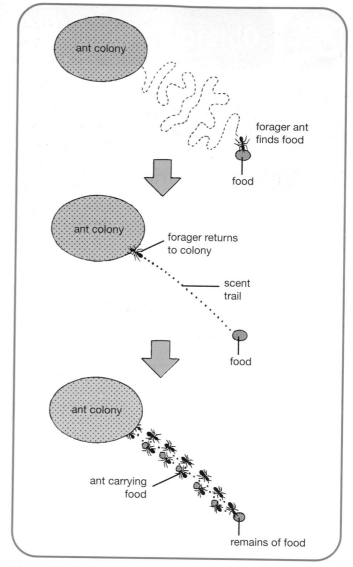

Figure 24.4 Foraging behaviour in ants

environment. These echoes are picked up by the bat's ears, enabling it to locate prey in the dark while at the same time avoiding harmful obstacles. So effective is **echolocation** that some bats can catch prey at a rate of one new victim every six seconds.

Whereas a cheetah depends on its **speed** (95 km/h over short distances) to catch its prey, a tiger uses **stealth** during the first part of the hunt and then captures its prey by leaping at it from a distance of up to six metres.

Figure 24.5 Long-eared bat

Economics of foraging behaviour

Time and energy are consumed by an animal during its search for food. If the energy gained from the food that it finds is less than the energy expended on the search, then the animal suffers a net loss of energy.

Such uneconomical behaviour would soon result in death if it were repeated on a regular basis. To be economical an animal must forage **optimally**. This means that it must consume those food items which will give it the best return for the time and energy spent. Foraging behaviour and subsequent choice of food items are therefore affected by the following factors.

Time

The search time is the name given to the length of time spent locating the food. The **pursuit time** is the time spent obtaining the food once located.

When the search time is short and the pursuit time long (e.g. a lion in constant sight of a herd of gazelles which are difficult to catch), then it is economical for the predator to be selective and wait for a chance to pick off an old or weak prey animal.

Where the search time is long and the pursuit time short (e.g. a bird searching at length through dense foliage and eventually finding a rich supply of insects which are easy to capture), then it is economical to be non-selective and eat as many prey items as possible since it may take a long time to find a similar haul.

Unproductive versus productive ecosystem

Consider a poor ecosystem where a number of different foods are thinly dispersed. The forager cannot afford to be too choosy because if it ate only the most desirable food items then too much time and energy would be spent on the search. On the other hand if it ate only the poor items, it would not gain enough energy. To be economical, it has to settle for a mixture of items with those of intermediate value giving the best net energy return for the time spent on the search (see Figure 24.6).

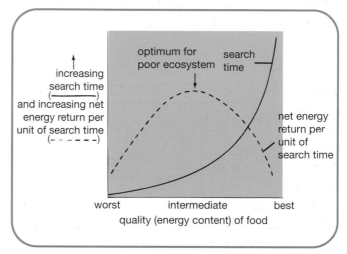

Figure 24.6 Effect of poor ecosystem on economics of foraging

In a rich ecosystem, search time is reduced and the forager can afford to be more selective since it can obtain all of its energy requirements from a few types of choice food items.

Risk versus good food supply

In some cases it is economical for a herbivore to settle for the food available in a poorer ecosystem if foraging in a food-rich area exposes it unduly to risk of attack by predators.

Size of prey

At first, **net energy gain** (benefit – cost) increases with increasing size of prey since larger items contain more energy than smaller ones. However an optimal prey size is eventually reached as shown in the graph in Figure 24.7.

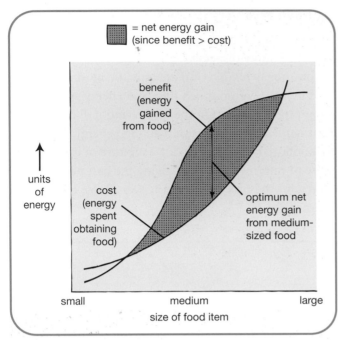

Figure 24.7 Effect of food size on economics of foraging

Beyond this point, net energy gain decreases because the very largest prey is the scarcest (hence involves a longer search time) and it tends to put up the best fight. It is not economical if the predator has to expend much energy subduing the prey.

Competition

Animals compete with one another for resources (such as food, water and shelter) if these are in short supply. Competition between individuals of different species is called **interspecific** competition whereas that between members of the same species is called **intraspecific** competition.

Interspecific competition

When two different species occupy the same ecological niche (see page 150), interspecific competition for a resource that is limited may become so fierce that one species ousts the other and the loser is forced to migrate or face extinction. This is called the competitive exclusion principle.

Paramecium

Paramecium (see page 3) is a unicellular organism. Two closely-related species of this animal are *Paramecium caudatum* and *Paramecium aurelia* (which is smaller in cell size).

Graph A in Figure 24.8 represents the growth of a population of *P. caudatum* cultured in favourable conditions and fed on bacteria. Graph B represents the growth of a population of *P. aurelia* cultured separately under exactly the same conditions. Each graph shows a normal population growth curve.

Testing Your Knowledge

1 **a) i)** Why is the random movement of *Planaria* an effective method of searching for food?
   ii) By what means does *Planaria* detect food during foraging?
   iii) In what way does *Planaria*'s movement change when it detects food? (3)

   b) i) What information does the duration of the dance performed by a honey bee on returning to the hive communicate to other bees about to go foraging?
   ii) Give TWO further pieces of information that the bee's dance indicates to the other workers. (3)

2 **a)** Copy and complete the following two equations which summarise the economics of foraging using the words: continued survival, death, gain, loss

   i) energy gained > energy expended = net _____ of energy and _____ .
   ii) energy gained < energy expended = net _____ of energy and _____ . (2)

   b) Which equation in a) would be described as uneconomic behaviour? (1)

3 **a)** What is the difference between *search time* and *pursuit time*? (2)

   b) Which food items give the forager the best net energy return in a poor ecosystem? (1)

   c) Explain why medium-sized prey often provides a predator with the greatest net energy gain. (2)

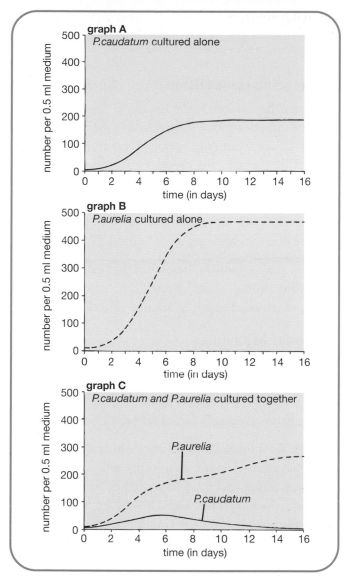

Figure 24.8 Effect of interspecific competition on population growth

Figure 24.9 Grey squirrel

Figure 24.10 Red squirrel

Graph C shows the growth curves that result when the two species are cultured together. During the first few days while food is still plentiful, both species increase in number in the absence of competition. However after about six days, the population of *P. caudatum* starts to decline whereas that of *P. aurelia* continues to rise. Eventually, as a result of interspecific competition, *P. caudatum* is completely ousted.

It is possible that *P. aurelia* is the successful competitor because it feeds more efficiently or survives on less food. However it is interesting to note that *P. aurelia* is also affected by the competition in that its population fails to reach as high a maximum size as it does when cultured alone.

Squirrels

The introduction of the North American grey squirrel (see Figure 24.9) to Britain has resulted in the widespread decline (almost to extinction) of the red squirrel (see Figure 24.10). Both types of squirrel occupy a similar ecological niche in the woodland ecosystem. The grey squirrel is thought to have become so widely distributed and successful because it competes aggressively for food and is able to make use of a wide variety of foodstuffs including acorns (despite their high tannin content).

The grey squirrel is therefore continuing to populate areas at the expense of the more timid red squirrel whose digestive system cannot cope with large quantities of acorns and other seeds rich in tannins.

Reduction in competition

Competition between members of two different species occupying a similar niche is sometimes reduced by the rivals reaching a 'compromise'. For example they might eat different foods, seek their food at different times of the day, nest in slightly different habitats and so on.

Cormorants

The common cormorant and the green cormorant are two species of sea bird which nest on cliffs and dive into the sea for fish (see Figure 24.11). They appear therefore to occupy exactly the same ecological niche. However one species has not forced the other out because competition is minimised by the common cormorant feeding mostly on flat fish and crustaceans from the sea bed while the green cormorant feeds further out to sea on eels and other fish swimming in the upper water.

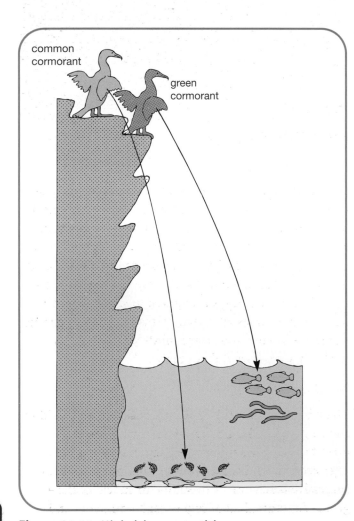

Figure 24.11 Minimising competition

It is highly likely, of course, that either type of cormorant, in the absence of the other, would extend its range.

Intraspecific competition

Whereas different species of an animal community can reduce competition by eating different foods, nesting in slightly different habitats, seeking food at different times of the day etc., the members of a population of the same species need exactly the **same resources**. Intraspecific competition is therefore even more intense than interspecific competition when there is a scarcity of some essential resource such as food.

Intraspecific competition regulates the size of the population of the species affected. The weaker members are weeded out by natural selection.

Territorial behaviour

Intraspecific competition often takes the form of **territoriality**. This is the name given to behaviour which involves competition between members of the same species (especially birds) for territories.

An animal's total **range** is the area which it covers during its lifetime. Within this range a male animal often establishes and inhabits a smaller area called its **territory** which contains enough food for himself and eventually a mate and their young.

He defends his territory fiercely using **social signals** (sign stimuli) as shown in Figures 24.12 and 24.13. He is most aggressive at the centre of his territory. The

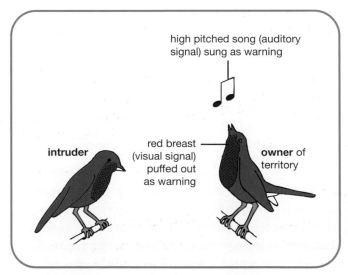

Figure 24.12 Territorial behaviour in robins

further he moves away from the centre, the less likely he is to attack intruders. Eventually during such a contest there comes a point when he is equally likely to fight or turn and flee. Such points of balance established by 'pendulum' fighting (alternating attack and escape) mark the **boundary** between the territories of two rivals.

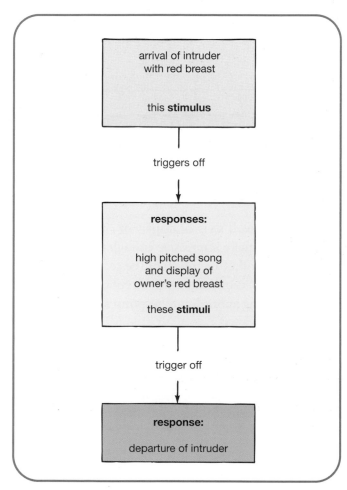

Figure 24.13 Flow chart of territorial behaviour

Unlike the members of a dominance hierarchy (see opposite) which live in a social group based on rank order, most territory holders are **solitary**. Each enjoys complete dominance within the boundaries of his own territory. Whereas a bird of prey will defend a full square mile or more, the needs of a robin are met by one small garden.

Advantages of territorial behaviour

Once territories have been established, aggression between neighbours is reduced to a minimum and energy is saved. Territorial behaviour spaces out the

population in relation to the available food supply. This ensures that there will be enough food for the number of young produced.

Red grouse

The red grouse (see Figure 24.14) lives on moorland and feeds on the shoot tips and flowers of heather plants. Young heather plants are more nutritious than older ones. Rather than compete directly for food, the male red grouse claims a territory large enough to provide food for his dependents during the breeding season.

Figure 24.14 Red grouse

Territorial size varies depending on the availability of food. Each enclosed space in Figure 24.15 represents a red grouse's territory on the same piece of moorland over a period of four years. During years 1 and 4, food was plentiful and the birds only needed to defend small territories. During years 2 and 3, the heather was poor and a larger territory was required to supply a bird's needs. When times were lean, intraspecific competition was more intense and weaker birds which failed to establish a territory did not breed.

Dominance hierarchy

Many higher animals live in large **social groups**. Within such a group, a **dominance hierarchy** is found to operate. This is a system where the individuals are organised into a graded order of rank resulting from aggressive behaviour between the members of the group. An individual of higher rank dominates and exerts control over others of lower rank.

second one which can peck all others except the first and so on down the line. This linear form of social organisation is called a **pecking order**.

Table 24.1 summarises the results from observing a group of hens over a period of time. Bird A dominates all of the others, bird B dominates all of the others except A and so on down the line to bird J which is dominated by all the other birds. A pecking order is an example of dominance hierarchy.

Mammals

Although not so clear cut, a similar system of social organisation exists amongst some mammals such as wolves and baboons. Because of his rank, the dominant individual has certain rights such as first choice of food, preferred sleeping places and available mates. This dominant male asserts his rank by employing social signals as shown in Figure 24.16.

The dominant wolf's visual display of **ritualised threat gestures** is normally impressive enough to assert his authority over other members of the social group. These in turn demonstrate their acceptance of his status by making **subordinate responses**.

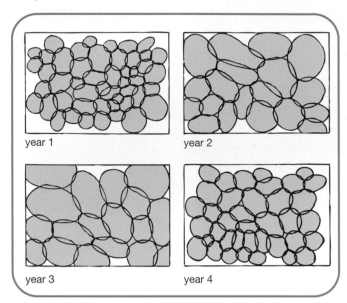

year 1 year 2

year 3 year 4

Figure 24.15 Sizes of red grouse territories

Birds

If a group of newly-hatched birds such as pigeons are kept together, one will soon emerge as the **dominant** member of the group. This bird is able to peck and intimidate all other members of the group without being attacked in return. It therefore gets first choice of any available food. Below this dominant bird there is a

		hen receiving pecks									
		A	**B**	**C**	**D**	**E**	**F**	**G**	**H**	**I**	**J**
h e n g i v i n g p e c k s	A		✓	✓	✓	✓	✓	✓	✓	✓	✓
	B			✓	✓	✓	✓	✓	✓	✓	✓
	C				✓	✓	✓	✓	✓	✓	✓
	D					✓	✓	✓	✓	✓	✓
	E						✓	✓	✓	✓	✓
	F							✓	✓	✓	✓
	G								✓	✓	✓
	H									✓	✓
	I										✓
	J										

Table 24.1 Pecking order for a group of 10 hens

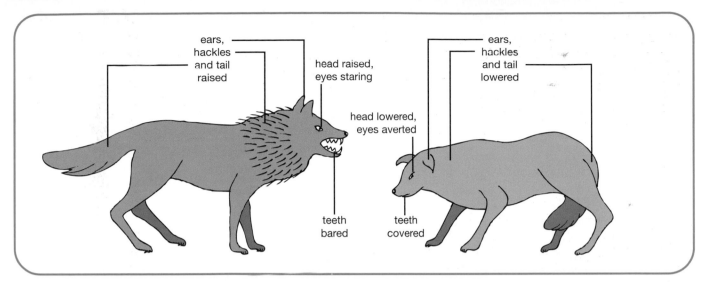

Figure 24.16 Dominant and subordinate responses

Advantages of dominance hierarchy

A system of dominance hierarchy increases a species' chance of survival since:

- aggression between its members becomes ritualised;

- real fighting is kept to a minimum;

- serious injury is normally avoided;

- energy is conserved;

- experienced leadership is guaranteed;

- the most powerful animals are most likely to pass their genes on to the next generation.

Co-operative hunting

Some predatory mammals such as killer whales, lions, wolves and wild dogs rely on co-operation between members of the social group to hunt their prey.

Killer whales hunt in packs called pods (see Figure 24.17) and employ various strategies. In a river mouth, a pod will sweep along in a line to catch migrating salmon. In coastal waters the same pod will encircle a shoal of herring and concentrate them into a seething mass. The whales then thrash the herring with their tails to stun them and gorge themselves on a catch of food that would be unavailable to a solitary predator.

The ambush strategy employed by lions involves some predators driving prey towards others that are hidden in cover and ready to pounce.

Figure 24.17 Pod of killer whales

Dogs and wolves, on the other hand, take turns at running down a solitary prey animal to the point of exhaustion and then attacking it (see Figure 24.18).

In the case of lions, wolves and dogs, the group of predators tends to concentrate its efforts on a prey animal that has become separated from the rest of the

Figure 24.18 Co-operative hunting

herd. This is often a young and inexperienced or old and infirm animal, making it an easy target.

Advantages of co-operative hunting

When a kill is achieved, all members of the predator group obtain food (some of which may be disgorged later by females to feed the young). Thus **co-operative hunting** benefits the subordinate animals as well as the dominant leader of the group. By working together, the animals are able to tackle large prey animals and, as a result, all gain more food than they would by foraging alone. Provided that the food reward gained by co-operative hunting exceeds that from foraging individually, the social group will continue to share food regardless of the fact that the dominant member(s) receives a much larger share than the subordinate ones.

Testing Your Knowledge

1 a) Explain the difference between the terms *intraspecific* and *interspecific* competition. (2)

   b) Explain what is meant by the term *competitive exclusion principle*. (2)

2 With reference to cormorants, describe the means by which competition between species can be kept to a minimum. Explain why this is of advantage to the species involved. (3)

3 a) Explain the difference between the terms *territory* and *territoriality*. (2)

   b) Briefly describe an example of territorial behaviour with reference to a named animal. (2)

   c) State TWO advantages of territoriality. (2)

4 Instead of fighting, wolves perform a ritual.

   a) Describe FOUR features of a dominant wolf's visual display when asserting his authority. (2)

   b) Describe the corresponding responses displayed by a subordinate animal. (2)

   c) i) What name is given to the type of social organisation that results from this behaviour pattern?

      ii) State TWO ways in which it is of advantage to the animals concerned? (3)

5 A pack of African wild dogs catch a large prey animal such as a wildebeest by running it down to the point of exhaustion. Give TWO advantages gained by the dogs from this form of co-operative hunting. (2)

Applying Your Knowledge

1 Copy and complete Table 24.2 which refers to some of the design features which apply to the planarian investigation described on page 205. (5)

2 Figure 24.19 shows an experiment set up to investigate the effect of a chemical extracted from the glands of ants on their food-searching behaviour.

feature of experimental design	reason
same number of animals of same species used and identical environmental conditions maintained in each dish	
	to ensure that the results are not based on only one animal whose behaviour might not be typical of the species in general
hungry animals used	
	to ensure that the animals are not simply attracted to any physical object
experiment repeated many times	

Table 24.2

and several moths (prey). Hungry blindfolded bats were released into chamber X and hungry bats with their ears plugged were released into Y.

a) State FOUR factors which would have to be kept constant between X and Y to make this a fair test. (4)

b) Identify the variable which was altered. (1)

c) Predict, with reasons, the outcome of the experiment. (2)

4 The graph in Figure 24.20 shows the economics of defending a territory of varying size by a certain species of bird.

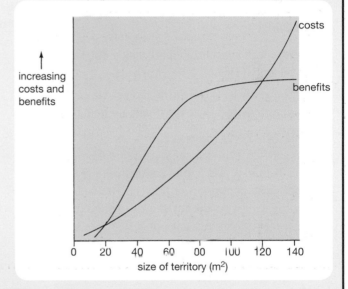

increasing costs and benefits

costs

benefits

size of territory (m²)

Figure 24.20

a) State the range of territory size that can be realistically defended by this species of bird. (1)

b) Suggest why the bird cannot survive in a territory of 10 m². (1)

c) Give ONE example of the high 'cost' that makes the defence of a territory of 140 m² unrealistic for this type of bird. (1)

d) What is the optimum size of territory for this species of bird? Explain why. (2)

5 Shore crabs feed on mussels by breaking open their shells and eating the soft inners. Figure 24.21 shows the results of offering crabs an unlimited choice of mussels of different sizes.

a) State the energy gained by a crab consuming:
   i) five mussels of length 18 mm per day;
   ii) three mussels of length 23 mm per day;
   iii) one mussel of length 27 mm per day. (3)

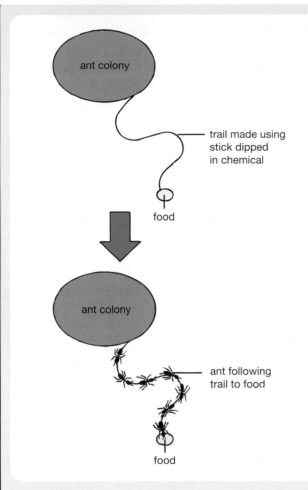

ant colony

trail made using stick dipped in chemical

food

ant colony

ant following trail to food

food

Figure 24.19

a) i) Describe the control that should have been set up.
   ii) Explain clearly why the control is necessary. (2)

b) What conclusion can be drawn from the experiment (assuming that the control remained unchanged)? (1)

c) What experiment (and control) could be set up to investigate whether the presence of food is necessary for ants to follow the chemical trail? (1)

d) The effect of the chemical is short-lived yet more and more ants emerge and follow the trail (even in darkness) provided that the food supply does not run out. How do they find their way? (1)

e) By what means are wasted journeys to locations that no longer contain food prevented? (1)

3 In an investigation involving bats, two environmental chambers, X and Y, were used which contained lengths of wire strung from ceiling to floor

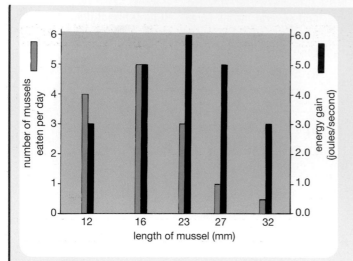

Figure 24.21

b) Which of these daily diets yields optimum energy? (1)

c) Suggest why tackling the largest size of mussel is uneconomical in terms of energy gain. (1)

6 Table 24.3 gives the results of a series of competition experiments involving two species of flour beetle, A and B. The beetles were reared together in containers of flour maintained at the environmental conditions shown in the table. The experiments were run until only one species of beetle ('the winner') remained in each container.

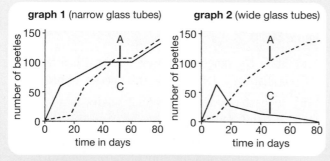

Figure 24.22

The graphs in Figure 24.22 refer to a further experiment using species A reared with a much smaller beetle, species C, in containers of flour in which a number of short lengths of glass tubing had been buried.

Graph 1 shows the results using glass tubing narrower in diameter than the body size of species A but wider than that of species C.

Graph 2 shows the results using glass tubing wider than the body sizes of both A and C.

a) From the table, state the most favourable environmental conditions for each species. (2)

b) Under what set of conditions was competition probably most intense? Explain your answer. (2)

c) In what way do the data in the table support the principle of competitive exclusion? (1)

d) Compare graphs 1 and 2 with respect to the effect of width of glass tubing on number of species C. Give a possible explanation for this difference. (2)

e) Is the competition illustrated by the above experiments intraspecific or interspecific? (1)

7 Five male zebra finches P, Q, R, S and T were kept together and observed over a period of several days. During this time, a record was kept of the results from 20 confrontations between each pair of birds. The bird which successfully dominated its rival in each contest was given a score of one point. The results are shown in Table 24.4.

a) Copy and complete the two right-hand columns in the table. (The first example has been done for you.) (1)

b) Which bird has the lowest status and is at the bottom of the pecking order? Explain your choice. (2)

environmental conditions	percentage number of experiments won by species:	
	A	B
dry and cold	100	0
dry and warm	85	15
dry and hot	88	12
wet and cold	69	31
wet and warm	16	84
wet and hot	0	100

Table 24.3

contest	score out of 20 (points)	winner	net number of contests won
T v Q	T 17, Q 3	T	14
T v R	T 3, R 17		
P v Q	P18, Q 2		
Q v R	Q 0, R 20		
Q v S	Q 8, S 12		
R v P	R 13, P 7		
P v T	P 14,T 6		
S v T	S 5,T 15		
R v S	R 19, S 1		
S v P	S 4, P 16		

Table 24.4

c) Which bird is at the top of this dominance hierarchy? Explain your choice. (2)

d) Give the pecking order of the five birds. (1)

8 Give an account of the ways in which animals obtain food under the following headings:

a) foraging behaviour and search patterns; (5)

b) economics of foraging behaviour. (5)

9 Write an essay on interspecific and intraspecific competition as shown by named animals in their pursuit of an adequate food supply. (10)

25 Obtaining food – plants

Sessility and mobility

Movement is a characteristic of all living things. An animal can normally move its whole body from place to place and is said therefore to be **mobile**. The movements of higher plants, on the other hand, are restricted to certain parts such as leaves turning towards the light. The plant cannot move about as a complete organism. It remains fixed to one position and is said to be immobile or **sessile**.

Such mobility and sessility have a bearing on the strategies employed by animals and plants to obtain food.

Heterotrophic nutrition

Animals are **heterotrophic**. This means that they need a ready-made organic substance to provide them with energy and materials for building cytoplasm. They are therefore dependent directly or indirectly on the synthetic activities of green plants.

In order to obtain food, most animals have to go foraging (see chapter 24) and therefore must be **mobile**. This need for mobility puts a limit on the maximum body size that an animal can attain and still function efficiently.

Autotrophic nutrition

Green plants are **autotrophic**. This means that they do not depend on an outside source of organic food but are able to synthesise all of their organic requirements from simple inorganic molecules. During **photosynthesis**, carbon dioxide and water are used to produce carbohydrates. Nitrate absorbed from the immediate environment provides the nitrogen needed for protein synthesis.

Since all the necessary chemicals are 'to hand', green plants produce their own food 'on site' and have no need to be mobile. This sessile existence allows many plants to attain a body size greatly in excess of any animal.

Leaf mosaic patterns

Natural selection favours a characteristic which increases the amount of light that a plant is able to absorb. A plant's leaves are therefore often arranged in a **mosaic pattern** (see Figure 25.1).

Figure 25.1 Leaf mosaic pattern

This allows the maximum surface area of leaf to receive light since overshadowing of one leaf by another is reduced to a minimum. The leaf mosaic pattern of beech leaves is so effective that almost no light passes through the leaf canopy of a beech wood and only a few shade-tolerant species can survive on the ground below.

Competition between plants

Plants growing in the same habitat compete with one another for factors such as light, water and soil nutrients if any one of these is in short supply (see Figure 25.2).

Investigating intraspecific competition

Plants of the same species have exactly the same growth requirements. When grown together, they will be in direct competition with one another if any resource is limiting and this competition will be intense.

The experiment shown in Figure 25.3 is set up to investigate the effect of density on germinating cress

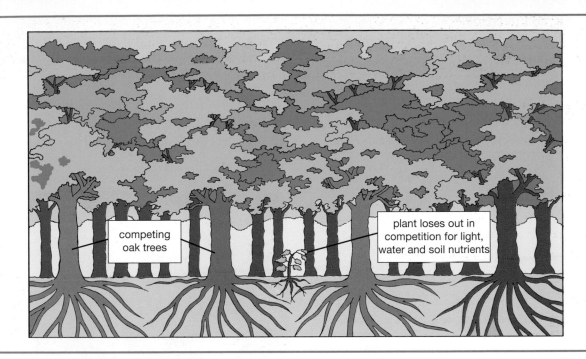

Figure 25.2 Intraspecific competition

seeds. Carton A with only 100 seeds represents low density of planting; carton B with 500 seeds represents high density.

After 5 days, almost all the seeds in A are found to have successfully germinated and grown into healthy seedlings. Although most of the seeds in B also germinate, many fail to grow into healthy plants and remain yellow and sickly. Of the seedlings that do grow successfully, most are found to be smaller than their counterparts in A.

From this experiment it is concluded that the spaced out plants in A grow well because each plant receives an adequate supply of each growth requirement.

In carton B where the seedlings are densely packed, much of the light that might have reached each plant is intercepted by the leaves of other plants so that on average each plant receives less light for photosynthesis. In addition their rooting systems become interwoven and may be competing for water and minerals.

If a plant is short of water, its stomata stay closed for a longer time and carbon dioxide uptake needed for photosynthesis is reduced. Thus intraspecific competition between densely-populated plants results in many individuals growing more slowly.

Commercial applications

Farmers use a drilling machine to space out crop seeds during planting. Horticulturalists mix tiny seeds with sand to sow them as thinly as possible and then thin out newly-germinated seedlings to reduce intraspecific competition.

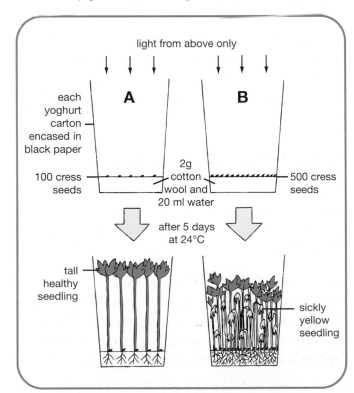

Figure 25.3 Investigating intraspecific competition

Interspecific competition

Plants of different species which occupy the same habitat often differ from one another in growth form (e.g. rooting depth, see Figure 25.4) and mineral requirements. Thus interspecific competition is normally less intense than intraspecific competition.

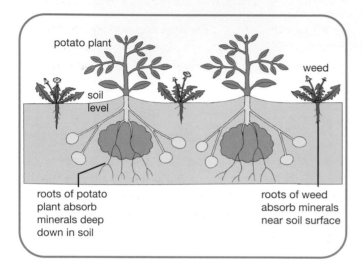

Figure 25.4 Interspecific competition

Nevertheless some species are able to become dominant at the expense of others. Conifer trees grown in closely packed plantations (see Figure 25.5) almost totally prevent the growth of other plants on the forest floor

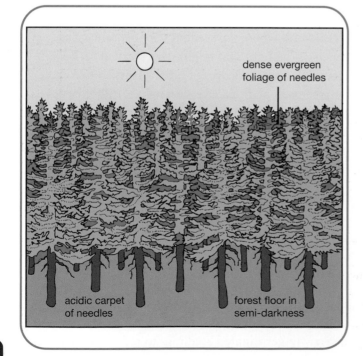

Figure 25.5 Competition by conifer trees

by cutting out the light. In addition, the carpet of needles on the ground produces an acidic soil low in minerals that further inhibits the growth of potential competitors.

Some plants such as *Rhododendron* produce toxic substances. These prevent other plants from growing in the soil beneath the *Rhododendron* plant (see Figure 25.6).

Figure 25.6 Clear zone under *Rhododendron* bush

Effect of grazing by herbivores on species diversity

Rabbits

Natural grassland normally contains a rich variety of plant species. Some are especially sturdy and show vigorous growth; others are more delicate.

Rabbits are relatively unselective grazers. Their effect on a grassland's plant community depends on the level of grazing pressure applied. At very low levels of grazing pressure, the aggressive dominant grasses are not held in check and tend to drive out the less vigorous plants. Species diversity amongst the plant community under these circumstances is low.

As grazing pressure increases (see Figure 25.7), more

and more of the competitive dominant grasses are eaten and kept in check. This tends to free resources and space for the smaller, more delicate plant varieties and promotes an **increase in diversity** of species. However at very high intensities of grazing pressure, species **diversity decreases** again as the dense population of rabbits struggle to find food and drive some of the more delicate species to extinction within that plant community.

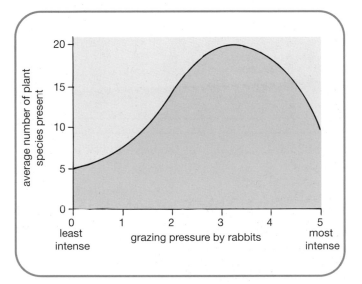

Figure 25.7 Effect of grazing on species diversity

Sheep

When grazing pressure is very intense on Scottish grassland that once was forest, such as Scots pine or birch, sheep eat any tree saplings that appear among the grass. Regeneration of a forest community by the process of succession (see chapter 36) is therefore prevented and **diversity** of plant species is **decreased**.

However when grazing pressure is kept very low on the same type of land, Scots pine and birch woodland communities are able to develop giving an **increase in species diversity**.

Compensation point

At a certain low level of light intensity, the rate of photosynthesis occurring in a leaf exactly equals the rate of respiration. This particular light intensity is called the plant's **compensation point**. It varies from species to species.

At compensation point there is no net gaseous exchange. Neither is there a net gain nor loss of food. Figure 25.8 compares the events occurring in a leaf at compensation point with those occurring in darkness and in light of high intensity.

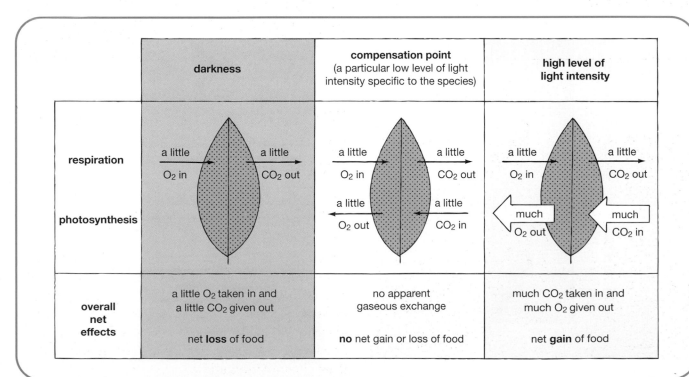

Figure 25.8 Comparison of compensation point with two extremes

During 24 hours of clear weather, compensation point in a plant occurs twice: shortly after dawn and shortly before nightfall. During the day, between these two compensation points, photosynthesis exceeds respiration and a store of carbohydrate builds up. Some of this is needed by the plant for use during the night and on dull days.

Sun and shade plants

Plants which thrive in brightly illuminated habitats and grow best when they are not shaded by neighbouring members of the community are called **sun** plants.

Plants which thrive in dimly lit habitats where they are shaded for long periods of the year by other members of the community are called **shade** plants (see Figure 25.9).

Consider the graph in Figure 25.10 which compares a sun and a shade plant with respect to their CO_2 output and uptake. Both plants give out CO_2 as a result of respiration in darkness. As the day dawns and light intensity gradually increases, each plant begins to photosynthesise until eventually photosynthetic rate equals and then exceeds respiration rate.

Figure 25.9 "...and she has the nerve to call herself a **sun** plant."

However, compared with the sun plant, the shade plant is found to take a **shorter** time to reach its compensation point because this occurs at a **lower** light intensity.

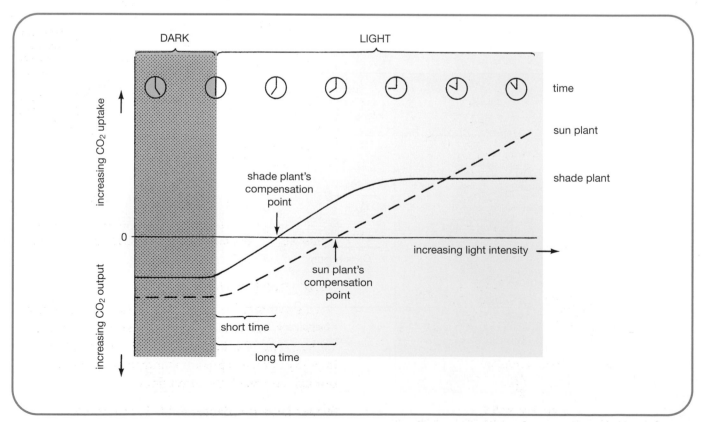

Figure 25.10 Compensation points of sun and shade plants

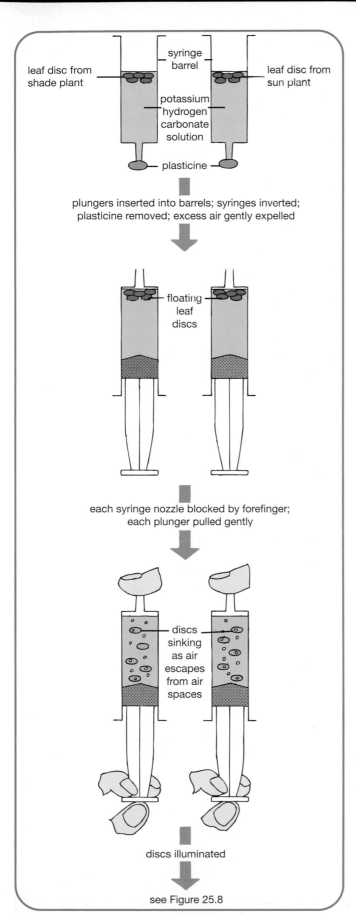

Figure 25.11 Setting up sun and shade plant investigation

Investigating the response of sun and shade plants to green light

In the experiment shown in Figure 25.11, leaf discs from sun and shade plants are immersed in a solution of sodium hydrogen carbonate to ensure that they receive an adequate supply of CO_2 for photosynthesis. Initially the discs **float** at the top of the liquid in the syringe barrel because air in their air spaces makes them less dense than the surrounding liquid. However when the syringe plunger is inserted and pulled down, air is extracted from their air spaces and the discs sink.

When photosynthesis occurs, oxygen is produced and collects in a leaf disc's air spaces. This makes the leaf disc become less dense once more and rise to the surface of the liquid.

When illuminated with **green light**, the shade plant's discs are found to rise to the surface (see Figure 25.12). The sun plant's discs fail to do so given the same length of time. It is therefore concluded that in green light, photosynthesis occurred (at a detectable level) only in the shade plant.

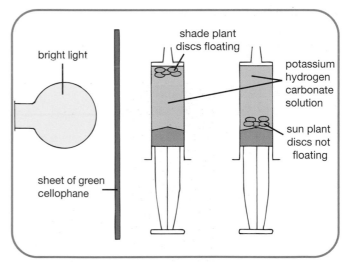

Figure 25.12 Response of sun and shade plants to green light

Adaptations of shade plants

During summer in a deciduous forest, the leaves of sun plants (e.g. oak and beech trees) absorb most of the sun's red and blue light and only allow light of low intensity (which is predominantly green) to pass through and reach the forest floor.

Shade plants are adapted to life in this poorly illuminated habitat in several ways:

Compared with sun plants, shade plants have a greater ability to make efficient use of light of **low intensity**.

Compared with sun plants, shade plants contain a richer supply of **carotenoid (yellow) pigments**

which enables them to absorb and make use of green (and blue-green) light for photosynthesis.

Shade plants produce their flowers and seeds **early** in the year before the dense leaf canopy above them reduces their rate of photosynthesis.

Testing Your Knowledge

1 a) Explain the meaning of the terms *mobility* and *sessility* with reference to living things. (2)

   b) Why is it essential to an animal such as a lion to be mobile? (1)

   c) How is a plant such as an oak tree able to survive without being mobile? (1)

2 The leaves of plants are often found to be arranged as a mosaic pattern. What is the benefit of this adaptation to the plant? (2)

3 a) Name TWO environmental factors for which neighbouring plants could be competing. (2)

   b) When a factor essential for plant growth is in short supply, is interspecific competition normally more or less intense than intraspecific competition? Explain why. (2)

4 What effect does moderate grazing pressure by rabbits have on the diversity of plant species on a piece of grassland. Explain your answer. (2)

5 a) What is meant by the term *compensation point*? (2)

   b) i) What is the difference between a sun and a shade plant?

      ii) Which type of plant is able to make maximum use of green light?

      iii) Why is this adaptation of survival value to the plant? (4)

Applying Your Knowledge

1 a) Copy and complete Table 25.1 which refers to some of the design features which apply to the cress seed investigation on page 219. (4)

   b) It could be argued that water was lost by evaporation from the two cartons at different rates and that this introduced a second variable factor. Suggest an improvement that could be made to the design to overcome this problem. (1)

2 A student set up an experiment to investigate the effect of competition on the growth of a population of cress seedlings. Figure 25.13 shows a simplified version of his set-up after 5 days in a seed propagator. Table 25.2 gives his results.

feature of experimental design	reason
equal mass of cotton wool and equal volume of water added to each carton	
yoghurt cartons are encased in black paper or painted black	
only recently purchased cress seeds are used	
experiment is repeated many times and results are pooled	

Table 25.1

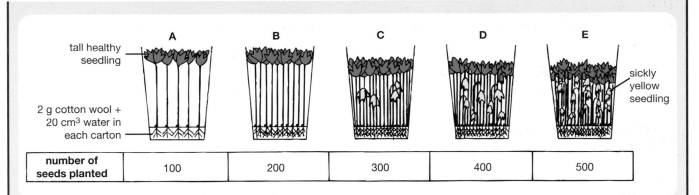

Figure 25.13

number of seeds planted	number of healthy seedlings with green leaves	% number of healthy seedlings with green leaves
100	87	
200	178	
300	204	
400	228	
500	225	

Table 25.2

a) State the means by which the factor under investigation was varied. (1)

b) State THREE factors that the diagram shows were kept constant. (3)

c) i) Copy and complete Table 25.2.
   ii) Why is it necessary to convert the results to percentages? (3)

d) Plot a line graph of number of seeds planted against percentage number of healthy seedlings with green leaves. (3)

e) Make a generalisation from your graph about the effect of competition on the growth of a population of cress seedlings. (1)

f) Suggest ONE factor other than water that cress seedlings in carton E could be competing for. (1)

g) Figure 25.14 is a partly completed diagram of an alternative method of investigating the effect of competition on cress seedlings. Make a simple diagram of plates B and C to show them set up and ready at the start of the experiment. (2)

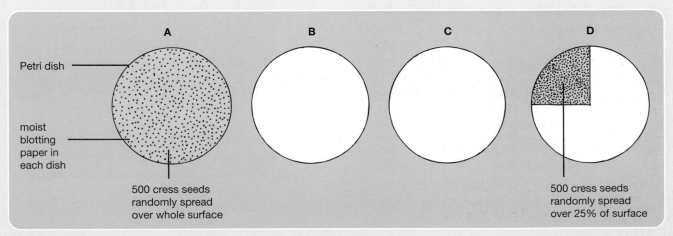

Figure 25.14

3   Table 25.3 gives the results of a plant competition experiment where five groups of cereal seed grains of the same species, but differing in number from one another, were grown in areas of uniform size and soil fertility.

number of grains planted per unit area	yield (grams of dry mass per unit area)
5	10
20	50
40	70
60	72
100	72

Table 25.3

   a) Plot the data as a line graph in the form of a curve. (2)

   b) i)   What relationship exists initially between number of grains planted and yield?
      ii) Why does this trend not continue indefinitely with increase in number of grains planted? (2)

   c) Apart from wastage of seed grain, what other disadvantage could result from planting too dense a crop? (1)

   d) Which of the following densities of planting would be likely to suffer most from interspecific competition?

      A 20   **B** 40   **C** 60   **D** 100 (1)

   e) From your graph state the optimum number of seed grains that should be planted per unit area. (1)

4   The experiment shown in Figure 25.15 was set up in an attempt to demonstrate that the weed, gold of pleasure, makes a chemical which inhibits growth of flax.

   a) In which pot does inhibition appear to have occurred? (1)

   b) Suggest which part of the weed plant could be making the inhibitor. Explain your answer. (2)

   c) Why was control pot A included in the experiment? (1)

   d) State TWO ways in which the experiment would need to be improved before the experimenter would be justified in drawing any valid conclusions. (2)

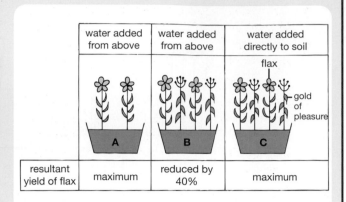

Figure 25.15

5   Many species of seaweed are found to grow in rocky pools at the seashore. Some of these algae are competitively dominant; others are more delicate and less aggressive. The periwinkle snail is a selective feeder and shows a preference for the dominant seaweeds. The graph in Figure 25.16 shows the effect of increasing population density of periwinkle snail on number of algal species present.

   a) The grazing pressure applied by a density of 50 periwinkles/m^2 is described as low. What effect does it have on the average number of algal species present in the rock pools? Explain your answer. (2)

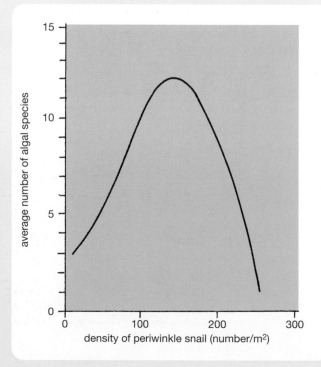

Figure 25.16

b) The grazing pressure applied by a density of 150 periwinkles/m² is described as moderate. What effect does it have on the average number of algal species present in the rock pools? Explain your answer. (2)

c) The grazing pressure applied by a density of 250 periwinkles/m² is described as very high. What effect does it have on the average number of algal species present in the rock pools? Explain your answer. (2)

6 Two similar squares of turf (X and Y) were cut out of a piece of grassland and examined. Each was found to possess the same community of 20 different plant species. The squares were kept in a greenhouse under identical conditions except that X was regularly cropped whereas Y was left uncut.

After three years one of the squares was found to possess 11 of the original species while the other still had all 20.

Identify which square was which and explain why. (3)

7 a) 'No plant can survive indefinitely in nature at the light intensity of its compensation point.' Is this statement true or false? Justify your answer. (2)

b) The graph in Figure 25.17 refers to a single plant during a period of 24 hours in summer.
   i) How long did the plant take to reach its compensation point in the morning?
   ii) At what two times did the plant reach its compensation point?
   iii) By how many hours were these two points separated during the day?
   iv) Predict how your answer to **iii)** would have differed if the graph had been drawn for the same plant in autumn. (4)

8 a) Copy and complete Table 25.4 which refers to some of the design features which apply to the

feature of experimental design	reason
	to ensure that CO_2 supply is not a limiting factor
lamp placed at an equal distance from the syringes in each treatment	
	to ensure that an additional variable factor is not introduced
experiment repeated and results pooled	
	to absorb all colours in spectrum of white light except green

Table 25.4

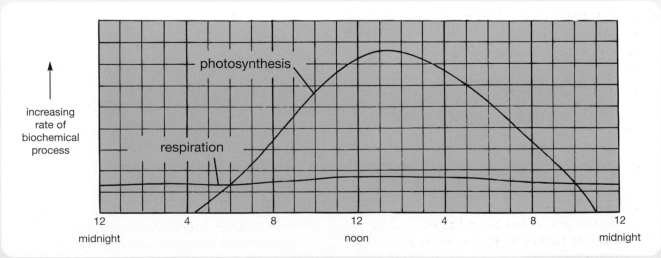

Figure 25.17

sun and shade plant investigation on page 223. (5)

b) In this investigation, it could be argued that light other than green may be able to reach the discs in Figure 25.12. How could this source of error be overcome? (2)

9 Figure 25.18 shows a separation of the leaf pigments from a sun and a shade plant using thin layer chromatography.

a) i) Which chromatogram shows the pigment separation from the shade plant?

ii) Explain your choice. (2)

b) Relate the difference between the two chromatograms to survival by a shade plant in a deciduous forest in summer. (2)

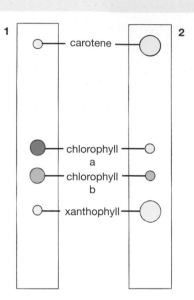

Figure 25.18

What You Should Know

Chapters 24–25

(See Table 25.5 for Word bank)

check	intensity	requirements
compensation	interspecific	respiration
co-operative	intraspecific	sessile
diversity	leadership	size
energy	light	social
food	materials	subordinate
foraging	nutrients	sun
hierarchy	optimally	territoriality
intense	rank	time

Table 25.5 Word bank for chapters 24–25

1 Many animals show distinct behaviour patterns when _____ for food.

2 Such behaviour tends to increase the animal's chance of gaining maximum net _____.

3 To be economical an animal must forage _____ with respect to search and pursuit _____, type of food selected and _____ of food selected.

4 _____ competition occurs between members of the same species; _____ competition occurs between members of different species.

5 Intraspecific competition for territories is called _____. It spaces out a population in relation to available _____ supply.

6 Dominance _____ amongst the members of a social group involves lower ranking individuals acknowledging the status of those with higher _____. They do this by showing _____ responses to the latter's threat displays. This behaviour conserves energy and ensures experienced _____.

7 _____ hunting benefits all the members of a _____ group since all the animals gain more food than they would hunting on their own.

8 Animals are mobile but plants are _____. This normally poses no problem for plants since their immediate environment provides all the raw _____ needed for survival.

9 Intraspecific competition exists amongst the members of a dense population of plants for _____, water and soil _____.

10 Interspecific competition between plants tends to be less _____ than intraspecific competition because different species often have different _____.

11 Moderate grazing of grassland by rabbits maintains species _____ since dominant plants are held in _____.

12 Compensation point is that low level of light _____ at which the rate of photosynthesis in a plant exactly equals the rate of _____. A shade plant is found to have its _____ point at a lower intensity of light than a _____ plant.

26 Coping with dangers

Animals

Avoidance behaviour and habituation

Consider the experiment shown in Figure 26.1. When the underside of the glass sheet is tapped sharply, the snail shows **avoidance behaviour** by retreating into its shell. This escape response is an example of **unlearned behaviour**. Snails are genetically programmed to respond to danger in this way. (Such a behavioural adaptation is of survival value and has therefore been favoured by natural selection during millions of years of evolution.)

When the snail re-emerges and the glass is tapped for a second time, the snail retreats as before. However, after many repeats, the snail eventually ceases to respond to the stimulus and therefore fails to show avoidance behaviour.

This is because the stimulus has proved, after many repeats, to be **harmless** and the animal has learned not to react to it. This type of **learned behaviour** is called **habituation**. It is beneficial because it prevents the animal from performing its escape response so often that it has no time or energy left to carry out essential activities such as obtaining food.

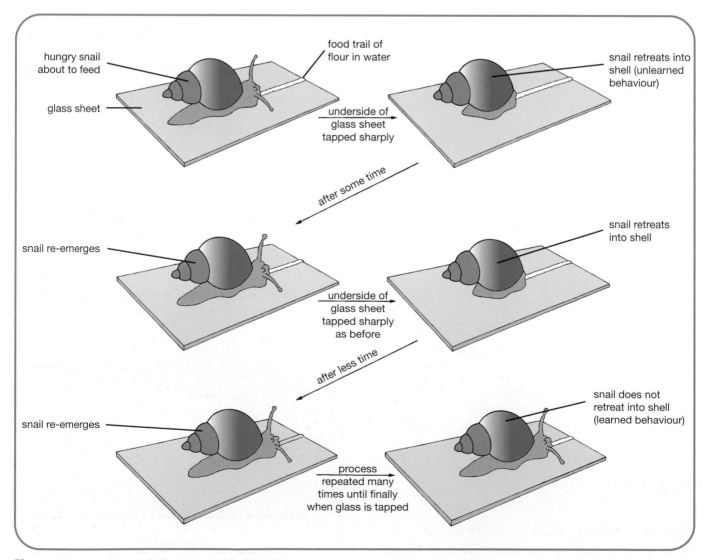

Figure 26.1 Avoidance behaviour and habituation

After a period lacking stimulation, the animal does respond to the stimulus as before and does show avoidance behaviour. It is important that habituation is short-lived because a long-term modification of the escape response would leave the animal open to danger.

In this experiment, hungry snails are used so that they will be **motivated** to re-emerge from their shells after being subjected to the stimulus. Repeat stimuli are applied as soon as the snail re-emerges to prevent it eating so much food that it loses its motivation.

Many snails of the same species are used and environmental factors such as light and temperature are kept constant to prevent a further variable factor being introduced into the experiment. The experiment is repeated many times to increase the reliability of the results.

Learning experiments in humans

Mirror drawing

Using her normal writing hand, the learner (see Figure 26.2) is timed as she joins up the dots forming the star outline while looking only at the mirror image.

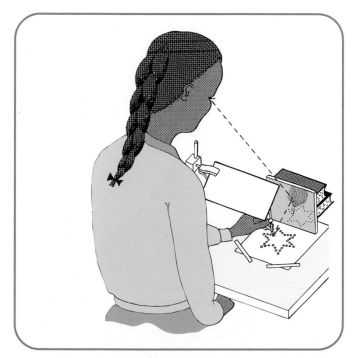

Figure 26.2 Mirror drawing

The learner is allowed ten trials to give her the opportunity to reach her best score. The same learner is used for each set of trials, the same writing hand is

used and a fresh copy of the same star outline is used each time to ensure that no second variable factor is introduced into the investigation.

At the end of a set of trials, the time required to perform the task successfully is found to have decreased to a minimum value. When graphed the results give a **learning curve** (see Figure 26.3). The curve never meets the x-axis but eventually runs parallel to it since a minimum time is needed to perform the task.

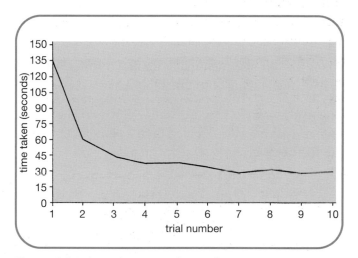

Figure 26.3 Learning curve (type 1)

Constructing words

The learner is allowed one minute in which to construct as many three-lettered words as possible using individual letter tiles. When the procedure is repeated several times, the learner is found to increase the score. When graphed the results give a second type of learning curve as shown in Figure 26.4.

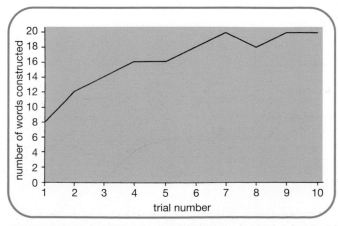

Figure 26.4 Learning curve (type 2)

Long-term modification

When each learning experiment is repeated several days later, the learner is found to have retained some of the learning. Now the performance at trial 1 is better than it was during the first experiment.

This shows that learning involves a long-term modification of the response made to a stimulus. In order to learn, an animal must be capable of remembering.

Learning to avoid danger

Toads

Toads feed on insects. A toad's natural instinct is to snap at any moving object that looks like an insect.

The toad shown in Figure 26.5 has previously been fed on harmless insects presented on the end of a thread but has never been presented with a bee.

When the toad rises to the bee bait, it receives a violent sting and spits the insect out. When offered a second bee, the toad ducks its head and refuses it. When offered a robber fly whose striped pattern closely resembles that of a bee, the toad refuses it but snaps at an insect that does not resemble the bee.

From this experiment it is concluded that the toad has learned that the yellow-striped pattern of the bee means danger and avoids it.

Birds

Newly-hatched ducklings and goslings quickly learn to follow the first large object that they meet if it moves and gives out sounds. The young birds are responding to the visual and auditory stimuli normally provided by their mother whom they learn to follow.

This form of learning, which can only occur during a brief period of early life, is called **imprinting** (see Figure 26.6). It is a behavioural adaptation of survival value since it provides the means by which a young bird avoids danger by staying close to and being protected by its mother.

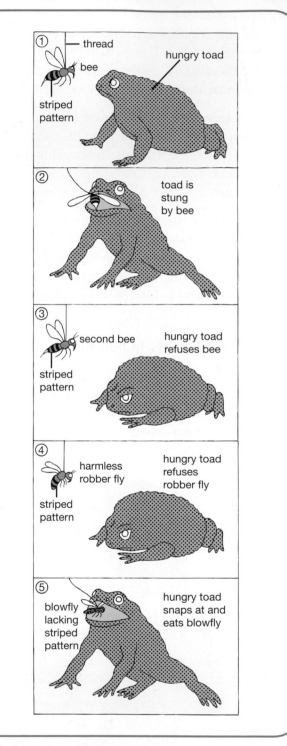

Figure 26.5 Learning to avoid danger

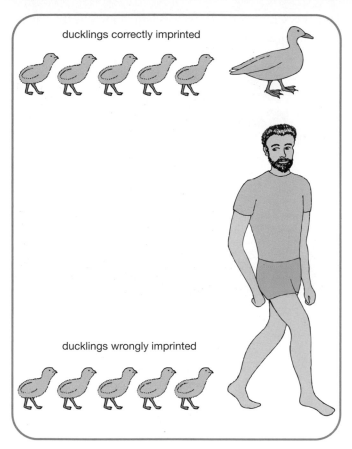

Figure 26.6 Imprinting

Humans

Children learn to avoid danger by being educated by, and imitating, their more experienced and knowledgeable elders.

In a science laboratory, danger is avoided by learning the meanings of the hazard signs (see Figure 26.7) and then applying appropriate precautions when using chemicals or materials bearing these signs.

Very young children are often unable to judge the speed and distance of approaching vehicles. Rather than learn

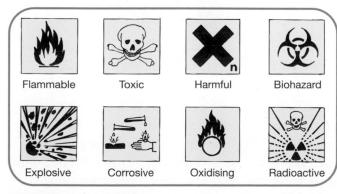

| Flammable | Toxic | Harmful | Biohazard |
| Explosive | Corrosive | Oxidising | Radioactive |

Figure 26.7 Hazard signs

'road sense' by trial and error, they are taught the Green Cross Code and are accompanied when crossing the road.

Intelligence

An animal's ability to learn depends on its intelligence. An intelligent animal is capable of reasoning and can solve a problem by applying previously learned concepts. Advanced animals can therefore figure out what constitutes a potentially dangerous situation and take steps to avoid it.

Testing Your Knowledge

1 a) i) Describe the type of avoidance behaviour that is shown by a snail.
      ii) Is this behaviour learned or unlearned? (2)
   b) i) Explain what is meant by the term *habituation* with reference to a snail.
      ii) Is this type of behaviour learned or unlearned? (3)

2 a) Why is it of survival value to a toad to learn to avoid insects with yellow and black stripes? (1)
   b) Explain why imprinting of young birds on the first noisy, moving object that they meet after hatching, normally helps to protect them from danger. (2)

3 a) Rewrite the following sentence using only the correct word from each choice in brackets.

      Learning in humans involves a (long-term/short-term) modification of the (stimulus/response) made to a (stimulus/response). (1)
   b) Give TWO examples of ways in which children learn to avoid danger. (2)

Individual mechanisms of defence
Active

In addition to 'tooth and claw', many other forms of active defence are found in the animal kingdom.

The skunk produces a foul-smelling secretion which it squirts at enemies. This causes great distress to the sprayed victim and allows the skunk to escape. Similarly many insects and reptiles defend themselves by injecting poison into their attackers (see Figure 26.8).

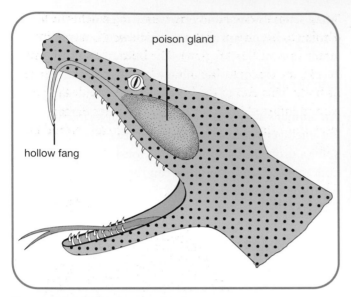

Figure 26.8 Snake's use of poison

26.10) resemble menacing eyes which are effective at startling predators.

Figure 26.9 Hedgehog's protective spines

Animals with long legs such as the antelope and ostrich defend themselves by **fleeing** from the enemy at top speed. Most birds take to the air when frightened. Many animals attempt to defend themselves by withdrawing from danger into **cover**. The octopus creates its own cover by ejecting an inky 'smoke screen' which aids escape.

Ground-nesting birds such as the avocet defend their young by performing **distraction displays**. Uttering plaintive cries, the adult lurches over the ground, helplessly flapping a 'broken' wing. Once the predator has been diverted from the nest, the adult suddenly flies off to safety.

A few animals attempt to defend themselves by **feigning death**. When threatened, the grass-snake often turns on its back, opens its mouth, lets its tongue loll and plays dead until the enemy goes away. Feigning death only works against predators that feed exclusively on freshly-killed prey and ignore already dead organisms that may have started to rot.

Passive

Many animals possess **protective coverings** such as the shell of the limpet, the armour plating of the armadillo, the bristles of certain caterpillars and the spines of the hedgehog (see Figure 26.9). All of these structural adaptations safeguard the animals from attacks by predators.

Markings on the wings of some moths (see Figure

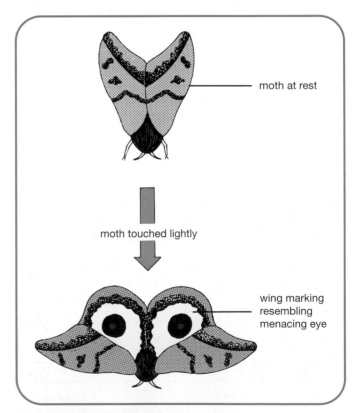

Figure 26.10 'Eye-spot' display in moth

Possession of **poisonous chemicals** in the body acts as an excellent defence when accompanied by warning colouration. The kokoi (poison-arrow) frog of the Amazon rainforest, for example, carries a deadly poison in its skin glands and advertises its presence using a brightly coloured coat (see Figure 26.11).

Figure 26.11 Poisonous kokoi frog's warning colouration

Camouflage is a further adaptation which provides prey animals with a means of passive defence. Some adult insects such as the stick insect closely resemble twigs; others look like leaves. Some harmless insects (e.g. the hoverfly) **mimic** the striped pattern of poisonous ones. (e.g. the stinging wasp) and are mistakenly ignored by predators.

Some animals have **patterns** which help them to blend with their surroundings. The flounder fish can even change the intensity of its colouring to resemble the background on which it is lying (see Figure 26.12).

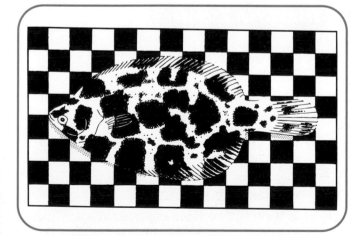

Figure 26.12 Camouflage in flounder

Some animals (e.g. trout) possess a two-tone colour pattern called **counter shading** which offers double camouflage. The dark upper half of the body viewed from above is almost invisible against the river bed background while the light underside viewed from below blends effectively with the sky.

Some animals defend themselves using a **deflection display** to deceive predators (which would normally target a prey's head region). The butterfly fish (Figure 26.13), for example, has a huge 'eye' spot at the rear of its body. This end of the body can better withstand a wound inflicted by a predator and deflects damage away from the vulnerable head end where it would be fatal.

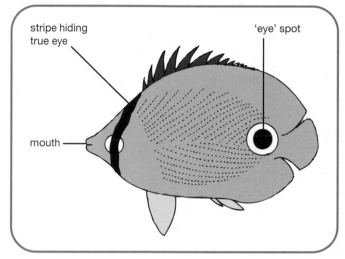

Figure 26.13 Deflection display in butterfly fish

Social mechanisms for defence

By staying together as a large group (e.g. school of fish, herd of mammals), many types of animals rely on the principle of 'safety in numbers' as a means of defence. Amongst a flock of birds, for example, there are many eyes constantly on the lookout for enemies. Following an **alarm call**, the bunching and swirling tactics adopted by the flock confuse the predator who finds it much more difficult to capture a member of a large unpredictable milling crowd than a solitary individual.

Defence is strengthened further by the members of a social group adopting a **specialised formation** as in the following examples.

Musk ox

Musk oxen (Figure 26.14) are native to Arctic regions of Canada and Greenland. Their natural environment is completely open and offers no scrub or woodland to use for concealment. Their natural enemy (apart from human beings) is the wolf. When threatened, a herd of musk oxen form a **protective group** with cows and calves in the centre and mature males at the outside

Figure 26.14 Social defence (musk oxen)

with their huge horns directed outwards. Individual wolves are gored and packs are driven off by a combined charge. This form of social defence is called mobbing.

Quail

Bobwhite quails roost in circles with their heads to the outside as shown in Figure 26.15. If disturbed, the circle acts as a defensive formation by 'exploding' in the predator's face. By the time their enemy has recovered from the confusion, the birds have flown away to safety.

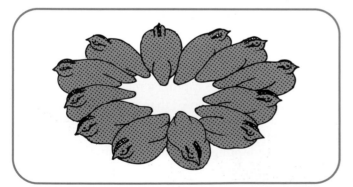

Figure 26.15 Social defence (quail)

Baboon

A strict social hierarchy is observed by the members of a troop of baboons. When they are on the march (see Figure 26.16), the dominant males stay in the centre close to the females with infants. Lower-ranking adult males and juveniles keep to the edge of the troop and raise the alarm if the group is threatened (e.g. by a leopard).

Testing Your Knowledge

1 Why is feigning death an especially risky mechanism of defence? (1)

2 Describe the mechanism used by ground-dwelling birds to defend their young from predators. (2)

3 **a)** Kokoi frogs live a relatively untroubled life on the floor of the Amazon rainforest. How do they defend themselves against potential predators? (2)

   b) Is the form of defence employed by kokoi frogs individual or social? Explain your answer. (1)

4 Explain how a trout's counter shading acts as a double camouflage. (2)

5 Musk oxen live in a completely open environment. How do they defend themselves against wolves? (2)

Figure 26.16 Social defence (baboon)

Plants

Unlike animals, plants are sessile. In the presence of hungry herbivores, plants cannot defend themselves actively by 'fight or flight'. Instead they employ various passive means of defence.

Structural defence mechanisms

Thorns

Hawthorn trees bear sharp thorns which are modified side branches (see Figure 26.17). These are able to cause serious damage to the delicate mouth lining of a browsing herbivore and protect the plant from future attacks.

Spines

In gorse, the leaves are reduced entirely to short spines (Figure 26.18) which are borne on modified branch stems. This presents an almost impenetrable barrier to hungry herbivorous mammals.

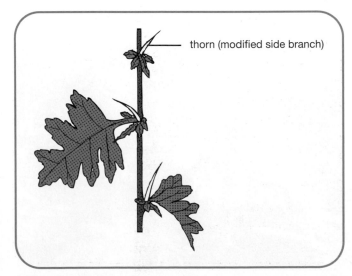

Figure 26.17 Thorn

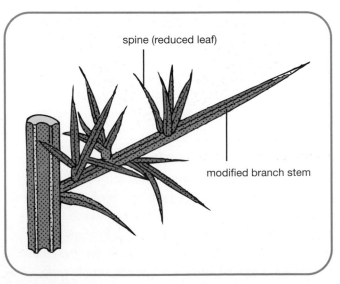

Figure 26.18 Spine

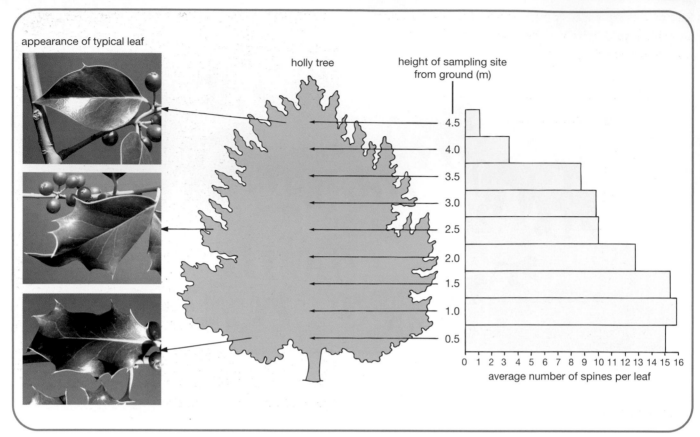

appearance of typical leaf

holly tree

height of sampling site from ground (m)

4.5
4.0
3.5
3.0
2.5
2.0
1.5
1.0
0.5

0 1 2 3 4 5 6 7 8 9 10 11 12 13 14 15 16
average number of spines per leaf

Figure 26.19 Relationship of spine number to height of holly leaf

The leaves of some cacti (see page 197) are completely reduced to spines. This helps to protect inner succulent tissues from thirsty desert animals.

Investigating the number of spines on holly leaves

In holly, the spines are restricted to the edges of the leaves. Figure 26.19 shows the results of an investigation into the relationship between number of spines present and height of leaf above the ground. Twenty leaves were collected at each sample site and on average the number of spines was found to decrease as the height of the leaf increased.

It is thought that natural selection has favoured those holly plants possessing spines concentrated on the lower leaves because this adaptation is of survival value. Leaves near the ground are within easy reach of many browsing herbivores and if they lacked spines they would be eaten. As height of leaf increases, the chance of attack is reduced and therefore the number of protective spines decreases.

Stings

The leaves of the nettle plant (see Figure 26.20) possess stinging hairs (see Figure 26.21). Each takes the form of a thin capillary tube ending in a spherical tip. When an animal touches a hair, its spherical tip breaks off

Figure 26.20 Nettle plant

leaving a sharp edge. This penetrates the animal's skin allowing liquid irritant to be injected and issue an impressive warning.

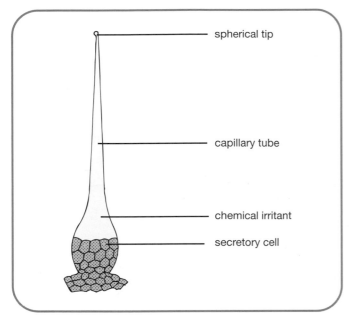

Figure 26.21 Nettle sting

Figure 26.22 Underground stems

Ability to tolerate grazing

The growing points (meristems) of most plant stems are located in buds well above ground level. If these are eaten by a herbivore during grazing, the plant may be unable to recover.

The **growth form** of some plants is specially adapted to survive conditions of continuous grazing. Most grasses, for example, have very **low growing points** and are therefore able to continue to send up new leaves despite the fact that their older ones are constantly being eaten. This enables grass to withstand a degree of grazing pressure that would be fatal to most other plants.

Some meadow grasses possess **underground stems** called rhizomes (see figure 26.22) that are capable of asexual reproduction. Others have **deep root systems** capable of regeneration. These organs, safely out of reach of grazing animals, ensure the plants' survival.

Plantain and dandelion have evolved a flat **rosette habit** (see Figure 26.23). Their leaves radiate out from a very short stem and often lie firmly pressed against the soil surface enabling them to escape the attention of grazing herbivores.

Figure 26.23 Rosette growth form

Testing Your Knowledge

structural adaptation	example of plant possessing this structure	how structure works as a defence mechanism
sting		
thorn		
spine		

Table 26.1

1 Copy and complete Table 26.1. (3)

2 **a)** What relationship exists between number of spines on a holly leaf and distance of leaf from the ground? (1)

   b) Suggest why holly trees have evolved in this way. (2)

3 Describe TWO adaptations possessed by grass plants which enable them to tolerate heavy grazing. (2)

Applying Your Knowledge

1 If an animal (e.g. a bird) is repeatedly exposed to a stimulus (e.g. a scarecrow) which fails to be dangerous, then eventually the animal fails to respond to the stimulus. Name this type of behaviour and explain why it is of advantage to the animal concerned. (3)

2 *Nereis* is a type of marine worm which lives in a tube. Its head emerges from the tube to feed but is quickly withdrawn in response to a variety of stimuli.

The graph in Figure 26.24 shows the results of an investigation into the effects of two stimuli tried at two-minute intervals on 40 worms living in glass tubes in a shallow tank of water.

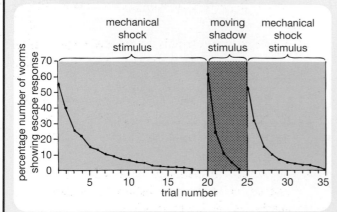

Figure 26.24

a) Why were as many as 40 worms used in the investigation? (1)

b) Suggest how the mechanical shock stimulus was applied simultaneously to the worms. (1)

c) Identify the unlearned escape response that *Nereis* employs to cope with such possible dangers. (1)

d) i) What evidence is there from the graph that the worms became habituated to both of the stimuli used in this investigation?

   ii) What evidence is there from the graph that one stimulus acts independently of the other? (2)

e) i) Which trial number's results best support the idea that habituation to mechanical shock is short-lived?

   ii) What is the survival value of this form of learned behaviour being short-lived? (2)

3 During his first day in a new job, an employee was shown how to assemble a type of machine component. Figure 26.25 charts his progress during his first eight days of employment.

a) What name is given to this type of graph? (1)

b) Suggest why the number of components assembled on day 2 was only slightly greater than on day 1. (1)

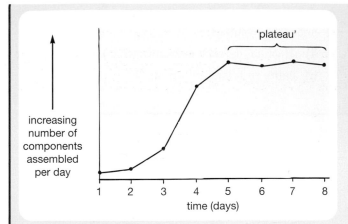

Figure 26.25

c) Between which two days did 'learning' occur at the fastest rate? Suggest why. (2)

d) Give ONE possible reason for the 'plateau'. (1)

4 The upper half of a pigeon's body is often darker than the lower half.

a) What name is given to this type of colour pattern? (1)

b) Describe its dual role in camouflaging the bird. (2)

c) Is this adaptation an individual or a social defence mechanism? (1)

5 Some caterpillars depend on camouflage for defence whereas others rely on bright warning colours.

To be effective, which group should be clustered together and which group widely scattered? For each, explain why. (3)

6 Read the passage and answer the questions that follow it.

The camouflage provided by cryptic colouration of body parts increases the chance of survival of many prey organisms. This affords little protection, however, to those organisms whose lifestyle demands much mobility. Many such animals are found, therefore, to have evolved an alternative means of protection – mimicry, the phenomenon of one species (especially an insect) closely resembling another.

Batesian mimicry involves an unequal partnership in which the mimic (an animal which is palatable to a predator) has come to closely resemble the model (an unpalatable, predator-resistant species). For example the conspicuous red and black colouration of the distasteful Monarch butterfly, which birds find especially noxious, is imitated by the harmless Viceroy butterfly.

Batesian mimicry is disadvantageous to predators since they are deceived into avoiding palatable prey or forced to run the risk of tackling distasteful/poisonous models by mistake. In addition, this relationship is also detrimental to the model whose security is undermined by any predator that eats a palatable mimic and is thereby encouraged to sample a similarly patterned model. Batesian mimicry only works, therefore, if the mimics are vastly outnumbered by the models; repeated stinging or poisoning of a predator by the models then renders the very rare encounter with a palatable mimic unprofitable.

Mullerian mimicry involves two or more predator-resistant species each of which has evolved a similar warning colouration, for example the yellow and black banded pattern shared by the stinging wasp and the distasteful cinnabar moth caterpillar. This form of mimicry is beneficial to both the predator (it only has to learn to avoid one warning pattern) and the prey (any losses suffered while the predator is still learning are shared by all the species involved).

a) To be successful why must cryptic colouration be accompanied by a high degree of immobility? (1)

b) From the passage name TWO predator-resistant features other than colouration which are often possessed by prey animals. (1)

c) Batesian mimicry involves mimic, model and predator. Which of these is the sole beneficiary? In what way does this form of mimicry operate to the disadvantage of the other two? (3)

d) Batesian mimicry sometimes involves warning sounds instead of colouring. Explain how this would work. (1)

e) Why is mutual resemblance an advantage to all of the prey species concerned in a case of Mullerian mimicry? (1)

f) Which form of mimicry involves the members of one species acting as 'sheep in wolves' clothing' in order to deceive the members of another species? (1)

g) In addition to sharing a similar warning colouration, a group of distasteful unrelated species vulnerable to attack by nocturnal predators are all found to emit a similar →

odour. Of what type of mimicry is this an example? Suggest how it operates. (2)

7 Give an account of individual and social mechanisms employed by animals to defend themselves against danger. (10)

8 Write an essay on how plants cope with danger under the headings:

a) structural defence mechanisms; (6)

b) ability to tolerate grazing. (4)

What You Should Know

Chapter 26

(See Table 26.2 for Word bank)

avoidance	group	remembered
camouflage	growth	short-term
danger	habituation	social
energy	individual	spines
grazing	long-term	structural

Table 26.2 Word bank for chapter 26

1 Some animals learn not to react to a stimulus if, after many repeats, it proves to be harmless. This is called _____.

2 Habituation prevents the animal wasting time and _____ on needless repeats of its _____ response.

3 Habituation only brings about a _____ modification of the escape response, otherwise the animal would be left open to _____.

4 In advanced animals, learning involves a _____ modification of the response made to a stimulus. Information is stored in the central nervous system and _____ for future use.

5 Animals show many _____ defence mechanisms such as _____ and protective coverings.

6 _____ mechanisms for defence are found amongst animals which remain together in a protective _____.

7 Some plants possess _____ adaptations for defence such as thorns and _____.

8 The _____ form of some plants enables them to tolerate _____ animals.

Control and Regulation

27 Plant Growth

Growth

Although **growth** of a multicellular organism is normally accompanied by an increase in body size, this criterion alone is inadequate to define growth. For example, a middle-aged person who gains several kilograms of fat round the waist and then sheds it by dieting is not showing true growth.

Growth is the **irreversible increase** in the **dry mass** of an organism and is normally accompanied by an increase in **cell number**. It involves the uptake of chemicals and the synthesis of new structures.

Fresh mass is a less reliable indicator of growth due to temporary fluctuations in an organism's water content.

Meristems

Growth occurs all over a developing animal's body. In plants, growth is restricted to regions called **meristems**. A meristem is a group of undifferentiated plant cells which are capable of dividing repeatedly.

Primary growth occurs at **apical** meristems; **secondary** growth occurs at **lateral** meristems.

Animals do not possess meristems.

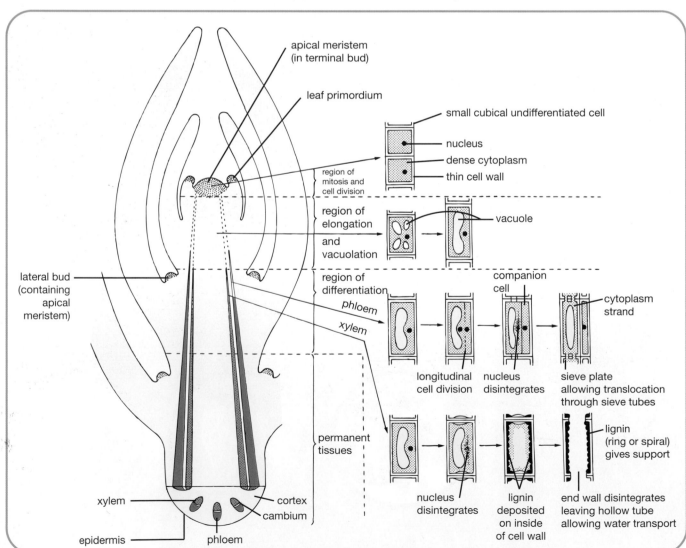

Figure 27.1 Growth of a shoot

Apical meristems

An apical meristem is found at a root tip and at a shoot tip (apex). Increase in length of a root or shoot depends on both the formation of new cells by the apical meristem and elongation of these new cells. Figure 27.1 shows how growth occurs at a shoot apex. Each new cell formed by mitosis and cell division in the meristem becomes **elongated, vacuolated** and finally **differentiated** (also see page 273). Figure 27.2 shows the apical meristem at a shoot tip.

Differentiation is the process by which an unspecialised cell becomes altered and adapted to perform a special function as part of a permanent tissue. Some cells, for example, become xylem vessels by developing into long hollow tubes supported by lignin; others become phloem tissue consisting of sieve tubes and companion cells.

Investigating growth in a young root

The experiment illustrated in Figure 27.3 shows how the same pattern of events is repeated in a growing root. When the young root is marked with ink at 1mm intervals and allowed to grow for a few days, the space between the tip and mark 1 is found to remain unaltered since this region consists of the protective root cap and the region of mitosis and cell division.

However marks 1, 2, 3 and 4 do become spread out since cells in this region are undergoing elongation and vacuolation. Above mark 4 there is no change in the spacing of the ink marks because here the cells are fully elongated and are undergoing differentiation (into xylem, phloem, root hairs etc.). Figure 27.4 shows the apical meristem at a root tip.

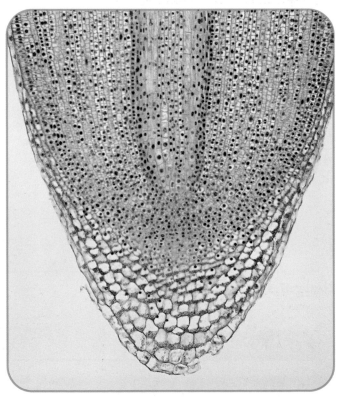

Figure 27.4 Apical meristem at root tip

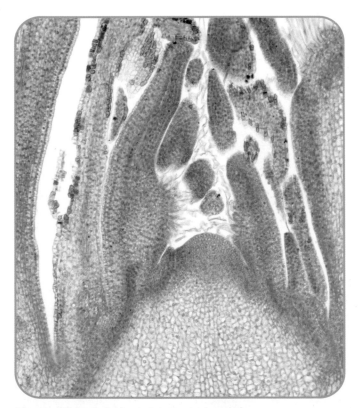

Figure 27.2 Apical meristem at shoot tip

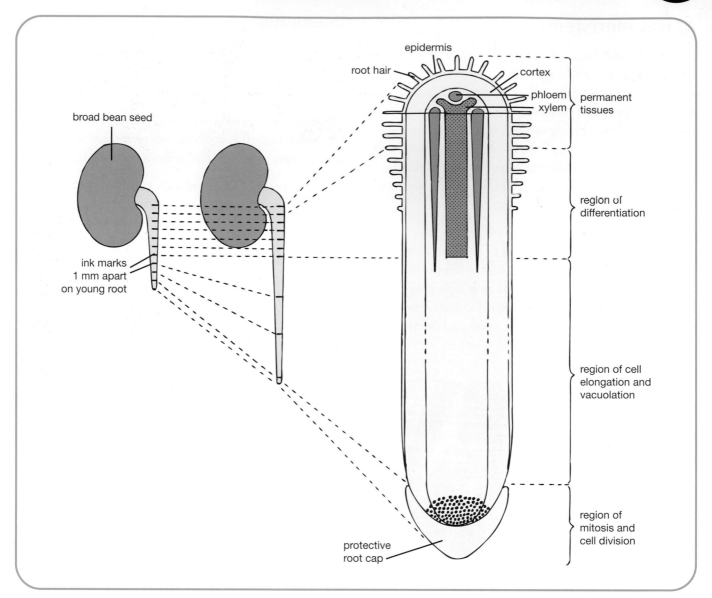

Figure 27.3 Growth of a root

1 a) Define the term *growth*. (2)

b) Why is gain in fresh mass an unreliable indicator of growth? (1)

2 a) What is a *meristem*? (1)

b) Identify TWO regions of a plant which possess an apical meristem. (2)

c) State the function of meristematic cells. (1)

3 State THREE differences between a xylem vessel and a meristematic cell. (3)

4 Give ONE way in which growth in an animal differs from growth in a plant. (1)

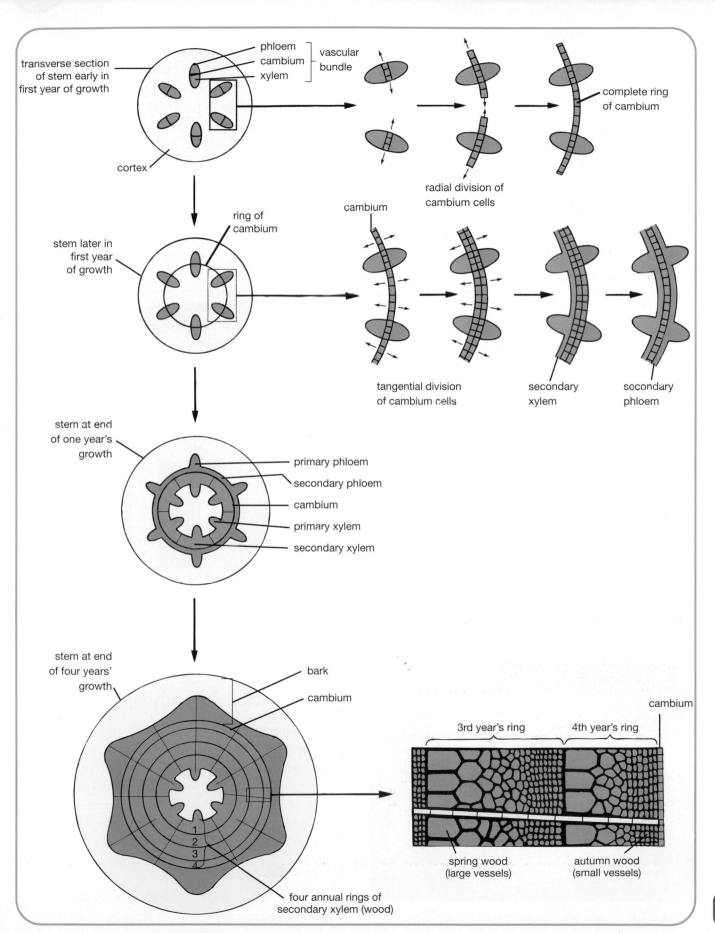

Figure 27.5 Secondary thickening in a woody stem

Lateral meristem

Whereas an annual plant dies after one year's growth, a perennial plant continues to grow year after year. As a perennial plant's overall size gradually increases, it requires more and more xylem tissue to provide it with additional support and transport of water. The development of this extra xylem each year makes the stem (and root) of a woody perennial (e.g. a tree) increase in thickness. This process is called **secondary growth** (**secondary thickening**). The secondary xylem is produced by a **lateral meristem** called **cambium**.

Cambium

Figure 27.5 shows the process of secondary thickening in a stem. A complete ring of **secondary xylem** (and a less obvious ring of secondary phloem) arise annually from the meristematic ring of cambium: xylem to the inside; phloem to the outside. (Differentiation is under genetic control, see chapter 29.)

Annual rings

In temperate regions, cambium is most active in spring and large xylem vessels (**spring wood**) are formed. These wide vessels are able to transport the large quantities of water and nutrients needed in spring for leaf growth and expansion. By the time autumn arrives, the cambium is less active and produces smaller xylem vessels (**autumn wood**). Cambium is inactive in winter.

This growth pattern is repeated annually allowing each year's ring of woody xylem to be clearly distinguished (see Figure 27.6) and the age of the tree to be easily calculated.

Figure 27.6 Annual rings of secondary thickening

The thickness of an annual ring is dependent upon the growth conditions that prevailed during that year. A wider ring indicates a year of mild temperatures during the spring and summer accompanied by plenty of sunshine and rainfall. A narrower ring indicates poor growing conditions caused by cold weather and/or drought and/or infestation of leaves by insects such as caterpillars.

Grain of timber

When timber is cut longitudinally through the vascular tissue, the rings show up as patterns called the **grain of the wood** (see Figures 27.7 and 27.8).

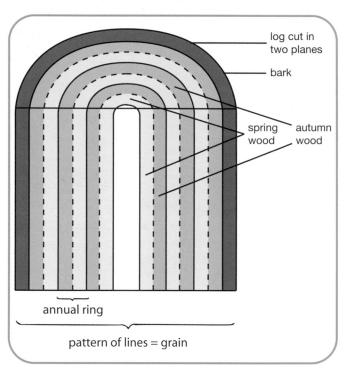

log cut in two planes

bark

spring wood autumn wood

annual ring

pattern of lines = grain

Figure 27.7 Grain of timber in a 3-year old

Regeneration in angiosperms

Regeneration is the process by which an organism **replaces** lost or damaged parts. Flowering plants (**angiosperms**) have extensive powers of regeneration. A root tip or a shoot tip containing an apical meristem or a piece of root or stem containing cambium, when cut from many types of plant, will develop roots and shoots and eventually regenerate the entire plant.

Commercial applications

Horticulturalists employ artificial methods of **vegetative propagation** (asexual reproduction) such as

Figure 27.8 Grain of timber

taking **cuttings** to increase their supply of a desirable type of plant. The success of this process rests solely on the ability of many types of angiosperm cuttings to regenerate their missing parts. Such regeneration is often promoted by the application of **rooting powders** containing growth substances (see page 280) to the cut surface.

Tissue cultures

Thousands of identical plants can be produced by growing **tissue cultures**. Early work in this field involved the use of carrot plants (see Figure 27.9). This technique now enables commercial growers to mass produce **clones** of plants such as pineapple, rose, orchid and oil palm.

Regeneration in mammals

Mammals have only limited regenerative powers by natural means. Almost without exception, regeneration is restricted to the healing of **wounds** (see Figure 27.10), the mending of broken **bones**, the replacement of **blood** after loss and the regeneration of damaged **liver**.

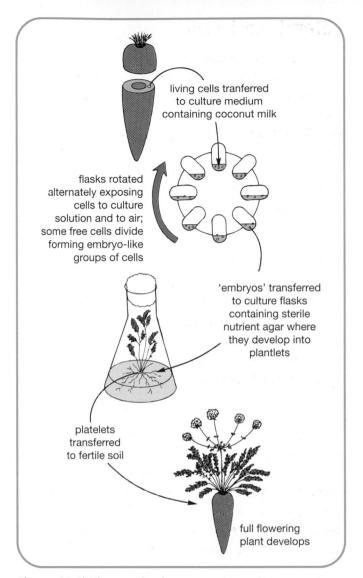

living cells tranferred to culture medium containing coconut milk

flasks rotated alternately exposing cells to culture solution and to air; some free cells divide forming embryo-like groups of cells

'embryos' transferred to culture flasks containing sterile nutrient agar where they develop into plantlets

platelets transferred to fertile soil

full flowering plant develops

Figure 27.9 Tissue culturing

So great is the liver's potential for regeneration that half of a liver can regain its full size within three months. This is made possible by the fact that liver cells are less highly differentiated than most other body cells and therefore the genes which control the synthesis of the essential growth factor can become 'switched on' when required. Cells affected in this way divide and behave like young cells until the liver has reached its full size. Then the genes become 'switched off' again. (The switching on and off of genes and the possible use of stem cells in the future are discussed more fully in chapter 29.)

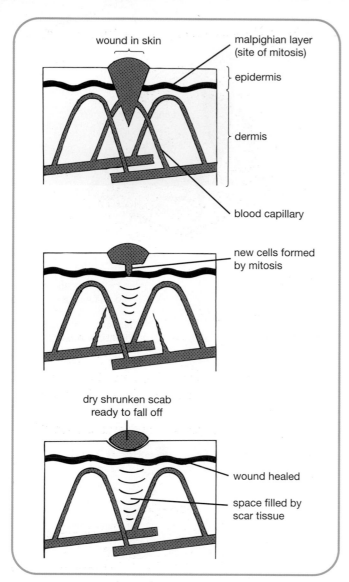

Figure 27.10 Healing of a wound

Testing Your Knowledge

1 a) What is the difference between an *annual* and a *perennial* plant? (1)

b) Why does a perennial plant need to become thicker every year? (1)

2 a) i) Name a tissue that is described as a lateral meristem.

ii) Where is this tissue found in a young stem which has not yet begun the process of secondary thickening? (2)

b) Which type of cells are formed to **i)** the outside; **ii)** the inside of this lateral meristem? (2)

3 a) Define the term *regeneration*. (2)

b) Name TWO techniques used by horticulturists to produce a clone of desirable plants. (2)

c) Give THREE examples of tissues that can be easily and rapidly regenerated by the human body. (3)

Applying Your Knowledge

1 a) Which TWO processes produce downward growth of a root? (2)

b) In an experiment, a young root was inked at 1mm intervals as shown in Figure 27.3. After 4 days of further growth, the spaces between the marks were measured and recorded as shown in Table 27.1.

i) Present the data as a bar graph.

ii) Draw TWO conclusions from the data. (4)

Table 27.1

space between:	distance (mm)
root tip and mark 1	1.0
marks 1 and 2	6.0
2 and 3	7.0
3 and 4	6.5
4 and 5	1.0
5 and 6	1.0
6 and 7	1.0
7 and 8	1.0
8 and 9	1.0
9 and 10	1.0

2 Figure 27.11 shows five stages in the development of a xylem vessel.

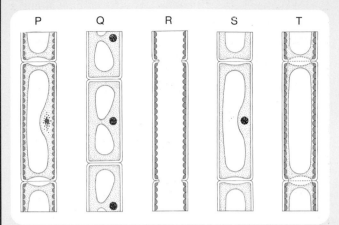

Figure 27.11

a) Arrange them in order starting with Q. (1)

b) Which stages contain cytoplasm. (1)

c) i) Which stage shows a fully differentiated xylem vessel?

ii) Identify the TWO features shown in the diagram that led to your choice.

iii) Relate each of these to a function performed by xylem. (5)

3 Figure 27.12 shows three early stages in the process of secondary thickening in part of a stem,

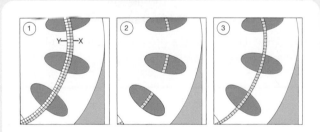

Figure 27.12

a) Arrange them into the correct order. (1)

b) Name tissue X. (1)

c) Between which two numbered stages did **i)** radial division; **ii)** tangential division of the cells in X occur? (2)

d) Name tissue Y. (1)

e) Explain why X is described as meristematic and Y is described as differentiated. (2)

4 Figure 27.13 shows a transverse section of a stem.

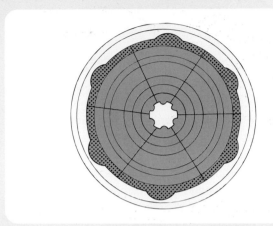

Figure 27.13

a) What age is the specimen? (1)

b) During which year were the conditions most suitable for the development of wood? (1)

c) i) Identify the year during which the tree's leaves may have been infested with insect larvae.

ii) Explain how you arrived at your answer. (2)

5 Figure 27.14 shows a log cut transversely and longitudinally.

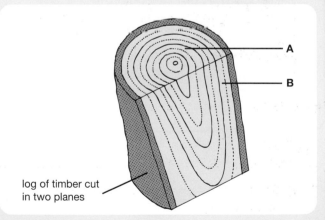

log of timber cut in two planes

Figure 27.14

a) Name A and B. (1)

b) From which tissue were A and B formed? (1)

c) Why are the vessels formed at B wider than those formed at A? (1)

d) What name is given to wood patterns such as those shown in this log's longitudinal plane? (1)

e) Suggest why trees growing near the equator do not contain distinct annual rings. (1)

owing:

eristems; (5)

g in a woody stem. (5)

and answer the questions based on

t-cloning laboratory, all members of the wo orce must wear sterile clothing. Gone are the traditional pots of soil found in the greenhouse and potting shed; in their place are countless rows of gleaming culture tubes. Here thousands of copies of one plant are produced by micropropagation. A plant with desirable characteristics (e.g. sweet fruit and disease resistance) is selected and tissue-cultured. One technique involves removing tiny pieces of bud tissue from the parent plant and growing them in sterile nutrient jelly in culture tubes. Each piece of tissue is kept in optimum conditions of light and temperature until it develops into a miniature plant. It is then checked carefully for disease before being supplied in a batch to a commercial grower who wishes to obtain a uniform crop (e.g. fruit that all ripen at the same time).

This seemingly ideal situation does, however, carry an element of risk. If a new disease-causing organism arrives that can attack the propagated plants, then the whole crop may be lost.

a) The flow diagram in Figure 27.15 is intended to give a summary of the procedure described in the passage. Copy it and complete the blanks. (2)

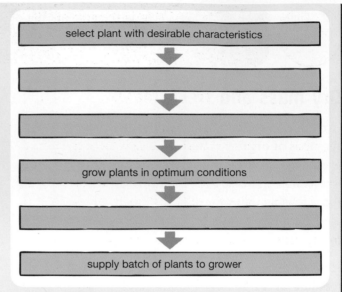

Figure 27.15

b) Name TWO environmental conditions given in the passage that would be kept at optimum levels to promote growth. (1)

c) State TWO precautions that are taken to prevent the young plants from becoming diseased. (2)

d) One of the benefits gained by the commercial grower is that there is no need to return again and again to pick the fruit crop. Identify the phrase that tells you why one visit to the crop at harvest time will be sufficient. (1)

e) What word given in the passage means 'producing a population of identical plants'? (1)

f) What could cause a grower to lose his entire crop? Explain why. (2)

28 Growth patterns

Dry mass and fresh mass

Growth occurs in an organism's body when the rate of synthesis of organic materials exceeds the rate of their breakdown. It is for this reason that growth involves an **irreversible increase in dry mass**. Gain in dry mass is a more reliable indicator of growth than gain in fresh mass because an organism's fresh mass varies from day to day depending on water availability.

However measuring the dry mass of an organism involves removing all of its water and this often proves to be impracticable. Therefore growth is usually investigated by measuring a variable factor such as fresh weight or height or cell number over a period of time.

Growth curve

Table 28.1 records the height of a plant shoot measured over a period of 16 days. When graphed, the results for total height of shoot give a **growth curve**. This is found to take the form of a **sigmoid (S-shaped) curve** as shown in Figure 28.1.

Careful study of this graph and the data in Table 28.1 reveals that during the first four days of growth, each daily increase in height was greater than that of the previous day. This phase is called the **period of accelerating growth**.

During days 4 to 8, the daily increase in height remained constant. This phase is called the **period of steady rapid growth**.

time from start (days)	total height of shoot (mm)	daily increase in height of shoot (mm)	growth phase
start	0	–	period of accelerating growth
1	1	1	period of accelerating growth
2	3	2	period of accelerating growth
3	7	4	period of accelerating growth
4	15	8	period of accelerating growth
5	23	8	period of steady rapid growth
6	31	8	period of steady rapid growth
7	39	8	period of steady rapid growth
8	47	8	period of steady rapid growth
9	51	4	period of decelerating growth
10	53	2	period of decelerating growth
11	54	1	period of decelerating growth
12	54	0	period of decelerating growth
13	54	0	period of no growth
14	54	0	period of no growth
15	54	0	period of no growth
16	54	0	period of no growth

Table 28.1 Growth of a plant shoot

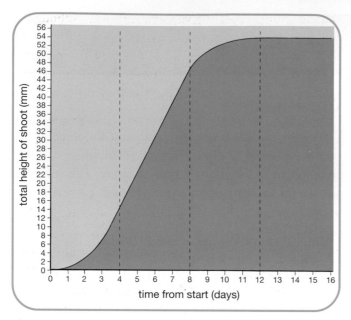

Figure 28.1 Graph of growth of height of plant shoot

During days 8 to 12, growth continued but each day's increase in height was less than that of the previous day. This phase is called the **period of decelerating growth**.

During days 12 to 16, the shoot failed to gain further height. It has now entered the **period of no growth**.

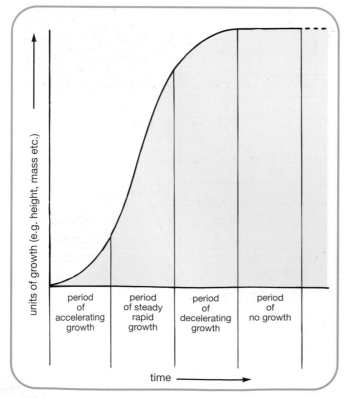

Figure 28.2 Sigmoid growth curve

This pattern of growth, with minor variations, is common to almost all living things. Figure 28.2 shows a generalised version of the sigmoid curve and its four phases.

Different growth patterns

The exact form that the sigmoid growth curve takes depends on the organism involved.

Annual plant

Figure 28.3 shows the growth curve for a pea plant. At first there is a **decrease** in mass during germination. This is due to food reserves being used for respiration to generate energy for growth. Soon the green leaves develop and photosynthesis begins. Once the rate of photosynthesis exceeds the rate of respiration, the plant **gains mass** and growth proceeds in the normal manner.

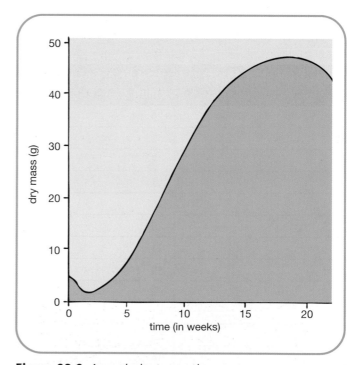

Figure 28.3 Annual plant growth curve

Annual plants only grow for a short time (less than a year) and then stop. Towards the end of the growing season the curve flattens out and then quickly goes into decline when the seeds are dispersed and the plant undergoes **senescence**.

Tree

Figure 28.4 shows the growth curve for a tree which is a perennial plant and continues to grow for many years.

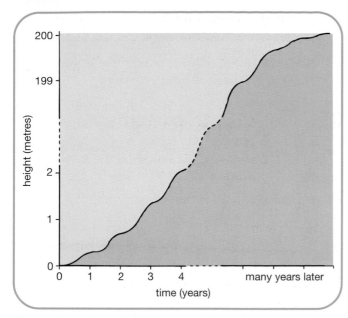

Figure 28.4 Perennial plant growth curve

Each year, spring and summer provide the best growing conditions. During winter, growth slows down and often comes to a complete halt. The growth pattern is therefore found to take the form of a series of **annual sigmoid curves** whose overall shape is, in turn, sigmoid.

In some trees growth occurs throughout the entire life of the plant and the graph never flattens out.

Human

Figure 28.5 shows the growth curve for a human. In this example, body mass has been graphed against time.

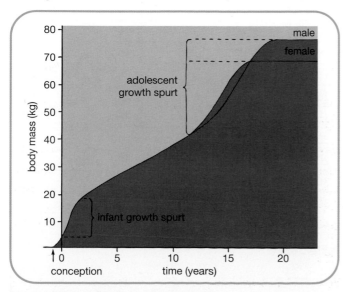

Figure 28.5 Human growth curves

Data for human height give a similar curve. Although basically sigmoid, it has two phases of rapid growth (called **growth spurts**) instead of one. The first occurs during the two years following birth; the second occurs at puberty after a long period of steady growth.

On average, the human male reaches puberty later and attains a larger adult size than the human female.

Insect

Figure 28.6 shows the growth pattern of an insect (e.g. locust). Insects possess a hard inelastic outer skin called an **exoskeleton** which prevents continuous increase in body size. This skin must be periodically shed to allow growth to occur. A new skin forms beneath the old one and when the latter is cast off during moulting (ecdysis), the insect quickly inflates its body before the new skin hardens. This produces an **abrupt increase** in body length giving the graph its step-like appearance.

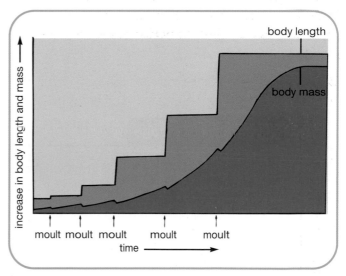

Figure 28.6 Insect growth 'curves'

Such **intermittent** growth only applies to measurements of body size (e.g. length). Although the body mass of an insect decreases slightly after each moult as the old skin is shed, it increases steadily between moults and, if graphed, gives a curve with a fairly sigmoid appearance.

Comparison of plant and animal growth

Table 28.2 summarises the growth differences between advanced plants and animals.

plant (e.g. angiosperm)	animal (e.g. mammal)
growth only occurs at meristems	growth occurs all over body
growth (increase in size) continues throughout life	growth (increase in size) stops on reaching adulthood
natural regenerative powers extensive	natural regenerative powers limited

Table 28.2 Comparison of plant and animal growth

Testing Your Knowledge

1 Name the FOUR periods of growth that make up the standard sigmoid growth curve. (4)

2 **a) i)** State TWO ways in which the growth curve of an annual plant differs from the normal sigmoid growth curve.
   ii) Explain why the annual plant's graph takes this shape. (4)

   b) Name the TWO phases of rapid growth shown by the human body. (2)

3 **a)** Identify a quantitative characteristic of a stick insect that would show an intermittent growth pattern. (1)

   b) Why would the sigmoid growth curve of a stick insect's body mass not be completely smooth? (1)

4 State THREE ways in which the growth of a rose bush differs from that of a rabbit. (3)

Applying Your Knowledge

1 **a) i)** Explain why measurements of fresh weight are unreliable indicators of growth.
   ii) Why then do scientists sometimes use such data instead of measuring changes in dry mass? (3)

   b) Suggest a suitable method of measuring growth in:
   i) a unicellular micro-organism (e.g. yeast);
   ii) an annual plant (e.g. sunflower);
   iii) a perennial plant (e.g. sycamore tree);
   iv) a mammal (e.g. pig);
   v) an insect (e.g. grasshopper). (5)

2 The data in Tables 28.3 and 28.4 refer to two different types of living organism.

time after hatching or germinating (days)	length (mm)
0.0	80
6.0	80
6.1	150
11.0	150
11.1	200
15.0	200
15.1	260
20.0	260
20.1	500
30.0	500
30.1	760
40.0	760

Table 28.3

time after hatching or germinating (weeks)	dry mass (g)
0	6
2	2
4	6
6	12
8	20
10	30
12	38
14	44
16	48
18	50
20	50
22	49
24	42

Table 28.4

age (years)	height (m)
birth	0.51
1	0.75
2	0.91
3	0.98
4	1.03
5	1.08
6	1.13
7	1.18
8	1.23
9	1.28
10	1.33
11	1.39
12	1.47
13	1.57
14	1.63
15	1.65
16	1.66
17	1.66
18	1.66
19	1.66
20	1.66

Table 28.5

a) Using two sheets of graph paper, plot a separate line graph of each set of data (with time on the x axis in both cases). (4)

b) Identify which graph depicts the growth curve of i) an annual plant; ii) an insect. (1)

c) Give TWO reasons for your choice of answer to b). (2)

3 The data in Table 28.5 refer to an average human female from birth to the age of 20 years.

a) Draw a line graph of the data. (3)

b) i) Which growth spurt is greater?
   ii) Explain how you arrived at your answer. (2)

c) In what TWO ways would the growth curve of an average male have differed from the graph that you have drawn? (2)

d) i) Predict TWO ways in which the growth curve of an average woman living in Britain 200 years ago would have differed from the graph that you have drawn.
   ii) Explain your answers. (4)

4 Figure 28.7 shows the growth rates of the human body and two types of organ.

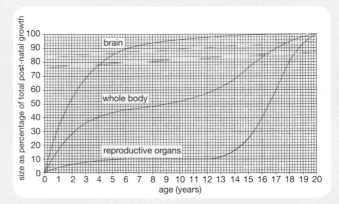

Figure 28.7

a) At age 2 years, what is the percentage of total post-natal growth shown by: i) reproductive organs? ii) whole body? iii) brain? (1)

b) At age 10 years, what is the percentage of total post-natal growth shown by: i) reproductive organs? ii) whole body? iii) brain? (1)

c) Which body part shows i) most; ii) least development during childhood?
   iii) Suggest why it is of survival value to humans to have this growth pattern rather than one where all the organs develop at the same rate as shown by many other mammals. (2)

257

d) i) Are the data graphed in Figure 28.7 based on averages for human males or females?

ii) Explain your choice of answer. (2)

e) During which 2-year period did general body growth occur at the fastest rate? (1)

f) At what age in years was each of the following body parts found to be at 50% of its post-natal final size? **i)** reproductive organs; **ii)** whole body; **iii)** brain. (1)

g) If the size as percentage of post-natal growth of the cranium had been graphed, which curve would it most closely resemble? Why? (1)

What You Should Know

Chapters 27–28
(See Table 28.6 for Word bank)

animals	flowering	primary
apical	height	regeneration
autumn	irreversible	rings
body	lateral	secondary
cambium	limited	sigmoid
differentiated	meristems	spring
dry	patterns	xylem

Table 28.6 Word bank for chapters 27–28

1 Growth is the _____ increase in the _____ mass of an organism.

2 In plants growth is restricted to regions called _____.

3 _____ plant growth occurs at _____ meristems (root and shoot tips) where newly formed cells become elongated, vacuolated and _____.

4 _____ growth occurs in a perennial plant at a _____ meristem called _____.

5 Cambial activity produces annual _____ of secondary xylem and phloem. Each ring of secondary _____ contains a region of large vessels called _____ wood and a region of smaller vessels called _____ wood.

6 _____ do not possess meristems. Instead growth occurs all over a developing animal's _____.

7 _____ plants (angiosperms) have extensive powers of _____ whereas mammals have only _____ powers.

8 Investigations into growth often involve measuring a variable factor such as fresh weight, _____, length etc.

9 A graph of the results normally takes the form of a _____ (S-shaped) growth curve.

10 Growth _____ vary from one type of organism to another.

29 Genetic control

Role of genes

To function, every living cell must possess a variety of proteins, some to form its basic structure, others to act as enzymes which control metabolic processes. These essential proteins are synthesised according to the base sequence encoded in the cell's DNA. A particular segment of DNA called a gene codes for each protein (or polypeptide). Thus the structure and function of a cell is determined and controlled by its genes.

Control of gene action

Some proteins are only required by a cell under certain circumstances. To prevent resources being wasted, the genes that code for these proteins are 'switched on' and 'switched off' as required.

Investigating the effect of β-galactosidase on lactose

Lactose is a sugar found in milk. Each molecule of lactose is composed of a molecule of glucose and a molecule of galactose (see Figure 29.1).

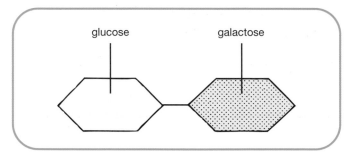

Figure 29.1 Molecule of lactose

The enzyme β-galactosidase can be immobilised in gel as shown in Figure 29.2 and then used in the experiment shown in Figure 29.3. Since milk is found to lack glucose at the start of the experiment but contain glucose after contact with β-galactosidase, it is concluded that the enzyme has promoted the breakdown of lactose to simple end products containing glucose. It is now known that β-galactosidase catalyses the following reaction:

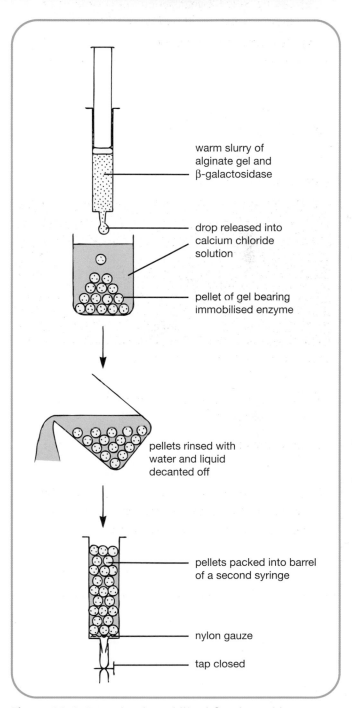

Figure 29.2 Preparing immobilised β-galactosidase

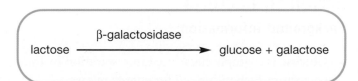

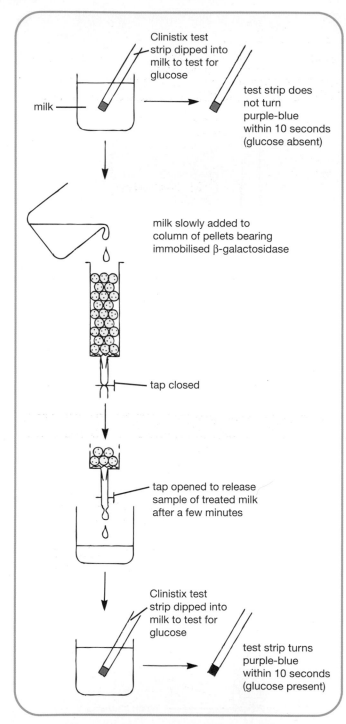

Figure 29.3 Using immobilised β-galactosidase

Jacob–Monod hypothesis of gene action in bacteria
Background information

- Glucose is a simple sugar used in respiration by the bacterium *Escherichia coli* for energy release.

- *E. coli* can only make use of the glucose in lactose if it is released from the galactose.

- Lactose is digested to glucose and galactose by the enzyme β-galactosidase.

- *E. coli*'s chromosome has a gene which codes for β-galactosidase.

- *E. coli* is found to produce β-galactosidase only when lactose is present in its nutrient medium and fails to do so when lactose is absent.

- Somehow the gene which codes for β-galactosidase is switched on in the presence of lactose and switched off in the absence of lactose.

- The process of switching on a gene, only when the enzyme that it codes for is needed, is called **enzyme induction**.

Operon

An **operon** (see Figure 29.4) consists of one or more **structural genes** (containing the DNA code for the enzyme in question) and a neighbouring **operator** gene which controls the structural gene(s). The operator gene is, in turn, affected by a **repressor** molecule coded for by a **regulator** gene situated further along the DNA chain.

Absence of lactose

Environments inhabited by *E. coli* normally contain glucose but not lactose. The bacterium would waste some of its resources if it made the lactose-digesting enzyme when no lactose was present to digest. Figure 29.4 shows how the bacterium is prevented from doing so by the repressor molecule combining with the operator gene. The structural gene remains switched off and its DNA is not transcribed.

Presence of lactose

When the bacterium finds itself in an environment containing lactose, the events shown in Figure 29.5 take place and lead to the induction of β-galactosidase.

Lactose (the **inducer**) prevents the repressor molecule from binding to the operator gene. The system is no longer blocked, the structural gene becomes switched on and production of β-galactosidase proceeds.

Within the figure 29.3:

- Clinistix test strip dipped into milk to test for glucose
- milk
- test strip does not turn purple-blue within 10 seconds (glucose absent)
- milk slowly added to column of pellets bearing immobilised β-galactosidase
- tap closed
- tap opened to release sample of treated milk after a few minutes
- Clinistix test strip dipped into milk to test for glucose
- test strip turns purple-blue within 10 seconds (glucose present)

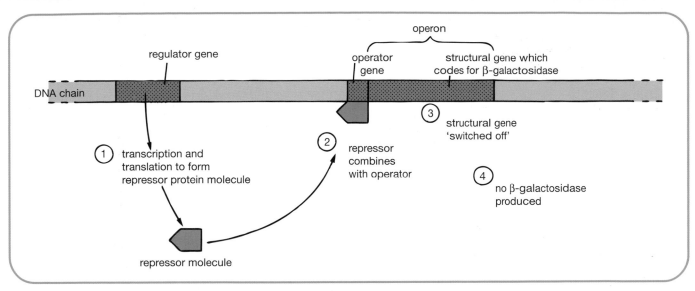

Figure 29.4 Effect of repressor in absence of lactose

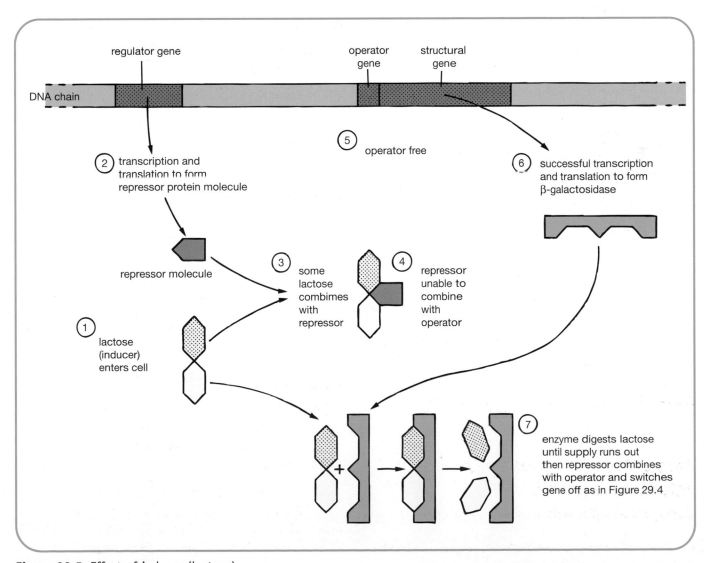

Figure 29.5 Effect of inducer (lactose)

When all the lactose has been digested, the repressor molecule becomes free again and combines with the operator as before. The structural gene becomes switched off and the waste of valuable resources (such as amino acids and energy from ATP) is prevented.

Hypothesis

This hypothesis of gene action was first put forward by two scientists called Jacob and Monod. It is now supported by extensive experimental evidence from work done using bacteria. However the mechanism by which genes are switched on and off in higher organisms such as humans is not yet fully understood.

Testing Your Knowledge

1 a) i) State TWO types of gene that make up an operon in a cell of *Escherichia coli*.
   ii) Which of these genes contains the DNA code for the enzyme β-galactosidase? (3)
   b) In the absence of lactose, what effect does the repressor molecule have on the system? (1)

2 a) To what molecule does lactose, when present in a cell of *E. coli*, become combined without being digested? (1)
   b) Can the structural gene be transcribed under these circumstances? Explain your answer. (1)

3 a) Give the word equation of the biochemical reaction that is catalysed by the enzyme β-galactosidase. (1)
   b) Describe the series of events that leads to the operon in *E. coli* being switched off again. (2)

Role of genes in control of cell metabolism

All the chemical processes that occur in a living organism and keep it alive are known collectively as its **metabolism.** Cell metabolism refers to the biochemical reactions that occur within a cell.

A **metabolic pathway** normally consists of several stages each of which involves the conversion of one metabolite to another during a breaking-down or building-up process.

Each stage in a metabolic pathway is controlled by an enzyme as shown in the imaginary example in Figure 29.6. Each enzyme is made of protein whose synthesis and structure is determined by genetic information held in a particular gene. However if a **fault** occurs in the genetic information due to a **mutation** (see chapter 16), the enzyme is not produced in its normal functional state. In the absence of the enzyme, some essential step in the pathway cannot proceed and an intermediate metabolite is often found to accumulate and lead to problems.

Normally the mutated allele of a gene is recessive to the normal allele and only shows its effect in the phenotype of a person who has inherited the mutated allele from both parents. Such a homozygous person suffers a disorder known as an **inborn error of metabolism.**

Phenylketonuria

Phenylalanine and tyrosine are two amino acids that humans obtain from protein in their diet. During normal metabolism excess phenylalanine is converted into tyrosine by an enzyme as shown in Figure 29.7.

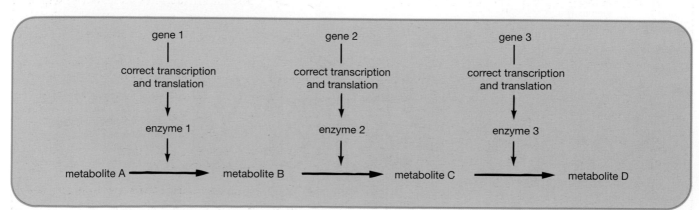

Figure 29.6 Genetic control of metabolic pathway

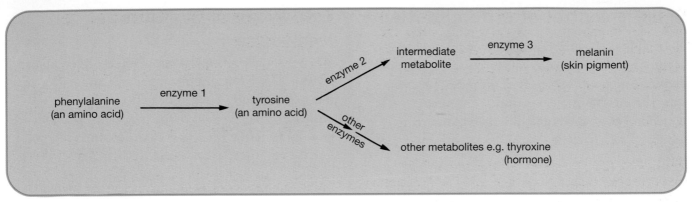

Figure 29.7 Normal fate of phenylalanine

Phenylketonuria (PKU) is a hereditary disorder caused by a genetic defect which disrupts this metabolic pathway. An affected person lacks the normal allele of the gene required to make enzyme 1 in Figure 29.7. Owing to this inborn **error of metabolism**, phenylalanine is no longer converted to tyrosine. Instead it undergoes alternative metabolic pathways producing **toxins** which affect the metabolism of brain cells and severely limit mental development in an untreated sufferer.

Although a sufferer of PKU has lighter skin pigmentation than normal, he or she is not an albino since some tyrosine is present in the diet for enzyme 2 to act upon.

Screening

PKU is a fairly common disorder. It occurs in about 1 in 10 000 of the UK population. In Britain all newborn babies are routinely **screened** for PKU by having their blood tested for the presence of excess phenylalanine within the first few days of life. If sufferers keep to a diet low in phenylalanine, especially during childhood when the brain is still developing, mental deficiency is prevented and other adverse effects are kept to a minimum.

It is possible that PKU may be cured in the future using genetically engineered cell implants.

Albinism

Owing to a genetic defect, albinos are unable to make enzyme 3 in the pathway shown in Figure 29.7 and therefore fail to synthesise **melanin**.

As a result an albino's skin is pink and fails to tan, the iris is pale blue or pink and the hair is pure white. However albinos suffer no medical problems provided that they avoid exposure to ultraviolet radiation.

Testing Your Knowledge

1 a) What is meant by the term *metabolism*? (1)

   b) Copy the following sentence and complete the blanks.

   A metabolic pathway normally consists of several _____ each of which involves the conversion of one _____ to another _____ during a breaking-down or _____ process. (2)

2 a) Of what substance is every enzyme composed? (1)

   b) Of what sub-units is every protein composed? (1)

   c) Where in the cell are the instructions for determining the specific sequence of amino acids in a protein kept? (1)

   d) If an enzyme which catalyses an essential step in a pathway is rendered non-functional following a mutation, what happens to the pathway? (1)

3 a) What happens to an excess of the amino acid phenylalanine in a normal human cell? (1)

   b) i)   Why is a sufferer of phenylketonuria (PKU) unable to deal with an excess of phenylalanine?

      ii) What happens to this excess phenylalanine? (2)

   c) What measures are taken nowadays to reduce the worst effects of PKU to a minimum? (2)

263

Genetic control of differentiation

Since all the cells in a multicellular organism have arisen from the original zygote by repeated mitotic division, every cell possesses all the genes for constructing the whole organism. However as development proceeds and **differentiation** occurs, groups of cells become **specialised** to perform different functions.

Within a particular type of differentiated cell, only certain genes continue to operate and code for certain proteins. In other types of differentiated cells, different groups of genes continue to operate and code for different proteins.

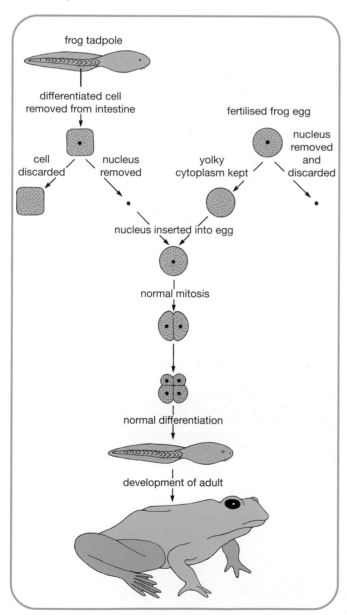

Figure 29.8 Reversibility of differentiation

Reversibility of differentiation

Amphibian

When the nucleus is taken out of a fully differentiated intestinal cell of an amphibian tadpole and transplanted into a fertilised egg from which the nucleus has been removed, the egg can successfully grow into a fertile adult (see Figure 29.8).

Dolly the sheep

In this famous experiment, the nucleus from an udder cell of a 6-year-old ewe was transferred into an egg from which the nucleus had been removed. The 'egg' was then implanted in the uterus of a surrogate mother and it developed into 'Dolly the sheep'. Dolly was therefore cloned from and genetically identical to the 6-year-old ewe whose udder cell supplied the original nucleus (also see Figure 29.9).

Both of these examples provide evidence that a differentiated cell contains all of the organism's genes and that genes which had been repressed (switched off) were able to be switched on again.

Two categories of genes

The genes present in all cells can be classified into two

"It just seems wrong to me. I don't think I could live with myself."

Figure 29.9 Cloning controversy

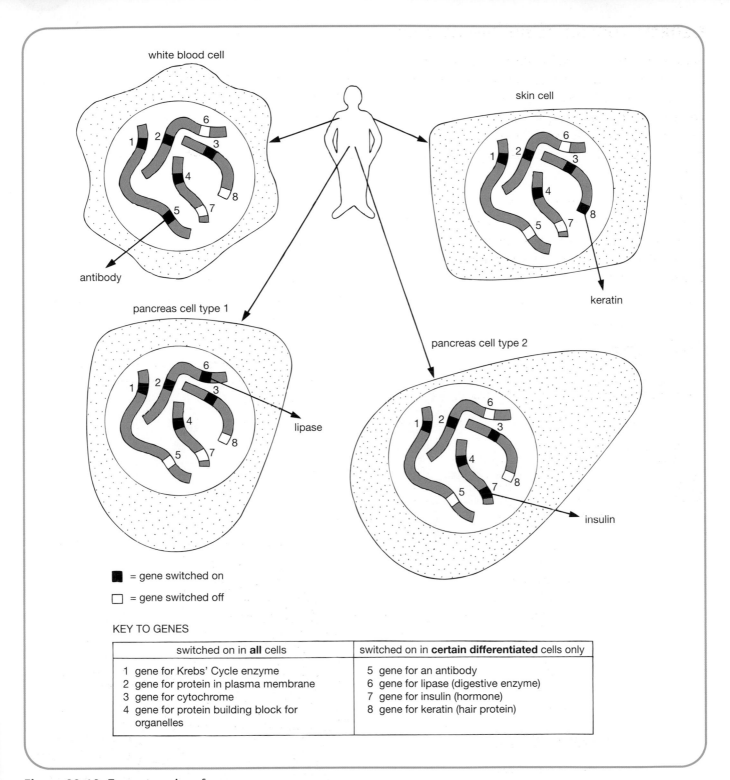

= gene switched on

= gene switched off

KEY TO GENES

switched on in **all** cells	switched on in **certain differentiated** cells only
1 gene for Krebs' Cycle enzyme 2 gene for protein in plasma membrane 3 gene for cytochrome 4 gene for protein building block for organelles	5 gene for an antibody 6 gene for lipase (digestive enzyme) 7 gene for insulin (hormone) 8 gene for keratin (hair protein)

Figure 29.10 Two categories of genes

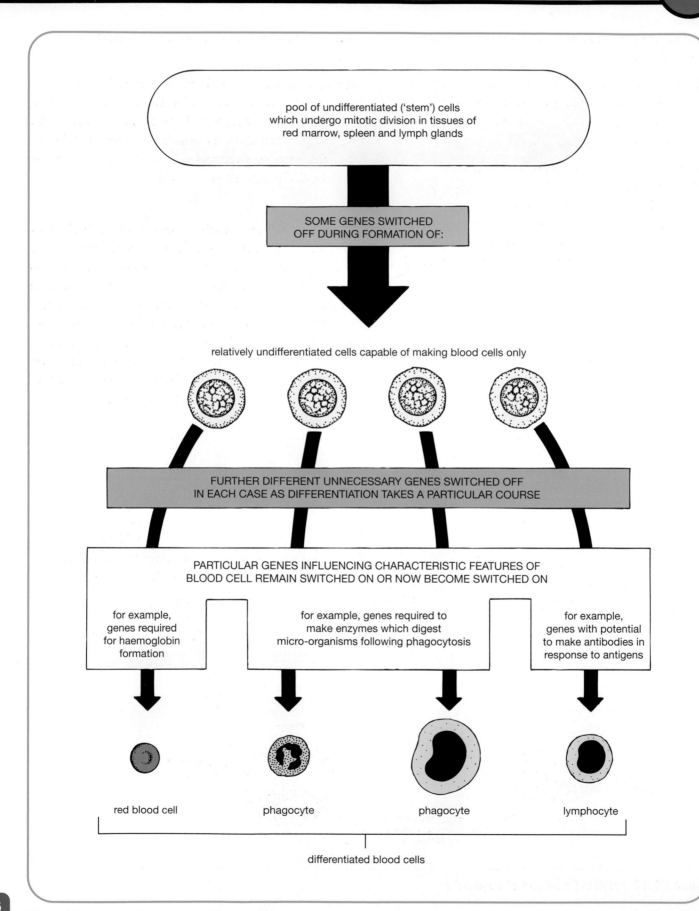

Figure 29.11 Genetic control of blood cells

distinct categories as shown in a simplified way in Figure 29.10.

The first group are expressed (switched on) in all cells and code for vital metabolites necessary for the biochemical processes basic to life (e.g. enzymes which control respiratory pathways).

The second group of genes are only switched on in a particular type of cell where they code for proteins characteristic of that cell type and no other (e.g. insulin made by pancreas cells). Many of these proteins also bring about **specific modifications** of the cell's structure and determine its characteristic features. The other genes not needed by the cell remain permanently switched off.

Genetic control of cell differentiation

Blood

Blood contains red blood cells and several types of white blood cell. The particular features possessed by a fully differentiated blood cell depend upon which of its genes are switched on and which were repressed during its differentiation from an undifferentiated ('stem') cell as shown in Figure 29.11.

Stem cells

A 6-day - old human embryo consists of a ball of about a hundred cells most of which are undifferentiated (i.e. none of their genes have been switched off). These cells can be extracted and cloned to form colonies. The **stem** cells produced can then be induced by chemical means to differentiate into a wide variety of human cell types (i.e. cells where certain genes have been switched off and others have been switched on). In the future it may be possible to use stem cells to repair damaged organs such as the heart or the brain (see Figure 29.12).

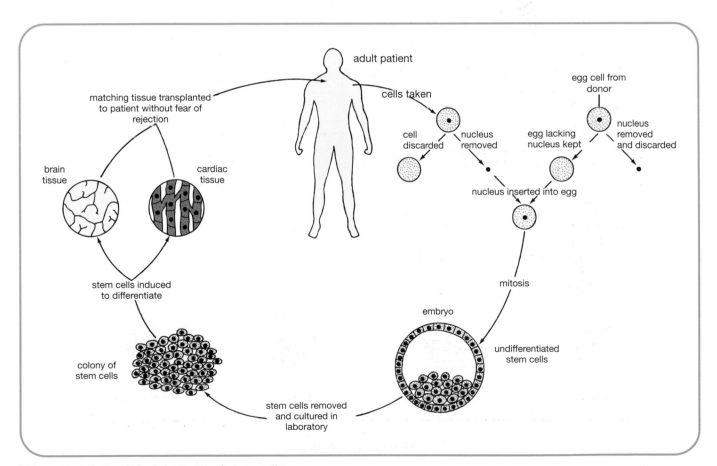

Figure 29.12 Possible future use of stem cells

Testing Your Knowledge

1 a) i) By what means does a zygote develop into a multicellular organism?
   ii) Explain the process of differentiation in terms of the 'switching on and off' of genes. (3)
   b) Give an example of an experiment that shows that differentiation can be reversed. (2)

2 a) Give an example of a type of gene that is switched on in all human cells capable of aerobic respiration. (1)
   b) i) Give an example of a type of gene that is switched on in a white blood cell only.
      ii) Give an example of a type of gene that is switched off in a white blood cell. (3)

Applying Your Knowledge

1 When a normal strain of the bacterium *Escherichia coli* is placed in growth medium containing glucose, it undergoes rapid cell division. When the same strain of *E. coli* is placed in growth medium lacking glucose but containing lactose, there is a delay before cell division occurs. Account for these observations. (2)

2 a) The following statements refer to events described by the Jacob–Monod hypothesis of gene action. Rewrite them so that each contains the correct word from the choice in brackets.
   i) The regulator gene produces the (repressor/inducer) molecule.
   ii) The inducer molecule combines with the (operator/repressor).
   iii) When the operator is free, the structural gene is switched (on/off). (3)

   b) i) Explain why *Escherichia coli* only produces the enzyme β-galactosidase when lactose is present in its food.
      ii) What is the benefit of this on/off mechanism to the bacterium? (3)

   c) Predict what will happen when lactose enters a cell of *E. coli* which has suffered a nucleotide pair deletion in the centre of the structural gene. (1)

   d) A different mutant strain of *E. coli* was found to produce β-galactosidase continuously whether lactose was present in its nutrient medium or not.
      i) Which of its genes had suffered a mutation?
      ii) Explain your answer. (2)

3 Figure 29.13 shows part of a metabolic pathway that occurs in humans. Each stage is controlled by an enzyme. Some of these stages have been given a letter.

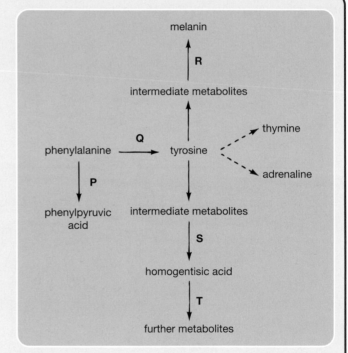

Figure 29.13

a) Explain how a mutation can lead to a blockage in such a pathway. (2)

b) Identify the letter that represents the point of blockage that leads to each of the following disorders: **i)** phenylketonuria; **ii)** albinism; **iii)** alkaptonuria (characterised by accumulation of homogentisic acid which is excreted in urine and turns black in light). (3)

c) Early methods of diagnosing PKU were based on the fact that phenylpyruvic acid could be detected in the urine of an affected baby a few weeks after birth. Suggest why the blood test now used is more satisfactory. (2)

4 The graph in Figure 29.14 shows the effect of a

phenylalanine meal on a normal person and on a sufferer of phenylketonuria (PKU).

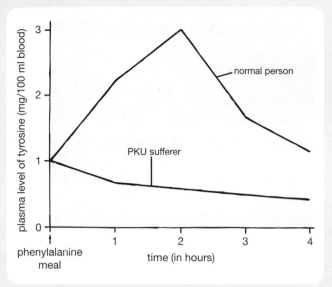

Figure 29.14

a) Explain the initial rise in level of tyrosine in the normal person. (1)

b) Why does the PKU sufferer not show a similar increase? (1)

c) Why does the level of tyrosine in the normal person fall after two hours? (1)

5 Figure 29.15 shows a cloning experiment done using an embryo from a sheep.

a) Give another term meaning 'fertilised egg'. (1)

b) What evidence is provided by this experiment to support the view that all of the organism's genes were still switched on at the 4-cell stage of embryo development? (2)

c) By what means could scientists investigate whether it is possible to reverse differentiation in this type of animal? (3)

d) Why are the lambs in Figure 29.15 described as a clone? (1)

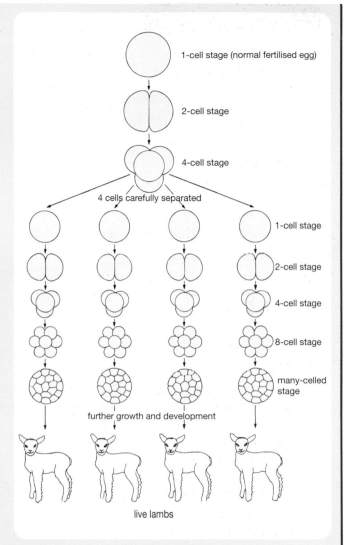

Figure 29.15

6 a) Give a brief account of the Jacob–Monod hypothesis of gene action. (6)

b) Write short notes on phenylketonuria. (4)

What You Should Know

Chapter 29

(See Table 29.1 for Word bank)

cell	Jacob–Monod	operator
differentiation	metabolism	pathway
enzyme	metabolites	PKU
error	mutated	proteins
functions	off	repressor
genes	on	resources
inducer	operate	reversal

Table 29.1 Word bank for chapter 29

1 The characteristic features of a cell are controlled by its _____.

2 To prevent _____ being wasted, some genes can be switched on and off as required.

3 The operon (_____) hypothesis states that a structural gene remains switched off while its _____ gene is combined with a _____ from a regulator gene. The structural gene becomes switched on and codes for its protein when an _____ prevents the repressor combining with the operator gene.

4 All the chemical processes that occur in a living organism are known collectively as its _____.

5 Each stage in a metabolic _____ is controlled by an enzyme.

6 Production of each _____ is controlled by a particular gene (or group of genes).

7 A _____ gene is unable to code the information needed to produce its enzyme. Lack of this enzyme may lead to an inborn _____ of metabolism such as _____.

8 During _____, cells become specialised to perform specific _____.

9 Within a differentiated cell, only certain genes continue to _____ and the other genes are switched _____.

10 In some cases genes that were switched off can be switched _____ again bringing about a _____ of differentiation.

11 Two categories of genes exist: the genes that code for vital _____ (and are switched on in all cells); the genes that code for _____ characteristic of a particular type of cell (and are switched on only in that type of _____).

Hormonal influences on growth

Animal hormones

In animals, endocrine glands produce **hormones** (chemical messengers) and secrete them directly into the bloodstream. Two of the human body's endocrine glands are shown in Figure 30.1.

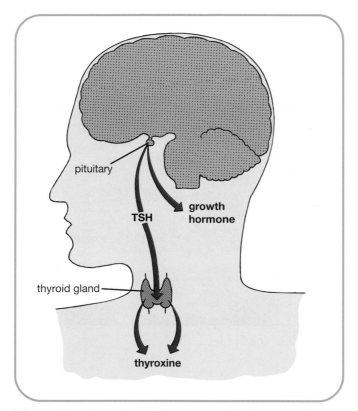

Figure 30.1 Secretion of hormones affecting growth

Role of the pituitary

The **pituitary gland** produces a variety of hormones. Two of these play a role in the control of growth and development.

Somatotrophin

Human **growth hormone** (somatotrophin) promotes growth by accelerating amino acid transport into the cells of both soft tissues and bones. This permits rapid synthesis of tissue proteins to occur. In particular, it promotes increase in length of long bones during the growing years.

Figure 30.2 shows the effect of over-production and under-production of human growth hormone during adolescence and over-production during adulthood.

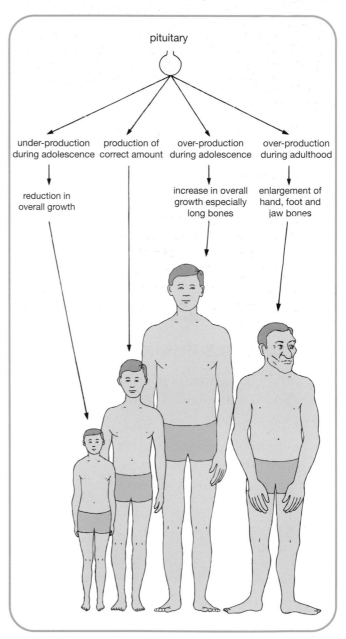

Figure 30.2 Effects of over- and under-production of somatotrophin

Thyroid-stimulating hormone

The pituitary gland also produces **thyroid-stimulating hormone** (TSH) which controls growth and activity of the thyroid gland. This latter endocrine gland responds to TSH by producing **thyroxine** which stimulates protein synthesis and is essential for proper development of the nervous system. In addition it regulates the body's metabolic processes and therefore affects growth and development in general. Thyroid deficiency during childhood results in retarded skeletal growth and poor mental development.

If the thyroid gland is inadequately stimulated by TSH then insufficient thyroid hormones are secreted and metabolic rate drops.

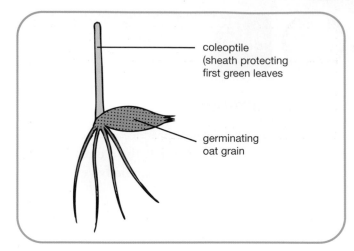

Figure 30.3 Oat seedling

Labels: coleoptile (sheath protecting first green leaves; germinating oat grain

Testing Your Knowledge

1 **a)** Name the growth hormone made by the human pituitary gland. (1)

   b) Briefly compare the effect of oversecretion of this hormone in a 15-year-old with oversecretion in a 50-year-old. (2)

2 **a)** Where in the body is thyroid-stimulating hormone (TSH) produced? (1)

   b) Explain why TSH is said to have an *indirect* influence on human growth. (2)

Plant growth substances

Plant growth substances (hormones) are chemicals which speed up, inhibit or otherwise affect growth and development of plants.

A hormone is produced at one site in a plant and transported to another site where, in low concentration, it brings about its effect.

Auxins

Early evidence for the existence of plant growth substances came from work done on the coleoptiles of oat seedlings (see Figures 30.3 and 30.4). From the series of experiments shown in Figure 30.5, the following conclusions were drawn:

● the shoot tip is essential for growth and produces a chemical;

● this chemical messenger diffuses down to lower regions of the shoot where it stimulates growth by making cells elongate;

Figure 30.4 Coleoptiles

● the growth substance is able to diffuse through agar (or gelatin) but not through metal.

The hormone involved is now known to be one of a group of plant growth substances called **auxins**. The most common auxin is **indole acetic acid (IAA)**.

Sites of IAA production

The auxin IAA is produced by the root and shoot tips and leaf meristems of plants. It is transported away from the tips (see Figure 30.6).

Movement of auxin over short distances, for example from cell to cell, occurs by **diffusion**. Long distance transport of auxin occurs in the plant's phloem tissue by **translocation**.

Auxins affect several aspects of plant growth as follows.

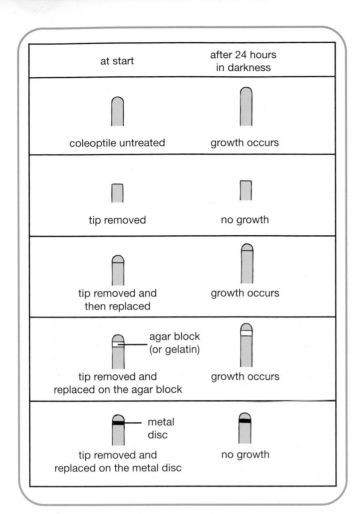

at start	after 24 hours in darkness
coleoptile untreated	growth occurs
tip removed	no growth
tip removed and then replaced	growth occurs
agar block (or gelatin) — tip removed and replaced on the agar block	growth occurs
metal disc — tip removed and replaced on the metal disc	no growth

Figure 30.5 Coleoptile experiment

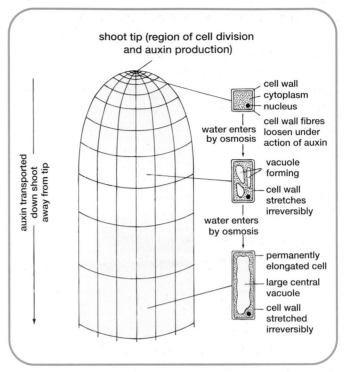

Figure 30.6 Cell elongation at a shoot tip

Effects of IAA at cell level
Cell division

IAA stimulates cell division (primary growth) in apical meristems. In perennial plants in spring, it is transported from apical meristems to cambium where it stimulates secondary growth.

Cell elongation

In apical and lateral meristems, IAA promotes cell elongation by increasing the plasticity of the cell walls, enabling them to stretch irreversibly when water enters by osmosis during vacuolation (see Figure 30.6).

Cell differentiation

IAA is necessary for the differentiation of newly formed unspecialised cells into specialised cells. This process occurs in the region of differentiation behind growing root and shoot tips and in the cells, formed by division of the cambium, which become xylem vessels during secondary thickening.

Effects at organ level
Investigating the effects of IAA on roots and shoots

The experiment shown in Figure 30.7 is set up to investigate the effects of specific concentrations of IAA on the growth of roots and shoots of cress seedlings.

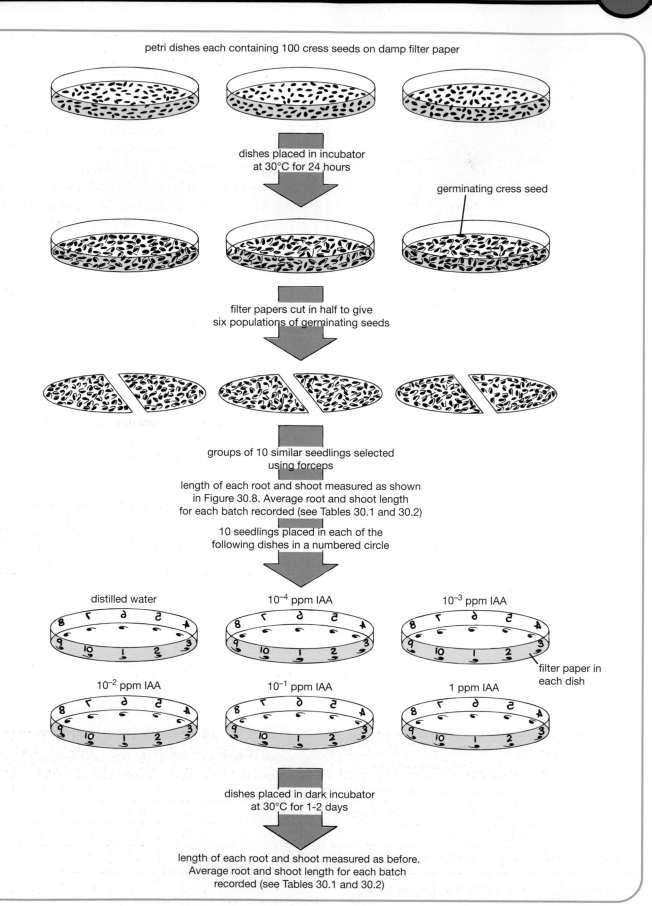

Figure 30.7 Investigating effects of IAA on roots and shoots

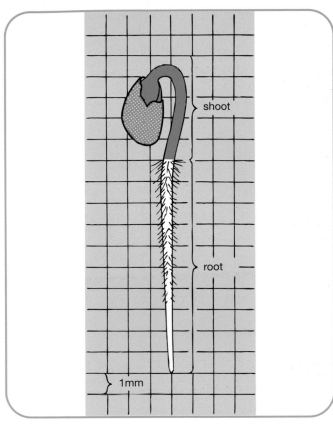

Figure 30.8 Cress seedling

The length of each root and shoot is measured before and after contact with IAA by placing it on a glass slide lying on graph paper as shown in Figure 30.8.

Concentrations of IAA

The solutions of IAA are made up by dissolving 0.1 g of IAA in 100 ml of water to give a stock solution containing 1000 parts per million (ppm) and then further diluting this to produce the five concentrations of IAA required. For example, 10 ml of a solution containing 10^3 ppm IAA added to 90 ml water produces a solution containing 10^2 ppm of IAA and so on.

The IAA solutions are kept in darkness when not in use since light destroys IAA.

Results

Tables 30.1 and 30.2 give typical sets of results. Since the seedlings in distilled water act as the control, an average increase in length greater than that found in water is regarded as a measure of stimulation of growth. An average increase in length less than that in water is regarded as inhibition of growth.

	IAA concentration (ppm)					
	0 (control)	10^{-4}	10^{-3}	10^{-2}	10^{-1}	1
average length of root before IAA treatment (mm)	10	10	10	10	10	10
average length of root after IAA treatment (mm)	20	22	21	19	15	11
average increase in length (mm)	10	12	11	9	5	1
average change in length relative to control (mm)	0	2	1	−1	−5	−9
% stimulation (+) or inhibition (−) of growth	0	+20	+10	−10	−50	−90

Table 30.1 Root results

	IAA concentration (ppm)					
	0 (control)	10^{-4}	10^{-3}	10^{-2}	10^{-1}	1
average length of shoot before IAA treatment (mm)	5	5	5	5	5	5
average length of shoot after IAA treatment (mm)	10	10	10.2	11	13	18
average increase in length (mm)	5	5	5.2	6	8	13
average change in length relative to control (mm)	0	0	0.2	1	3	8
% stimulation (+) or inhibition (−) of growth	0	0	+4	+20	+60	+160

Table 30.2 Shoot results

The effect of IAA concentration on a root or shoot is therefore quantified by subtracting the average increase in length of the control from the average increase in length due to IAA (where negative values indicate inhibition).

Each result is then expressed as a percentage stimulation or inhibition of growth using the formula:

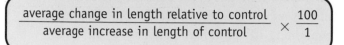

$$\frac{\text{average change in length relative to control}}{\text{average increase in length of control}} \times \frac{100}{1}$$

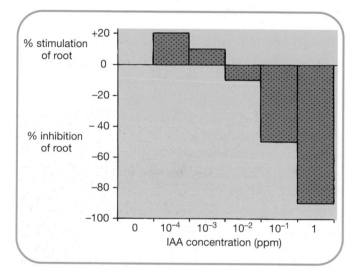

Figure 30.9 Bar graph of root results

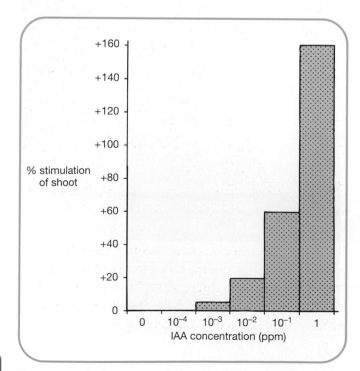

Figure 30.10 Bar graph of shoot results

Figures 30.9 and 30.10 show bar graphs of the results. Under ideal circumstances and using an even wider range of IAA concentrations, results are obtained which can be presented as two line graphs (see Figure 30.11).

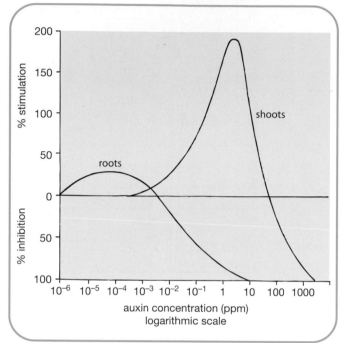

Figure 30.11 Effect of IAA on roots and shoots

Conclusions

From these results it is concluded that very low concentrations of auxin which stimulate elongation of roots, have little or no effect on shoots. On the other hand, higher concentrations of auxin which stimulate shoots, inhibit the elongation of roots. At very high concentrations of auxin, shoot growth is inhibited.

Table 30.3 lists the reasons for adopting certain techniques and precautions during the experiment.

Growth curvature effects

The experiment in Figure 30.12 shows that auxin will diffuse from a shoot tip into an agar block and then from the agar into a cut coleoptile where the cells resume elongation.

When an agar block containing auxin is placed asymmetrically on a decapitated coleoptile (see Figure 30.13), the shoot bends because the side below the agar block receives a higher concentration of growth substance causing greater cell elongation on that side.

design feature or precaution	reason
300 seeds germinated at start	to allow 60 similar seedlings to be selected
large number (10) seedlings used in each dish	to increase the reliability of the results
IAA solutions kept in dark when not in use	to prevent light destroying IAA
separate syringes used to add IAA solutions to dishes	to avoid contamination of one IAA concentration by another
all factors kept equal except IAA concentration	to ensure that IAA concentration is the only variable factor involved in the experiment

Table 30.3 Design techniques

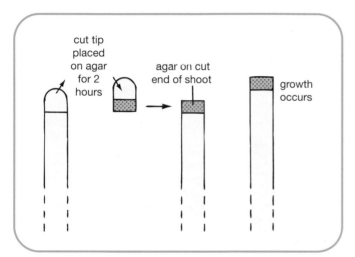

Figure 30.12 Use of agar block

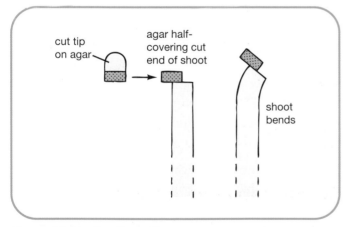

Figure 30.13 Bending effects

Within limits (see graph in Figure 30.11) the higher the concentration of auxin, the greater the curvature produced.

Phototropism

This is the name given to a **directional growth movement** by a plant organ in response to light from one direction (see Figure 30.14).

Figure 30.14 Positive phototropism

Oat coleoptiles exhibit **positive phototropism** by bending towards a unidirectional source of light as shown by shoot X in Figure 30.15. Since shoot Y fails to bend, it is concluded that the shoot tip is the region responsible for detecting unilateral light.

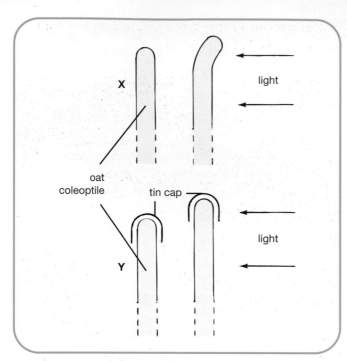

Figure 30.15 Response to light from one side

Mechanism of phototropism

The experiment shown in Figure 30.16 shows that a unidirectional light source causes an unequal distribution of hormone to occur in the shoot tip. A

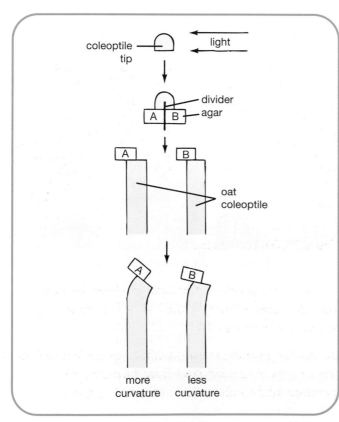

Figure 30.16 Mechanism of phototropism

higher concentration of auxin is present in the non-illuminated side than in the illuminated side.

These findings can be used to explain the mechanism of positive phototropism. If the shaded side of a shoot contains more auxin, then more cell elongation will occur on that side making the shoot bend towards the light as shown in Figure 30.17.

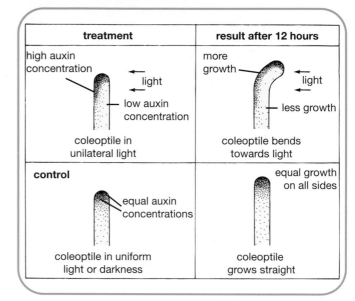

Figure 30.17 Hormonal explanation of phototropism

Apical dominance

Very high concentrations of auxin inhibit growth (see Figure 30.11). In many plants, the apical bud is able to inhibit the development of lateral (side) buds further down the stem by producing a sufficiently high concentration of auxin which is translocated down the stem's phloem tissue to the side buds.

Such auxin-controlled **apical dominance** is demonstrated by the experiment shown in Figure 30.18. In the absence of the apical bud (or an auxin-containing substitute), hormone concentration drops to a level which no longer inhibits growth of lateral buds. These now develop into side branches.

The degree of apical dominance exerted is found to vary from species to species. In some plants (e.g. sunflower) it is very strong and lateral buds normally remain dormant. In other plants (e.g. tomato), it is much weaker and the lateral branches normally develop giving an extensively branched plant.

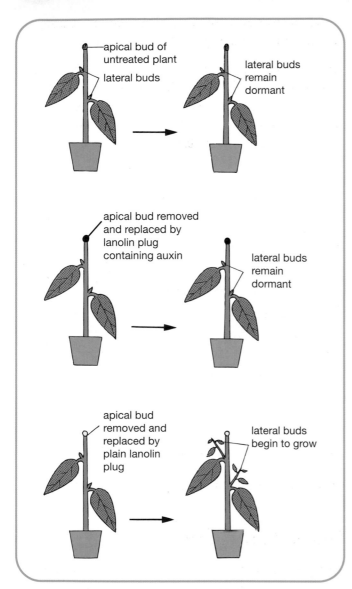

Figure 30.18 Apical dominance

Gardeners often remove apical buds to promote the development of side branches in plants which would otherwise become too tall and spindly.

Fruit formation

Following fertilisation in a flower, auxin promotes the formation of the **fruit coat** from the ovary wall. The coat may become soft and succulent, as in a grape for example.

Leaf abscission

Abscission is the separation of leaves or fruit from a plant. Prior to leaf fall (see Figure 30.19) or fruit fall, auxin concentration drops and a thin **abscission layer**

of cells is formed at the base of the leaf stalk or fruit stalk as shown in Figure 30.20.

Figure 30.19 Leaf fall

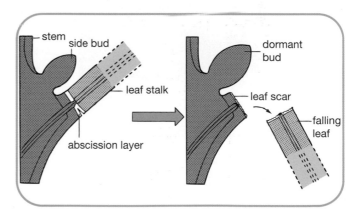

Figure 30.20 Leaf abscission

The walls of the cells in this layer gradually become weakened until eventually the leaf or fruit stalk snaps and the leaf or fruit falls.

During the growing season, auxin from the leaf or fruit prevents the abscission layer from forming and therefore inhibits abscission of the leaf or fruit.

Commercial applications of auxins

Seedless fruit

Fruit development without fertilisation can be induced artificially by treating unfertilised flowers with auxin. This process produces a plentiful supply of seedless fruit that is ripe for harvesting at the same time.

Delaying abscission of fruit

Fruit growers often spray crops with synthetic auxin towards the end of the growing season to delay the formation of the abscission layer in the fruit stalks. This prevents the fruit dropping early before they are fully ripe.

Rooting powders

When rooting powder which contains synthetic auxin is applied to the cut end of a stem, it stimulates the formation of roots at the cut end (see Figure 30.21). This makes the plant easy to propagate.

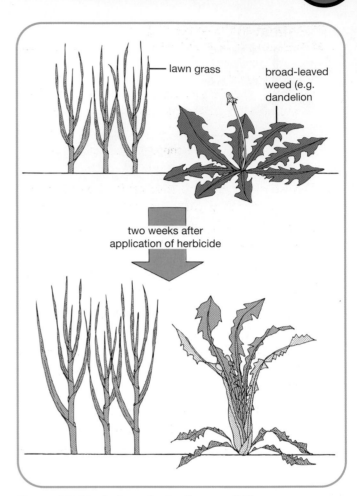

Figure 30.22 Action of selective weedkiller

Figure 30.21 Roots from a cut stem

Herbicides

Synthetic auxins are used as **selective weedkillers** (herbicides). They work by stimulating the plant's rate of growth and metabolism to such an extent that the plant exhausts its food reserves and dies of starvation.

Selective weedkillers are especially effective on lawns since weeds with broad leaves (see Figure 30.22) absorb a lot of the chemical and die, whereas narrow-leaved grass plants absorb little of the chemical and are hardly affected. Also see Figure 30.23.

Figure 30.23 "Oh no, Daisy's committing herbicide!"

Testing Your Knowledge

1 a) What is a *coleoptile*? (1)

   b) Which part of the coleoptile produces a chemical essential for growth and development of the shoot? (1)

   c) Which region of the shoot is affected by the chemical? (1)

   d) What effect does the chemical have on the cells in this region of the shoot? (1)

   e) To which group of plant growth substances does the chemical belong? (1)

2 Rewrite the following sentences choosing the correct word from each choice in brackets.

   Very low concentrations of auxin stimulate elongation of (roots/shoots) but have little effect on (roots/shoots). Higher concentrations of auxin which stimulate (roots/shoots), inhibit elongation of (roots/shoots). (2)

3 Most cereal crops are narrow-leaved. The weeds that compete with them tend to be broad-leaved.

   a) What type of chemical could be used in an attempt to destroy only the weeds?

   b) Explain how this method works. (3)

4 a) What effect does an auxin have on the ovary wall in a flower following fertilisation? (1)

   b) By what means can seedless grapes be produced? (1)

5 a) What name is given to the inhibition of side buds by a high concentration of auxin made by the apical bud on the same plant shoot? (1)

   b) What happens to the side buds if the apical bud is cut off and replaced with a plug of plain lanolin paste? (1)

   c) What happens to the side buds if the apical bud is cut off and replaced with a plug of lanolin paste containing a high concentration of auxin? (1)

Gibberellins

In the 1920s, a Japanese farmer discovered that some of his rice plants which were infested with a fungus had grown abnormally tall. This growth effect was found to have been caused by a growth substance (hormone) secreted by the fungus. Since the fungus was called *Gibberella*, the hormone was called gibberellin.

Many gibberellins are now known to exist. They make up a second group of plant substances which occur naturally in low concentrations in most plants. The most common one is gibberellic acid (GA).

Like auxins, gibberellins stimulate cell division and elongation in stems. However they play no part in phototropic movements or bending of shoots. Gibberellins affect several other aspects of growth and development as follows.

Effect of gibberellic acid (GA) on dwarf pea seedlings

Look at the experiment shown in Figure 30.24. The dwarf pea seedlings in pot A which receive gibberellic acid reach the same height as the normal tall variety in pot C. This is due to an increase in length (not number) of their internodes.

The dwarf plants in the control (pot B) show that this effect is not due to the application of lanolin alone.

It is concluded from this experiment that gibberellin is needed by a plant to make its internodes increase in length. Tall varieties are able to manufacture sufficient gibberellin to make them grow to full height. The dwarf condition (which is inherited) is caused by a shortage of gibberellin due to a genetic deficiency.

Effect of gibberellic acid (GA) on germinating barley grains

Figure 30.25 shows the internal structure of a soaked barley grain. It indicates the plane in which it should be cut to give an embryo 'half' and an endosperm 'half'. Several of these 'halves' are needed for the experiment shown in Figure 30.26.

The results of this experiment show that digestion of starch (in the starch agar) has only occurred in plate A

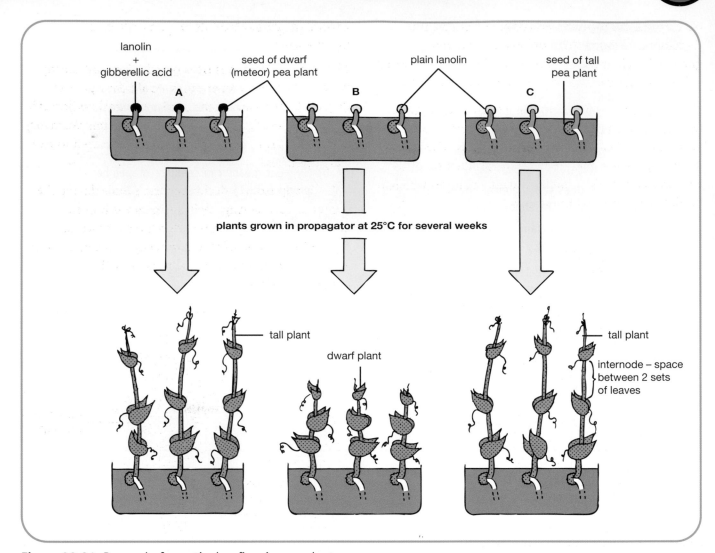

Figure 30.24 Reversal of genetic dwarfism in pea plants

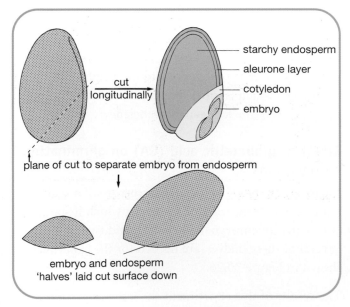

Figure 30.25 Barley embryo and endosperm

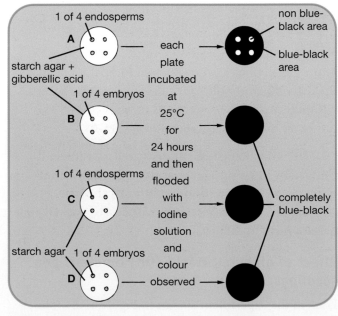

Figure 30.26 Effect of GA on barley endosperm

beneath the endosperm parts of barley grains. It is therefore concluded that both endosperm and gibberellic acid must be present for the starch-digesting enzyme, α-amylase, to be produced in a barley grain.

Under normal circumstances in a soaked cereal grain, the hormone gibberellin is made by the embryo and passes up to the **aleurone layer** (see Figure 30.27). Here it acts at gene level and induces the production of α-amylase. This digests starch to sugar (maltose) which is needed for growth by the seedling.

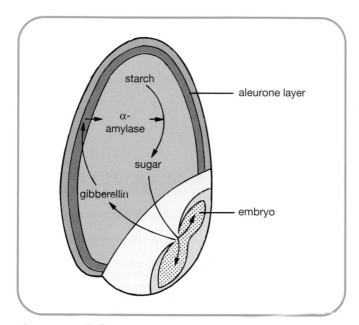

Figure 30.27 Induction of α-amylase by GA

By this means, gibberellin **breaks the dormancy** of many types of seed and induces them to germinate when water is present in the environment and temperature conditions are suitable for growth.

Model of hormone action

By definition, a hormone is a chemical substance produced at one site and transported to another site where a low concentration of it brings about an effect.

Gibberellin is produced at one site (the embryo) and is active at another site (the aleurone layer) where, in low concentration, it exerts its effect (induction of α-amylase). Hence this sequence of events provides a model of hormone action.

Effect of gibberellic acid (GA) on bud dormancy

The buds of deciduous trees remain dormant during winter when low temperatures would damage new delicate tissues and organs. Under natural conditions in spring, GA produced by the plant breaks this dormancy and allows the buds to open ready to develop into new leafy branches.

If GA is applied to a deciduous tree's buds during the winter or early spring, their dormancy is broken artificially and they begin to develop as shown in Figure 30.28. Similarly the dormancy of the buds in the 'eyes' of a potato tuber can be broken by the application of GA.

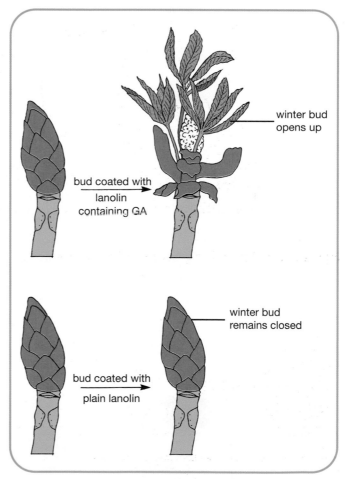

Figure 30.28 Effect of GA on winter buds

Testing Your Knowledge

1 a) What effect does gibberellic acid (GA) have on stem length when applied to the shoots of dwarf pea plants?
   b) Is this effect due to increase in number or length of internodes? (2)

2 a) i) Name the TWO regions of a barley grain separated by the cotyledon.
     ii) Which of these contains a store of starch? (3)
   b) i) Which layer of the endosperm can be induced to make a starch-digesting enzyme?
     ii) Name this enzyme. (2)
   c) i) What plant growth substance is needed to induce production of the enzyme?
     ii) Under normal circumstances, which part of the grain makes the growth substance? (2)

   d) i) When the enzyme acts on its substrate, what end product is formed?
     ii) Which part of the grain uses this product for growth? (2)
   e) If the experiment shown in Figure 30.26 is not properly set up then a small amount of digestion may also be found to occur in a plate other than A. Suggest a possible source of error responsible for this result. (2)

3 Copy and complete Table 30.4 by placing ticks in the appropriate columns to indicate the roles played by gibberellin (G) and auxin (A). (7)

role	G only	A only	G and A
induction of α-amylase production in cereal grains			
stimulation of cell division and elongation			
reversal of genetic dwarfism			
stimulation of root formation			
prevention of leaf abscission			
promotion of phototropic growth movements			
breaking of dormancy of winter buds			

Table 30.4

Applying Your Knowledge

1 The graph in Figure 30.29 summarises the results of an experiment in which the rate of increase in length of an oat coleoptile was followed over a period of 10 hours.
   a) Name the growth substance made by the tip of an oat coleoptile. (1)
   b) i) At approximately what time was the coleoptile in the graph decapitated?
     ii) Explain your answer. (2)

   c) Suggest why a slight increase in coleoptile length still occurred after decapitation. (1)
   d) Name TWO possible courses of action that could have been taken at 17.00 hours to bring about the change in rate of increase in length of coleoptile that followed. (2)

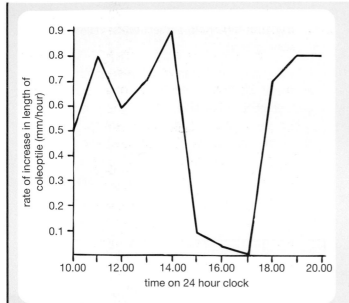

Figure 30.29

2 Table 30.5 shows the results from an investigation into the effect of different concentrations of auxin on roots and shoots of cress seedlings. In this investigation 10 lengths of root and 10 lengths of shoot were immersed in each of the auxin solutions and kept at 30°C in darkness.

After 24 hours averages were calculated and the results in Table 30.5 obtained.

a) Present the data as two line graphs sharing the same axes. (4)

b) Which concentration of auxin produced the greatest stimulation of **i)** roots; **ii)** shoots? (2)

c) What effect did a molar concentration of 10^{-1} have on the growth of **i)** roots; **ii)** shoots? (2)

d) Which concentration of auxin had the effect of inhibiting growth of both roots and shoots? (1)

e) Why were 10 lengths of each type of plant organ used and averages calculated? (1)

f) **i)** Why were the dishes kept in darkness during the experiment?

   ii) Identify TWO other variables that must be kept constant during the experiment. (3)

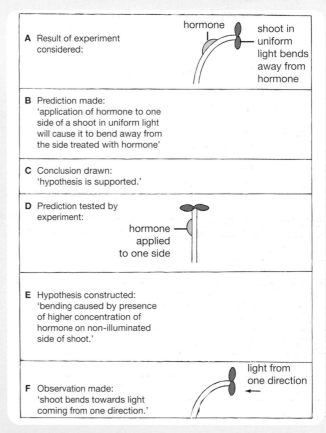

Figure 30.30

molar concentration of auxin applied	average length of organ after 24 hours (in mm)	
	root	shoot
0	20.0	20.0
10^{-5}	22.5	20.0
10^{-4}	23.5	20.0
10^{-3}	22.0	20.5
10^{-2}	18.5	22.5
10^{-1}	14.0	27.5
1	11.0	35.0
10^{1}	10.0	30.0
10^{2}	10.0	16.0

Table 30.5

3 Figure 30.30 shows six stages involved in constructing and testing a hypothesis. Arrange the stages into the correct order starting with F.

4 In an experiment, four groups of ten French bean plants were treated as shown in Table 30.6.

group	treatment
W	apical buds removed and no further treatment given
X	apical buds replaced with lanolin plugs containing auxin
Y	apical buds replaced with plain lanolin plugs
Z	apical buds replaced with lanolin plugs containing gibberellin

Table 30.6

The lengths of each plant's side buds and shoots were measured at the start and at 2-day intervals. The average total length of side buds and shoots per plant was calculated for each group and recorded in Table 30.7.

a) Draw line graphs of the data on the same sheet of graph paper. (2)

b) i) In which group of plants did apical dominance occur?

ii) Explain how you arrived at your answer. (2)

c) What is the purpose of including group Y in the experiment? (1)

d) From the data it was concluded that application of one of the hormones stimulated growth of side shoots.

i) Identify the hormone.

ii) Explain why the data justify drawing such a conclusion.

iii) Calculate the percentage stimulation brought about by this hormone at day 8. (3)

e) Suggest why the result recorded for group X at day 10 differs from earlier results recorded for the same group. (1)

5 Leaves X and Y in Figure 30.31(i) were approaching an age at which they would be lost naturally from each plant by abscission. However one of these leaves was treated with auxin. Figure 30.31(ii) shows the appearance of the plants three weeks later.

a) Which leaf was treated with auxin? (1)

b) Give TWO reasons for your answer. (2)

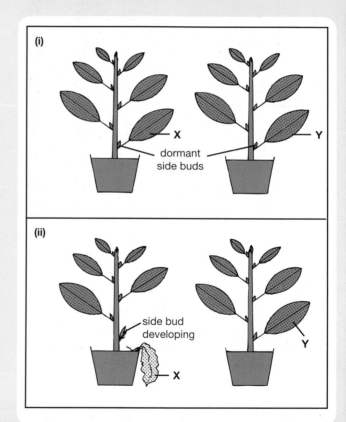

Figure 30.31

average total length of side buds and shoots per plant (mm)		time in days from start of experiment					
		start	2	4	6	8	10
	W	4	4	10	35	60	96
	X	4	4	4	4	6	20
	Y	4	4	11	34	60	98
	Z	4	4	16	56	102	146

Table 30.7

6 The bar graph in Figure 30.32 shows the results of an investigation into the effect of gibberellic acid on the growth of dwarf pea plants. The growth substance was applied in lanolin paste to the tip of each shoot 2 days after it had emerged from the germinating seed. The height of each shoot was measured after 7 and after 22 days of growth.

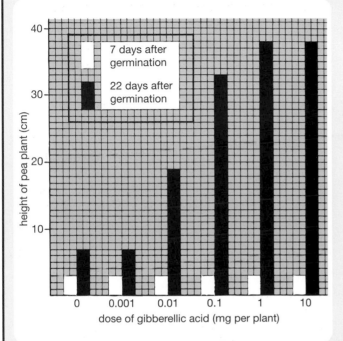

Figure 30.32

a) Make a generalisation about the effect of increasing concentration of gibberellic acid on the height of dwarf pea plants after 22 days of growth. (1)

b) i) What was the lowest dose of GA needed to give the maximum effect?
   ii) Suggest how you could find out whether plants showing this maximum effect are equal in height to the normal tall variety of this strain of pea plant. (2)

c) i) Which dose of GA had no effect?
   ii) Suggest why. (2)

d) Which treatment was the control? (1)

e) Give a possible explanation for the difference between the results at 7 days and at 22 days. (1)

7 In an experiment, barley embryo and endosperm 'halves' were prepared as shown in Figure 30.25. These were placed, cut surface down, on plates of starch agar. In some cases gibberellic acid (GA) had also been added to the agar as indicated in Table 30.8.

After 24 hours each plate was flooded with iodine solution to locate starch-free areas which indicated digestion. The results are given in Table 30.8.

a) What conclusion can be drawn from plates 1 and 4 about the effect of GA on an endosperm's ability to digest starch? (1)

b) What conclusion can be drawn from plates 2 and 5 about the effect of GA on an embryo's ability to digest starch? (1)

c) Compare the contents of plates 3 and 4 and suggest the source of the GA supply in a barley grain under natural conditions of germination. (1)

d) What effect does GA have on endosperm tissue that subsequently allows successful digestion of starch to occur? (1)

e) Why is it of survival value to a seed grain that its dormancy is broken by the above mechanism only when water enters the seed grain and the temperature is above freezing point? (1)

starch agar plate	presence or absence of GA in agar	part(s) of barley grain put on agar	digestion (✓) or no digestion (✗) of starch after 24 hours
1	absent	endosperms	✗
2	absent	embryos	✗
3	absent	endosperms and embryos	✓
4	present	endosperms	✓
5	present	embryos	✗
6	present	endosperms and embryos	✓

Table 30.8

8 Give an account of:

  a) the role of the pituitary gland in the control of human growth and development; (4)

  b) the commercial applications of auxins. (6)

What You Should Know

Chapter 30

(See Table 30.9 for Word bank)

α-amylase	elongation	phototropic
abscission	gibberellic acid	root
auxin	gibberellins	side
bend	herbicides	somatotrophin
cuttings	higher	stems
division	IAA	thyroid
dominance	inhibit	thyroid-stimulating
dormancy	internode	thyroxine
dwarfism	narrow	weeds

Table 30.9 Word bank for chapter 30

1 The pituitary gland produces both _____ (which promotes human growth) and _____ hormone (TSH).

2 The _____ gland responds to TSH by producing _____ which controls metabolic processes.

3 The auxins make up a group of plant growth substances (hormones). The most common _____ is called indole acetic acid (_____).

4 IAA is produced at root and shoot tips and promotes cell _____, cell _____ and differentiation of cells.

5 Low concentrations of auxin which stimulate _____ elongation have little effect on shoots.

_____ concentrations of auxin which stimulate shoot elongation _____ root elongation.

6 Unequal distribution of auxin causes shoots to _____ producing _____ movements.

7 Apical buds exert apical _____ by producing high concentrations of auxin which inhibit growth of _____ buds.

8 Auxin prevents leaf and fruit _____ and promotes fruit formation.

9 Synthetic auxins act as _____ by disrupting growth of broad-leaved _____ more than grass-like plants which absorb less through their _____ leaves.

10 Synthetic auxins stimulate formation of roots on stem _____.

11 The _____ make up a second group of plant growth substances. The most common gibberellin is called _____ (GA).

12 GA promotes cell division and elongation in _____.

13 GA reverses genetic _____ by promoting stem _____ elongation.

14 GA induces production of _____ in cereal grains and breaks _____ in winter buds.

Effects of chemicals on growth

Macro-elements in plants

In addition to the large amounts of carbon (C), hydrogen (H) and oxygen (O) needed for the manufacture of carbohydrate during photosynthesis, a green plant requires small but appreciable quantities of several other chemical elements for healthy growth.

Four of these **macro-elements** are **nitrogen** (N), **phosphorus** (P), **potassium** (K) and **magnesium** (Mg). They are present in fertile soil and are absorbed by the plant's root hairs.

Water culture experiments

The importance of each macro-element to a plant is investigated by setting up a series of **water culture** experiments where one element at a time is omitted from the mineral solution. The control experiment contains all the elements necessary for growth.

The glass beaker used in the water culture experiment shown in Figure 31.1 is initially rinsed with concentrated nitric acid to remove traces of mineral elements. It is surrounded by an opaque cover (e.g. aluminium foil) to keep out light and prevent the growth of algae. (If these were allowed to grow, a second variable factor would be introduced into the experiment making the results invalid.) Several

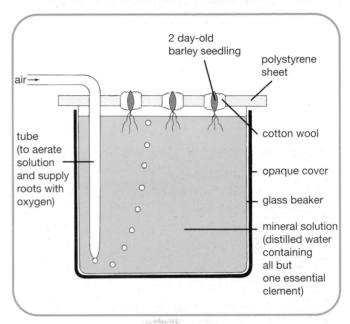

Figure 31.1 Water culture experiment

seedlings are used in each beaker to improve the reliability of the results.

After several weeks' growth, each plant lacking an essential macro-element is found to display certain **deficiency symptoms** when compared with the control (see Figure 31.2). These symptoms often indicate the role normally played by the macro-element in the plant's metabolism as summarised in Table 31.1.

element omitted	symptoms of deficiency	reason for deficiency symptom (role of element)
nitrogen	overall growth reduced, leaves **chlorotic** (pale green or yellow), leaf bases red, roots long and thin	required for formation of amino acids, proteins and nucleic acids
phosphorus	overall growth reduced, leaf bases red	required for formation of ATP and nucleic acids
potassium	overall growth reduced, early death of older leaves	required for important role in transport of molecules across membranes
magnesium	overall growth reduced, leaves chlorotic	required for chlorophyll formation

Table 31.1 Effects of macro-element deficiency in plants

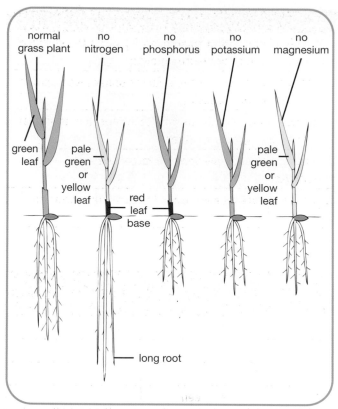

Figure 31.2 Effects of macro-element deficiency in plants

Investigating the effect of nitrate concentration on algal growth

Different masses of nitrate fertiliser are added to five flasks containing a solution of complete medium, as shown in Figure 31.3. A further flask receives no fertiliser and acts as the control.

A sample of unicellular algae from a pond or fish tank is shaken up vigorously in 20 cm³ of complete medium solution and then a 3 cm³ inoculum of algal suspension is added to each flask using a syringe. The flasks are kept in constant illumination in a warm room and distilled water is added as required to compensate for loss by evaporation and maintain the solution in each flask at its starting level.

After four weeks the appearance of the plant population in each flask is noted and then the contents are filtered. Each filter paper is dried in a warm oven to constant mass and the dry mass of algae from each flask calculated by subtracting the weight of an unused filter paper. The experiment is repeated several times and the class results pooled to allow averages to be calculated.

Table 31.2 summarises the reasons for adopting certain design features during the investigation. Table 31.3 gives a specimen set of results. The effect of increasing concentration of nitrate fertiliser on dry mass of algae is represented by a graph of the results in Figure 31.4 which shows the line of best fit.

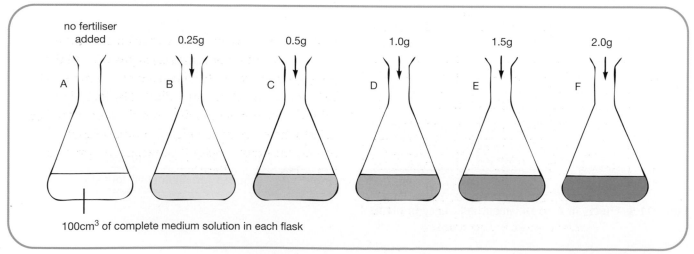

Figure 31.3 Investigating the effect of nitrate concentration on algal growth

design feature	reason
use of complete medium solution	to supply plants with a small quantity of all the chemical nutrients required for healthy growth
use of control flask A	to check that the one variable factor under investigation is responsible for the results
algal sample shaken vigorously before taking each inoculum	to disperse algae evenly so that each inoculum is similar in density and variety of algae
light intensity, temperature and volume of liquid kept equal for all flasks	to ensure that only one variable factor is being investigated
experiment repeated, results pooled and averages calculated	to obtain a more reliable result for each concentration of nitrate

Table 31.2 Design features

flask	mass of nitrate fertiliser added (g)	colour of solution in flask		dry mass of algae (g)
		at start	after 4 weeks	
A	0.0	faint green	yellow-green	0.04
B	0.25	faint green	dark green	0.09
C	0.5	faint green	dark green	0.14
D	1.0	faint green	very dark green	0.16
E	1.5	faint green	very dark green	0.18
F	2.0	faint green	very dark green	0.18

Table 31.3 Algal growth results

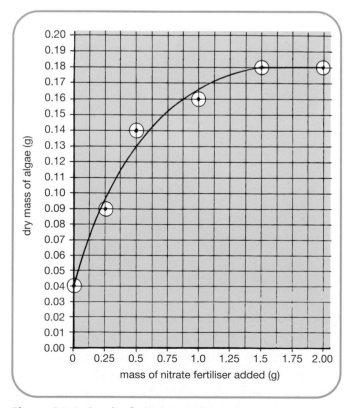

Figure 31.4 Graph of algal growth results

Discussion of results

At the end of the investigation the population of algae in the control flask is found to be the smallest (as indicated by dry mass) and least healthy (as indicated by colour). It is concluded that the low concentration of nitrate present in complete medium solution acts as the **limiting factor** and prevents a population explosion in flask A.

In the other flasks, increasing enrichment of the solution by nitrate promotes algal growth and the population increases (as indicated by dry mass). The cells are also healthier (as indicated by their dark colour). The population does not show uncontrolled growth, however. At the higher concentrations of nitrate, the graph levels off showing that some other factor has become limiting.

Lead

Effect of lead on activity of catechol oxidase

When a potato, an apple or a banana is peeled, its inner light-coloured tissues turn brown if left exposed to oxygen. This effect is brought about by the enzyme **catechol oxidase** which is present in the cells and catalyses the following reaction:

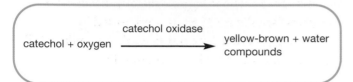

catechol + oxygen $\xrightarrow{\text{catechol oxidase}}$ yellow-brown + water compounds

In the experiment shown in Figure 31.5, the substrate **catechol** is added to each test tube. The enzyme catechol oxidase is present in the liquid released from the ground-up potato tissue.

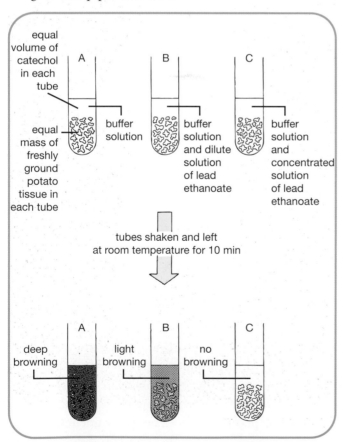

Figure 31.5 Effect of lead on catechol oxidase

Buffer solution is used to keep the contents of the tubes at the same pH. If buffer solution were not used, it is possible that the contents of the tubes might differ from one another in pH. It could then be claimed that the results obtained were due to this difference in pH and not to the action of lead ethanoate on the enzyme.

The contents of tube A turn deep brown in the absence of lead ethanoate. The contents of tube B turn light brown in the presence of dilute lead ethanoate solution. The contents of tube C fail to turn brown in the presence of a high concentration of lead ethanoate.

It is therefore concluded that **lead inhibits the activity** of the enzyme catechol oxidase. Similarly lead can have a harmful effect on the human body by inhibiting enzymes that control the metabolic pathways involved in essential processes such as respiration and growth.

Iron and calcium in animals

Unlike lead, **iron** and **calcium** are essential elements needed for healthy growth and development.

Iron

This element is an essential constituent of many **enzymes** (e.g. catalase). It is needed to make **cytochrome** (the hydrogen carrier molecule in respiration) and it forms part of the haem group in the respiratory pigment **haemoglobin** (see Figure 31.6).

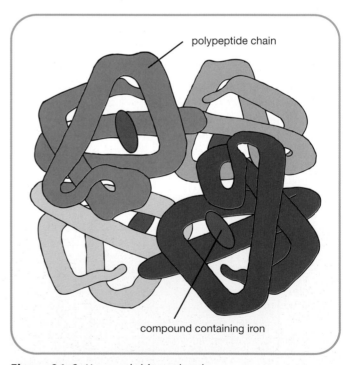

Figure 31.6 Haemoglobin molecule

A small quantity of iron is lost every day by humans in dead skin cells, bile and urine and therefore needs to be replaced. Growth and menstruation make further demands as shown in Table 31.4.

factors creating demand for iron	average mass of iron required to satisfy demand (mg/day)
basic loss in skin cells, bile and urine	0.5–1.0
growth: 0–1 year	0.7–0.8
1–11 years	0.3
adolescence	0.5
normal menstruation	0.5–1.6
pregnancy	6.0

Table 31.4 Iron demands

It is therefore essential that growing children and pregnant women consume a diet containing large quantities of iron. Many foodstuffs such as meat, eggs, fish, green vegetables and cereals are rich in iron. However iron present in vegetables and cereals is less well absorbed than that from animal sources. Vegetarians must be careful in their choice of diet if they are to avoid deficiency and anaemia.

Calcium

This element is required by animals to form calcium phosphate which becomes built into the hard parts of **bones** and **teeth**. Calcium carbonate is required by many invertebrates to form their **shells** (see Figure 31.7).

In humans, calcium is also needed for the **clotting of blood** and the **contraction of muscle.** The main sources of calcium are milk, cheese and green vegetables.

Figure 31.7 Role of calcium

Vitamin D deficiency

Vitamin D is essential in humans to promote the absorption of calcium and phosphate from the intestine and their uptake by bone.

Deficiency of vitamin D in young children leads to the formation of soft abnormal bones. The long bones of the legs tend to bend under the weight of the body making the child bow-legged. This condition is known as **rickets** (see Figure 31.8). It is prevented by consuming foods such as cod-liver oil, egg yolk or full cream milk which are rich in vitamin D.

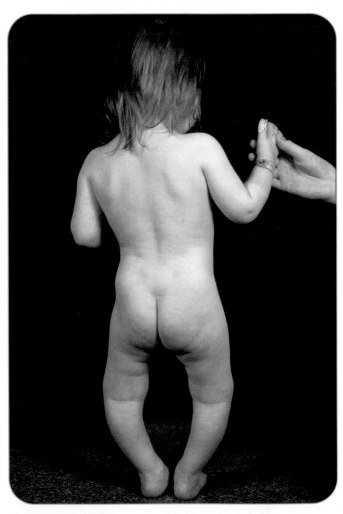

Figure 31.8 Rickets

Rickets is almost unknown in tropical countries because vitamin D is formed in the skin when it is exposed to the ultraviolet rays present in sunshine. The condition is more likely to be found amongst poorly-fed children in cities of northern latitudes where sufficient ultraviolet radiation fails to penetrate the cloud cover and layer of atmospheric pollution.

Effect of drugs on fetal development

Thalidomide

In the 1950s the drug **thalidomide** was developed and prescribed to pregnant women in order to counteract the feelings of nausea (morning sickness) that many experience. However if the thalidomide was taken during a certain critical period, very early in the pregnancy, then the limbs of the fetus failed to develop properly.

As a result the baby's hands developed attached to the shoulders and the feet to the hip joints. In addition malformation of eyes, ears and heart occurred in some cases together with mental subnormality and epilepsy.

This tragedy demonstrated the danger of taking drugs during pregnancy. Thalidomide has been withdrawn from use in the UK. However it is still used as a treatment for leprosy in some countries and may be consumed by pregnant women unfamiliar with its side effects.

Alcohol

A higher incidence of spontaneous abortion is found to occur amongst women who drink **alcohol** in excess during pregnancy. In those women who do not suffer a miscarriage but who do continue to drink heavily, some of the alcohol crosses the placenta and causes the blood vessels in the umbilical cord to collapse temporarily (see Figure 31.9). During these periods the fetus fails to receive an adequate **oxygen supply** vital for the proper development of growing tissues such as the brain.

In extreme cases of alcohol abuse, the fetus suffers several harmful effects known collectively as the **fetal alcohol syndrome**. These include:

- pre- and post-natal growth retardation;
- facial abnormalities;
- heart defects;
- development of abnormal joints and limbs;
- mental retardation.

Nicotine

Nicotine is a drug present in tobacco plants which stimulates the central nervous system and increases

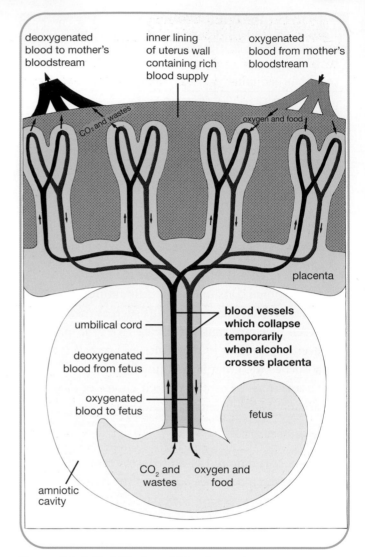

Figure 31.9 Umbilical cord blood vessels

pulse rate and blood pressure. In addition to nicotine, tobacco smoke contains **carbon monoxide** (CO) and **tar** (which has been shown to contain up to 30 different harmful substances).

Smoking during pregnancy has been shown to retard the growth of unborn children. CO reduces the concentration of oxygen that the blood can carry and nicotine in the bloodstream prevents adequate glucose reaching fetal tissues including brain cells.

Evidence suggests that babies born to heavy smokers are smaller and lighter in birth weight and that they do not develop intellectually at the same rate as children of non-smokers. Fetal development is also adversely affected if the mother, herself a non-smoker, lives with smokers and is constantly subjected to a smoky environment.

Testing Your Knowledge

1 **a) i)** Name an element other than nitrogen which, if in deficient supply to a plant, brings about chlorosis.

   ii) Relate this element's role in the plant's metabolism to chlorosis.

   iii) Chlorosis rarely becomes apparent until a few weeks after germination. Suggest why. (3)

   b) Name TWO elements (apart from carbon, hydrogen and oxygen) needed by a plant to make DNA. (2)

   c) Give ONE reason why a plant needs potassium. (1)

   d) What control should be set up in a water culture experiment? (1)

2 **a)** Name a chemical element that is an essential constituent of many enzymes in the human body. (1)

   b) Name the chemical element that has an inhibitory effect on the action of many enzymes. (1)

3 Give THREE reasons why the human body needs calcium in the diet. (3)

4 State ONE effect that each of the following drugs can have on fetal development. **a)** thalidomide; **b)** nicotine; **c)** alcohol. (3)

Applying Your Knowledge

1 The apparatus shown in Figure 31.10 was one of a set of water cultures used to investigate the requirement by cuttings of *Tradescantia* (wandering sailor) for certain chemical elements. Each of the gas jars was thoroughly cleaned in hot soapy water and rinsed under the tap before use.

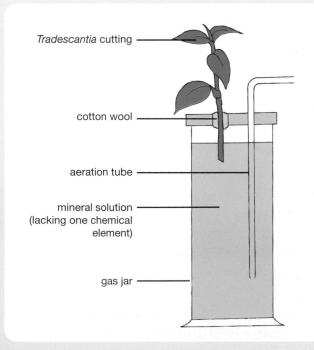

Figure 31.10

   a) i) Identify TWO sources of error contained in the procedure adopted during the setting up of this investigation.

   ii) Explain your answers. (4)

   b) Assume that the two sources of error have been eliminated in an improved version of the experiment. Compare Figure 31.10 with Figure 31.1 and state which set-up would be likely to give more reliable results. Justify your answer. (1)

   c) After six weeks, the *Tradescantia* plants in an improved version of the experiment were examined for deficiency symptoms.
   i) What is meant by the term *deficiency symptom*?
   ii) Which leaves would be most likely to show them? Why? (3)

   d) Predict with reasons what would have happened to a *Tradescantia* plant if distilled water had been used in the gas jar instead of mineral solution. (2)

2 In an investigation into the effects of different levels of minerals on the growth of the roots of radish plants, a student set up seven versions of the apparatus shown in Figure 31.11. She varied the number of slow-release pellets used per carton as indicated in Table 31.5. After six weeks of growth, she weighed each radish root and then pooled her ➔

In figure labels:
- *Tradescantia* cutting
- cotton wool
- aeration tube
- mineral solution (lacking one chemical element)
- gas jar

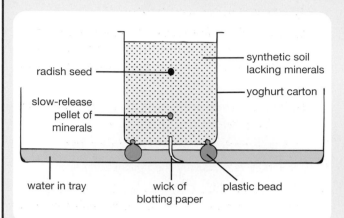

Figure 31.11

results with those of four other students who had carried out the same investigation. The results are shown in Table 31.5.

number of slow-release mineral pellets used	average fresh mass of radish root (g)
2	10.3
4	12.9
6	16.2
8	17.7
10	19.1
12	18.9
14	19.0

Table 31.5

a) i) Plot the results as seven points on a sheet of graph paper and then draw a line of best fit.

  ii) Draw a conclusion from the graph. (4)

b) Suggest a control that should have been set up in this investigation. (1)

c) Why was mineral-free soil used in this experiment? (1)

d) Why were slow-release mineral pellets used? (1)

e) What procedure was carried out to increase the reliability of the results? (1)

f) The students used fresh mass of radish roots as their results.

  i) Why would it have been better to have used dry mass?

  ii) Describe how accurate measurements of dry mass could be obtained. (3)

3 Although the iron content of porridge oats is 4 mg/100 g, only 10% of this iron is absorbed from the gut into the bloodstream.

a) If a woman in late pregnancy needs a daily uptake of 6 mg of iron, how much porridge would she have to eat to meet this iron demand? (1)

b) Since this would be unrealistic, name THREE other foods that she could also consume in order to gain iron. (3)

4 Figure 31.12 shows the daily turnover of calcium in a human adult who has consumed 900 mg. Bone is continually being remodelled by the processes of bone formation and resorption which are equal in an adult whose bones have reached their final size.

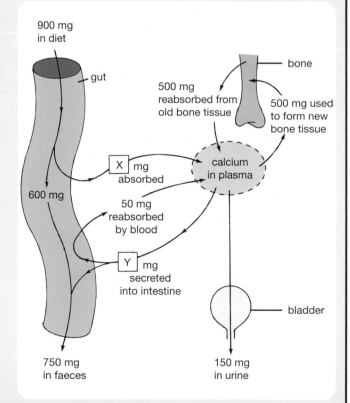

Figure 31.12

a) State the figures that should have been inserted in boxes X and Y. (2)

b) Assume that a 13-year-old girl also consumes 900 mg of calcium each day. Suggest a way in which her overall calcium turnover would differ from that of an adult. (1)

5 During the 1960s, many children of Asian immigrants who had settled in British cities were found to be suffering from rickets.

a) In what vitamin must the children's diet have been deficient? (1)

b) Why did the children's parents who had been fed a similar diet during their childhood in Asia not suffer from rickets in their youth? (1)

c) Describe an obvious physical feature displayed by a 2-year-old child suffering rickets. (1)

d) Name a food rich in the vitamin needed to cure rickets. (1).

e) Briefly describe the role played by this vitamin in the body. (2)

6 The five test tubes shown in Figure 31.13 were set up to investigate the effect of increasing concentration of lead ethanoate on the activity of catechol oxidase. The tubes were kept in a warm room for 20 minutes and then a colorimeter was used to measure the percentage transmission of light through each tube. (This means the percentage of light able to pass through the contents of the test tube.) The results are given in Table 31.6.

a) Present the results as a graph by plotting the points and then drawing a line of best fit. (3)

b) Estimate the percentage transmission that would have been obtained using 2.5% lead ethanoate solution. (1)

c) What was the source of the enzyme in this experiment? (1)

d) Why was buffer solution used? (1)

e) Which of the five tubes would have been i) the lightest; ii) the darkest in appearance to the naked eye at the end of the experiment? Explain your answer. (2)

f) Why is no tube completely colourless at the end of the experiment? (1)

g) What conclusion can be drawn from this experiment? (1)

7 The graphs in Figure 31.14 show the percentage number of babies of different birth weights born in a hospital over a period of one year. Some of the mothers were smokers and some were non-smokers.

a) i) What percentage of babies born to mothers who smoked, weighed 3.0 kg or less?
   ii) What percentage of babies born to mothers who did not smoke, weighed 3.0 kg or less? (2)

b) i) What percentage of babies born to mothers who smoked, weighed 3.1 kg or more?
   ii) What percentage of babies born to mothers who did not smoke, weighed 3.1 kg or more? (2)

c) Draw a conclusion from this data about the effect of nicotine on birth weight. (1)

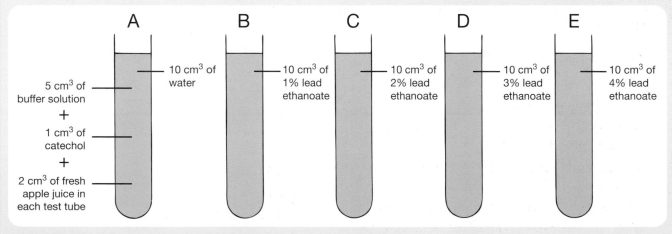

Figure 31.13

	test tube				
	A	**B**	**C**	**D**	**E**
concentration of lead ethanoate solution added (%)	0	1	2	3	4
transmission of light (%)	36	49	60	71	84

Table 31.6

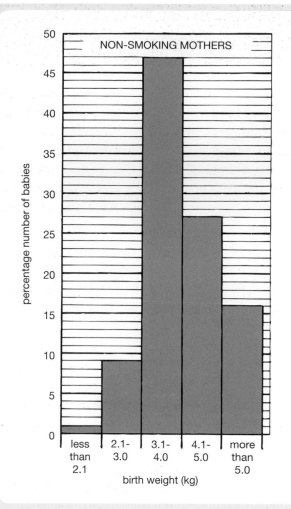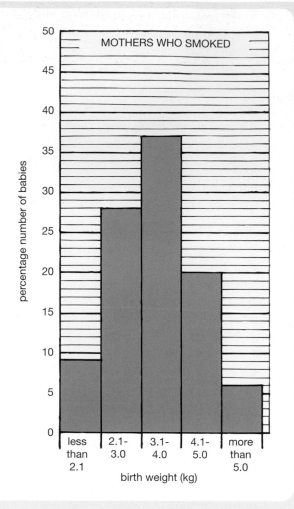

Figure 31.14

8 Give an account of:

a) the possible effects of alcohol and nicotine on the development of the human fetus; (6)

b) the requirement for iron in the human diet. (4)

Effect of light on growth

Vegetative shoot

A **vegetative** shoot is one which does not bear flowers. To investigate the effect of light on growth and development of a leafy shoot, two sets of broad bean seedlings arc grown in identical environmental conditions except that one is kept in **daylight** and the other in **darkness**.

From a comparison of the results (see Figure 32.1), it is concluded that light is necessary for the development of fully expanded green leaves and short strong stem internodes.

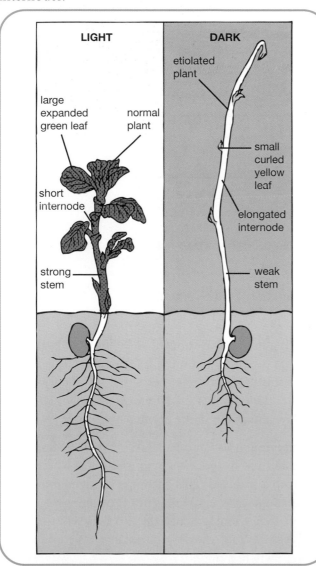

Figure 32.1 Etiolation

Plants grown in darkness and possessing small curled yellow leaves and long weak internodes are said to be **etiolated**. Microscopic examination reveals that increased cell elongation accounts for the formation of these long internodes. It is thought that the cells of an illuminated plant do not become as elongated because they receive less auxin (since some has been destroyed by light).

In terms of evolution, etiolation is of survival value to a plant since the chance of some leaves eventually reaching light is increased by the plant expending most of its resources on the development of long thin stems.

Phototropism

The shoots of green plants normally show **positive phototropism** by growing towards a unidirectional source of light. An explanation of the mechanism of phototropism, involving the hormone IAA, is given in chapter 30.

Growth of a shoot towards light is of survival value to the green plant because it increases the chance of the shoot being exposed to the light energy needed for photosynthesis.

Effect of light on flowering
Photoperiodism

Many species of plant stop producing vegetative (leaf) buds and start producing flower buds at a certain time of the year. This occurs in response to a change in the period of illumination (**photoperiod**) to which they are exposed each day. Such a response to a photoperiod is called **photoperiodism**. Three distinct types of plant exist.

Long day (short night) plants

These plants only flower when the number of hours of light to which they are exposed each day is **above** a certain critical level (i.e. the number of hours of darkness is below the critical level).

In spinach (*Spinacea oleracea*), shown in Figure 32.2, the plant must receive at least 13 hours of light in order

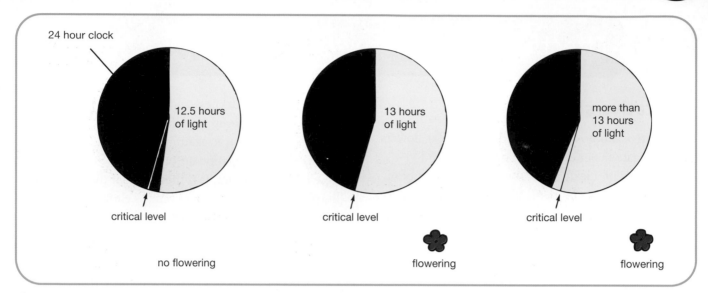

Figure 32.2 Photoperiodism in spinach, a long day (short night) plant

species of long day (short night) plant	critical duration of light (hours)	species of short day (long night) plant	critical duration of of darkness (hours)
Dill	11	*Bryophyllum*	12
Italian ryegrass	11	*Chrysanthemum*	9
Red clover	12	Cocklebur	9
Spinach	13	Strawberry	14
Winter wheat	12	Winter rye	12

Table 32.1 Critical periods of light and darkness

to flower. The length of the critical period of illumination varies amongst different species of long day (short night) plants (see Table 32.1). In some cases it is found to be less than 12 hours.

Short day (long night) plants

These plants only flower when the number of hours of light to which they are exposed each day is **below** a certain critical level (i.e. the number of hours of darkness is above the critical level).

In strawberry (*Fragaria chiloensis*), shown in Figure 32.3, the plant must receive at least 14 hours of darkness in order to flower. The duration of the critical period of darkness varies amongst different species of short day (long night) plants (see Table 32.1). In some cases it is found to be less than 12 hours.

Day neutral plants

These are plants in which flowering is not dependent

upon photoperiod. Examples include celery, geranium and tomato.

Mechanism of the response

Hormone

Flowering in plants is thought to be initiated by a hormone (or combination of hormones). The experiment in Figure 32.4 shows that it is the leaf that perceives the stimulus and produces sufficient hormone to induce flowering in the entire plant.

The experiment in Figure 32.5 shows that the hormone is **translocated** in the phloem to meristems where flowering is induced. This change is brought about by the hormone causing those genes controlling vegetative features to become switched off and those controlling floral features (e.g. ovaries, stamens, petals etc.) to become switched on.

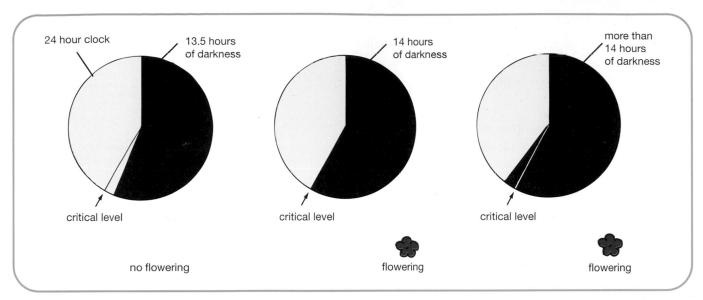

Figure 32.3 Photoperiodism in strawberry, a short day (long night) plant

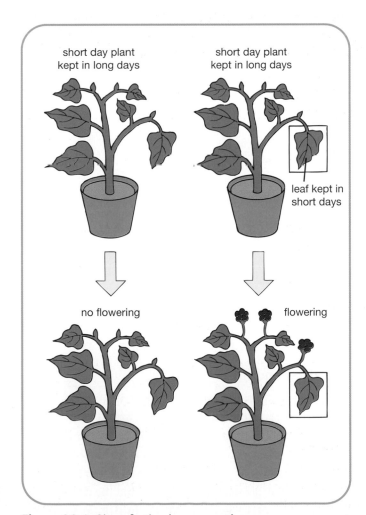

Figure 32.4 Site of stimulus perception

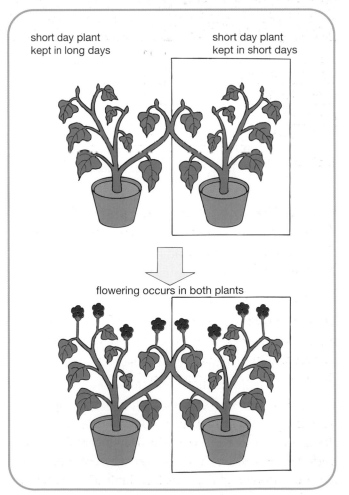

Figure 32.5 Transport of hormone

Biological significance

By responding to a photoperiod of particular length, all members of a species produce their flowers at the same time of year. This allows **cross-pollination** to occur, often on a large scale.

Cross-pollination is of advantage to the species because it results in the production of **increased variation** amongst the offspring. When these are subjected to natural selection, only the fittest survive. This increases the species' long-term chance of survival.

Effect of light on timing of breeding in animals

Many birds and mammals are **seasonal breeders**. Their gonads (testes and ovaries) become active only at a certain time of the year. These changes are triggered by the arrival of daily photoperiods of a certain critical length.

In birds, the reproductive activity of **long day breeders** is stimulated by the increasing daylengths that occur in spring. Hormones are secreted which promote the enlargement and activity of gonadal tissues and the production of sex cells as shown in Figure 32.6. The diagram also shows how a similar series of events occurs in many small mammals, such as the rabbit, which are long day breeders.

In larger mammals (e.g. sheep and deer) which require a longer period of gestation, a **short day breeding cycle** occurs. Gonadal activity and reproductive behaviour are triggered by the decreasing photoperiods which occur in autumn.

Biological significance

In each case, seasonal breeding behaviour is timed so that the offspring will be born in spring during favourable environmental conditions. Several months of plentiful food and mild temperatures during summer will allow the young animals to grow and become strong before winter and unfavourable conditions return.

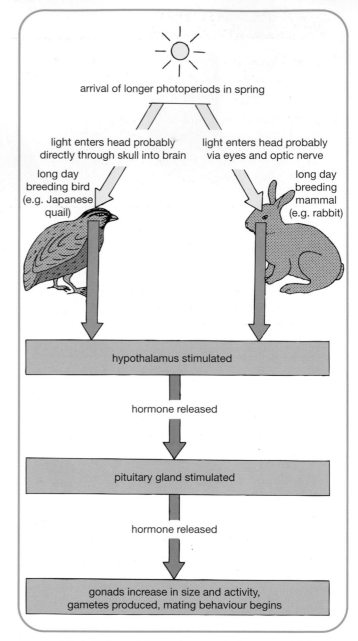

Figure 32.6 Effect of long photoperiods on gonadal activity

Migration and hibernation

The onset of shorter daylengths also triggers a series of events that lead to the **migration** of certain birds (e.g. swallow) and the **hibernation** of certain mammals (e.g. hedgehog) in autumn. These forms of behaviour are further examples of photoperiodism. They are of biological significance since they enable the animal to survive the winter when conditions in Britain would be severe and food would be scarce.

Testing Your Knowledge

1 State TWO ways in which the **i)** leaves; **ii)** stem internodes of an etiolated plant differ from those of a plant grown in light. (4)

2 **a)** What is meant by the term *positive phototropism*? (1)

   b) What is the survival value of positive phototropism to a plant? (1)

3 **a)** What is meant by the term *photoperiodism*? (1)

   b) Explain the difference between a long day and a short day plant. (2)

4 **a)** Amongst mammals, what is the difference between a long day and a short day breeder? (2)

   b) At which time of the year does each type of breeder produce its young? (1)

Applying Your Knowledge

1 **a)** Growth may be defined as **i)** an increase in cell number or **ii)** an increase in dry mass. Briefly discuss whether or not an etiolated bean seedling shows real growth. (2)

   b) i) Predict the ultimate fate of an etiolated bean seedling kept permanently in darkness.
   ii) Explain your answer. (2)

   c) With reference to **i)** existing leafy shoots and **ii)** new leafy shoots, predict the effect of transferring a healthy plant into total darkness for a few weeks. (2)

2 *Chrysanthemums* are short day plants which normally flower in autumn.

   a) Which part of the plant must be exposed to the critical photoperiod for flowering to occur? (1)

   b) Describe the procedure that should be adopted by a horticulturalist in order to produce a crop of *Chrysanthemum* plants in full bloom at Christmas time. (2)

3 Healthy young specimens of four unrelated species of flowering plants (A–D) were subjected to a range of different light and dark treatments. The results are recorded in Table 32.2 (where + = flowering and – = no flowering).

   a) State the photoperiodic group to which each species of flowering plant A–D belongs. Where appropriate, give the critical number of hours of light or darkness that the species needs in order to flower. (4)

   b) Which treatment in Table 32.2 results in flower production by all four species of plant? (1)

4 Under natural conditions in northern latitudes, ferrets do not breed from September to March. However they come 'on heat' in April regardless of how cold the weather is.

   a) Suggest which environmental factor triggers gonadal activity in ferrets. (1)

   b) If this is the case, are ferrets long day or short day breeders? (1)

	treatment					
	1	2	3	4	5	6
hours of light	15	14	13	12	11	10
hours of darkness	9	10	11	12	13	14
species A	–	–	+	+	+	+
species B	+	+	+	+	–	–
species C	–	–	–	+	+	+
species D	+	+	+	+	+	+

Table 32.2

c) Describe a possible effect of exposing ferrets to artificially prolonged photoperiods during midwinter. (1)

d) Why is daylength a more reliable indicator of time of year to an animal than temperature? (1)

5 The graph in figure 32.7 shows how photoperiod and mass of testes in a species of hare vary throughout the year.

a) State the type of relationship that exists between photoperiod and mass of testes. (1)

b) Predict the effect on mass of testes of keeping this type of hare in artificially short days. (1)

c) i) Name the THREE months of the year when male hares would be most likely to mate with females.

   ii) Explain why the timing of mating activity increases the chance of survival of the species given that the length of the gestation period in hares is 31 days. (3)

6 Give an account of photoperiodism in plants and animals including reference to its biological significance. (10)

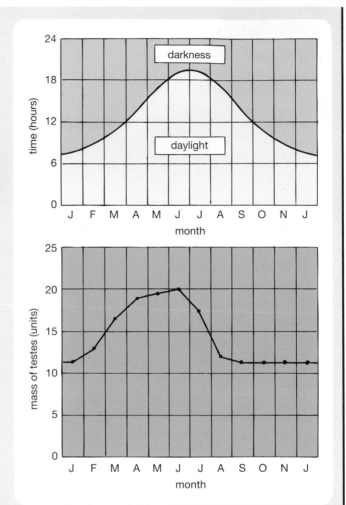

Figure 32.7

What You Should Know

Chapters 31–32

(See Table 32.3 for Word bank)

above	intelligence	photoperiods
calcium	lead	phototropism
darkness	light	rickets
deficiency	long	short
elongated	magnesium	spring
etiolated	nicotine	teeth
fetus	phosphorus	thalidomide
haemoglobin	photoperiodism	vitamin D

Table 32.3 Word bank for chapters 31–32

1 Nitrogen, _____, potassium and _____ are four macro-elements needed by plants for healthy growth.

2 A plant grown in a water culture experiment lacking one essential macro-element displays certain _____ symptoms.

3 _____ inhibits the action of certain enzymes. It is poisonous to humans and may have a detrimental effect on the _____ of children.

4 Iron is essential to humans for the formation of _____.

5 _____ is essential to humans for the formation of healthy bones and _____.

6   _____ is needed by humans to promote the uptake of calcium ions by bone and prevent _____.

7   Drugs such as _____, alcohol and _____ harm a developing human _____.

8   A plant grown in darkness develops a weak _____ stem and small curled leaves and is said to be _____.

9   Green plants show positive _____ by growing towards a source of _____ from one direction.

10   In order to flower, long day plants require a number of hours of light _____ a certain critical level.

Short day plants need a critical number of hours of _____. Such a response by an organism to a photoperiod of a certain length is called _____.

11   Seasonal gonadal activity in many birds and mammals is stimulated by the arrival of daily _____ of a certain length.

12   Small mammals with short gestation periods tend to be _____ day breeders. Large mammals with long gestation periods tend to be _____ day breeders.

13   Breeding is timed so that the young are born in _____ when the conditions are favourable.

33 Physiological homeostasis

Internal environment

A human being is a multicellular organism; the body consists of a community of cells. Since each type of cell is specialised to perform a particular function, the different parts of the body are dependent upon one another for survival. The millions of cells that make up the community and the tissue fluid that bathes them are collectively known as the **internal environment**.

Need for control

For the human body to function efficiently and work as an integrated whole, the state of the internal environment must be maintained within tolerable limits as shown in Figure 33.1. These features (and many others) of the internal environment are controlled by homeostasis.

Physiological homeostasis

Physiological homeostasis is the maintenance of the body's internal environment within certain tolerable limits despite changes in the body's external environment (or changes in the body's rate of activity).

Principle of negative feedback control

When some factor affecting the body's internal environment deviates from its normal optimum level (called the **norm** or **set point**) this change in the factor is detected by **receptors**. These send out nerve or hormonal messages which are received by **effectors**.

The effectors then bring about certain responses which counteract the original deviation from the norm and return the system to its set point. This corrective mechanism is called **negative feedback control** (see Figure 33.2). It provides the stable environmental conditions needed by the body's community of living cells to function efficiently.

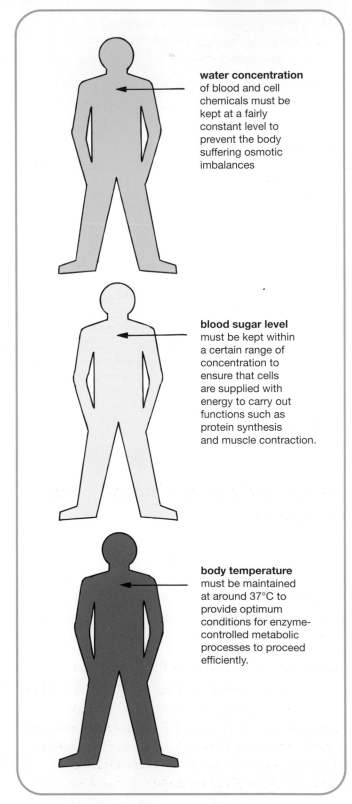

water concentration of blood and cell chemicals must be kept at a fairly constant level to prevent the body suffering osmotic imbalances

blood sugar level must be kept within a certain range of concentration to ensure that cells are supplied with energy to carry out functions such as protein synthesis and muscle contraction.

body temperature must be maintained at around 37°C to provide optimum conditions for enzyme-controlled metabolic processes to proceed efficiently.

Figure 33.1 Maintenance of tolerable limits

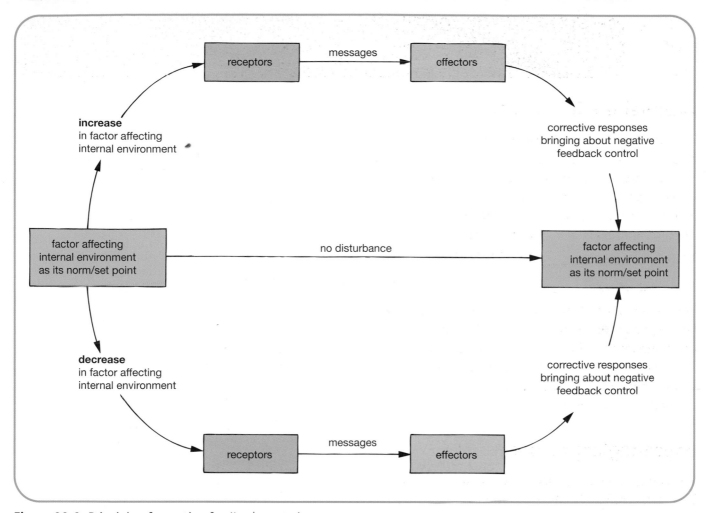

Figure 33.2 Principle of negative feedback control

Homeostasis in the human body

Osmoregulation

This is the means by which the body maintains its concentration of water, salts and ions at the correct level. Osmoregulation is under homeostatic control. The kidneys act as the effectors in the system by responding to hormonal messages and bringing about negative feedback control.

Control of water concentration of blood

If the water concentration of the blood decreases (due to increased sweating, lack of drinking water or consumption of salty food) then **osmoreceptors** in the **hypothalamus** (see Figure 33.3) are stimulated. These trigger the release of much **ADH** (**antidiuretic hormone**) by the **pituitary gland** into the bloodstream.

On arriving in the kidneys, ADH increases the permeability of the tubules and collecting ducts to water. As a result more water is reabsorbed into the bloodstream by osmosis. This increases its water concentration until it is returned to normal.

If the water concentration of the blood becomes too high, less ADH is released and the kidney tubules and collecting ducts become less permeable to water.

Less water is reabsorbed by osmosis and the blood's water concentration soon drops to around its normal level.

This process is summarised in Figure 33.4.

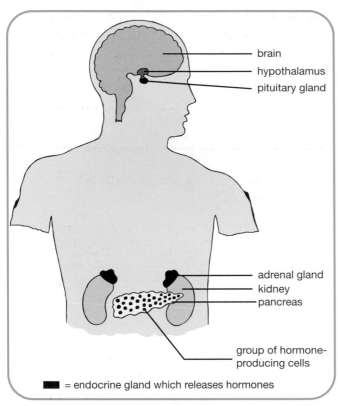

Figure 33.3 Location of three endocrine glands

Testing Your Knowledge

1 a) What is meant by the term *physiological homeostasis*. (2)

   b) i) Outline the principle of negative feedback control.

      ii) Why is such control of advantage to an organism? (5)

2 a) i) Where in the human body are the osmoreceptors found that respond to a decrease in water concentration of the blood?

      ii) Which part of the body releases an increased concentration of ADH under these circumstances? (2)

   b) By what means does ADH bring about its effect so that water is conserved by the body? (2)

3 Compare the relative volume and concentrations of urine produced when the water concentration of the blood is i) above the set point; ii) below the set point. (2)

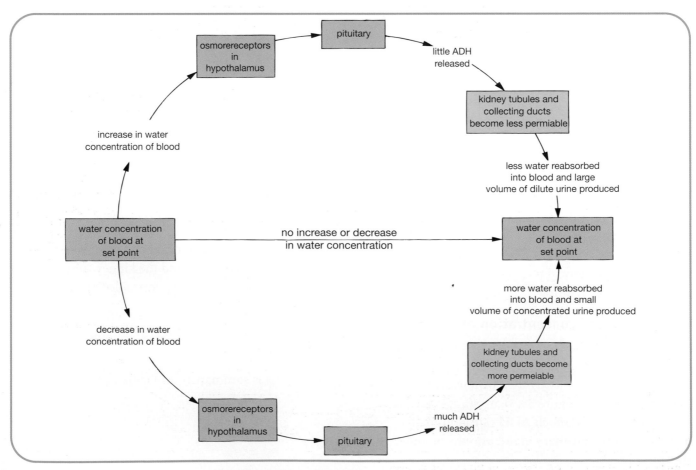

Figure 33.4 Homeostatic control of blood water concentration

Control of blood sugar level

All living cells in the human body need a continuous supply of energy. Most of this energy is released by the oxidation of glucose (see page 29). Cells are therefore constantly using up the glucose present in the bloodstream (i.e. **blood sugar**).

However the body obtains supplies of glucose only on those occasions when food is eaten. To guarantee that a regular supply of glucose is available for use by cells regardless of when and how often food is consumed, the body employs a homeostatic mechanism.

Liver as a storehouse

About one hundred grams of glucose are stored as **glycogen** in the **liver**. Glucose can be added to or removed from this reservoir of stored carbohydrate depending on shifts of supply and demand.

Insulin and glucagon

A rise in blood sugar level to above its set point (e.g.

following a meal) is detected by cells in the **pancreas** (see Figure 33.3) which produce **insulin**. This hormone is transported in the bloodstream to the liver where it activates an enzyme which catalyses the reaction:

This brings the blood sugar concentration down to around its normal level.

If the blood sugar level drops below its set point (e.g. between meals or during the night), different cells in the pancreas detect this change and release **glucagon**. This second hormone is transported to the liver and activates a different enzyme which catalyses the reaction:

$$glycogen \longrightarrow glucose$$

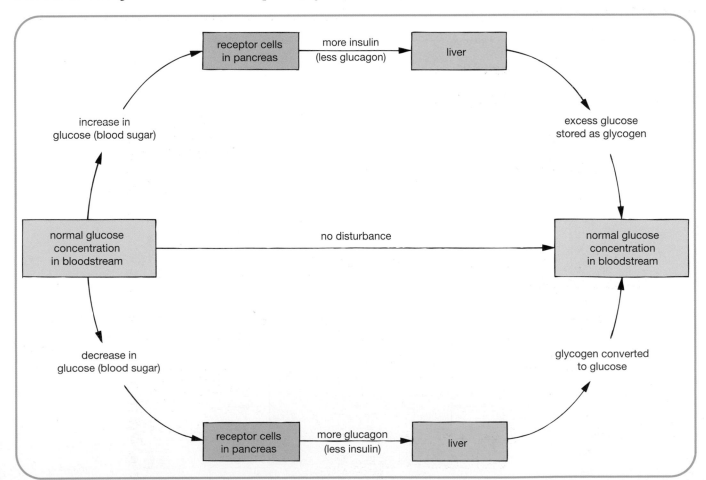

Figure 33.5 Homeostatic control of blood sugar level

The blood sugar concentration therefore rises to around its normal level. Figure 33.5 gives a summary of this homeostatic mechanism.

Adrenaline

During an **emergency** when the body needs additional supplies of glucose to provide energy quickly for 'fight or flight', the **adrenal glands** (see Figure 33.3) secrete more of the hormone **adrenaline** into the bloodstream.

Adrenaline overrides the normal homeostatic control of blood sugar level by inhibiting the secretion of insulin and promoting the breakdown of glycogen to glucose.

Once the crisis is over, secretion of adrenaline is reduced to a minimum and blood sugar level is returned to normal by the appropriate corrective mechanism involving insulin or glucagon.

Alternative diagram

Every factor under homeostatic control can be represented by the type of diagram shown in Figure 33.2. However it must be kept in mind that when a factor deviates from its norm and is then returned to this set point by negative feedback control, it often overshoots the mark. This triggers the reverse set of corrective mechanisms.

To illustrate that a factor which is in a state of dynamic equilibrium is constantly wavering on either side of its set point, homeostasis is often represented as two interrelated circuits (see Figure 33.6).

Diabetes mellitus

Some people suffer a disorder known as *diabetes mellitus.* Some (or all) of their insulin-secreting pancreas cells are non-functional. Since sufferers produce insufficient (or no) insulin, the concentration of glucose in their blood rises to 10–30 millimoles per litre compared with the normal concentration of 5 millimoles per litre.

The glomerular filtrate formed in the kidneys of a diabetic is so rich in glucose that much of it is not reabsorbed into the bloodstream but is instead excreted in **urine**.

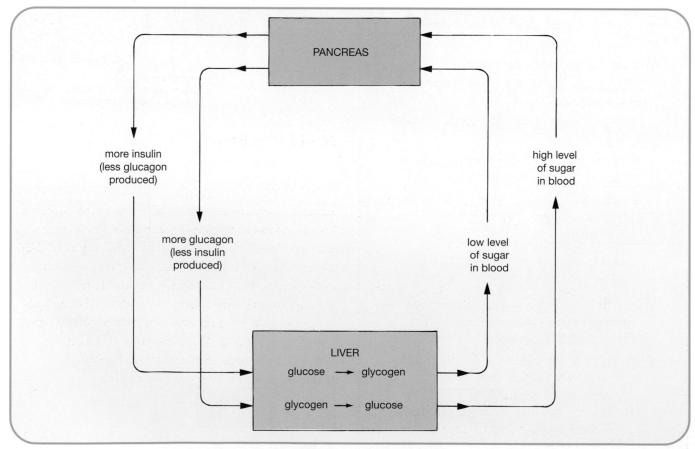

Figure 33.6 Alternative diagram of blood sugar control

In the absence of insulin, cells are unable to use glucose efficiently and fat stores become depleted leading to loss in weight and wasting of tissues.

Whereas *diabetes mellitus* used to be a fatal disorder, it is now successfully treated by regular injections of insulin and a controlled diet.

Testing Your Knowledge

1 Hormones are chemical messengers released directly into the bloodstream by endocrine glands.

a) Copy Table 33.1 and complete the central column. (4)

hormone	endocrine gland from which hormone originates	letter(s) indicating effect(s) of hormone
adrenaline		
insulin		
ADH		
glucagon		

Table 33.1

b) Complete the right hand column by using a selection of one or more answers from the following list:

A promotes conversion of excess glucose to glycogen

B increases permeability of kidney collecting ducts

C promotes conversion of glycogen to glucose

D decreases blood sugar level

E prepares the body to cope with an emergency (4)

2 a) With reference to concentration of sugar in the bloodstream, state the circumstances that lead to: **i)** glycogen being converted to glucose; **ii)** glucose being converted to glycogen. (2)

b) i) In the homeostatic control of blood sugar level, which organ is the receptor and which is the effector?

ii) Suggest why such a corrective mechanism is described as a form of *negative* feedback control. (4)

Control of body temperature

Ectotherm

An **ectotherm** (see Figure 33.7) is an animal which is **unable** to regulate its body temperature (by physiological means). All invertebrates, fish, amphibians and reptiles are ectotherms and their body temperature normally varies directly with that of the external environment. They obtain most of their body heat from the surrounding environment. Certain ectotherms can regulate their body temperature to some extent by **behavioural** means. Lizards, for example, bask in sunshine to gain energy and raise their body temperature.

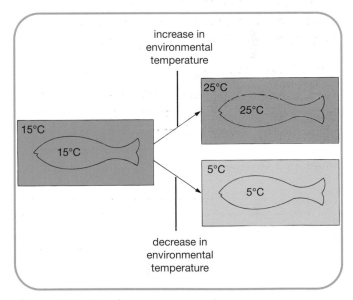

Figure 33.7 Ectotherm

Endotherm

An **endotherm** (see Figure 33.8) is an animal which is **able** to maintain its body temperature at a relatively constant level independent of the temperature of the external environment. All birds and mammals are endotherms. They have a **high metabolic rate** which generates much heat energy. Regulation of their body temperature is brought about by homeostatic control. The graph in Figure 33.9 summarises the effect of increase in external temperature on the body temperature of the two types of animal.

and cold **thermoreceptors** in the skin. These convey information to it about the surface temperature of the body.

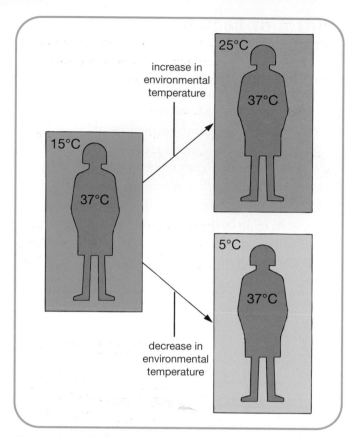

Figure 33.8 Endotherm

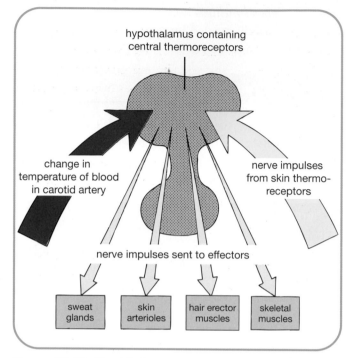

Figure 33.10 Hypothalamus as temperature-monitoring centre

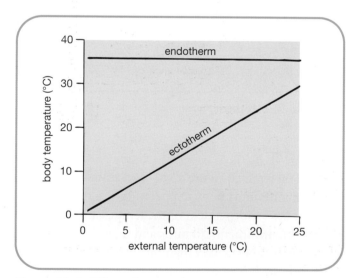

Figure 33.9 Effect of external temperature on body temperature

Role of hypothalamus

In addition to playing many other roles, the hypothalamus is the body's temperature-monitoring centre (see Figure 33.19). It acts as a **thermostat** and is sensitive to nerve impulses that it receives from heat

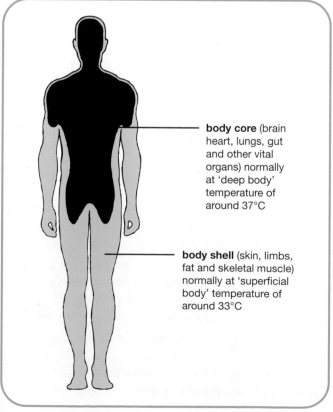

Figure 33.11 Body core and body shell

In addition, the hypothalamus itself possesses **central thermoreceptors**. These are sensitive to changes in temperature of blood which in turn reflect changes in the temperature of the **body core** (see Figure 33.11).

The thermo-regulatory centre in the hypothalamus responds to this information by sending appropriate nerve impulses to effectors. These trigger corrective feedback mechanisms and return the body temperature to its normal level (set point).

Role of skin

The skin plays a leading role in temperature regulation. In response to nerve impulses from the hypothalamus, the skin acts as an effector.

Correction of overheating

The skin helps to correct overheating of the body by employing the following mechanisms which promote heat loss.

Vasodilation

Arterioles leading to skin become **dilated** (see Figure 33.12). This allows a large volume of blood to flow through the capillaries near the skin surface. From here, the blood is able to lose heat by **radiation**.

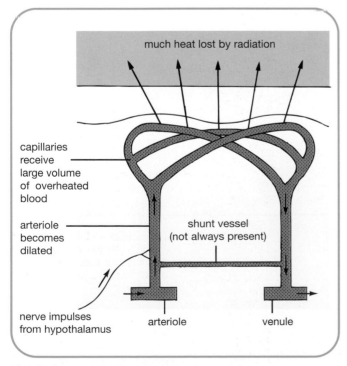

Figure 33.12 Vasodilation in skin

Increase in rate of sweating

Heat energy from the body is used to convert the water in sweat to **water vapour** and by this means brings about a lowering of body temperature.

Correction of overcooling

The skin helps to correct overcooling of the body by employing the following mechanisms which reduce heat loss.

Vasoconstriction

Arterioles leading to the skin become **constricted** (see Figure 33.13). This allows only a small volume of blood to flow to the surface capillaries. Little heat is therefore lost by radiation.

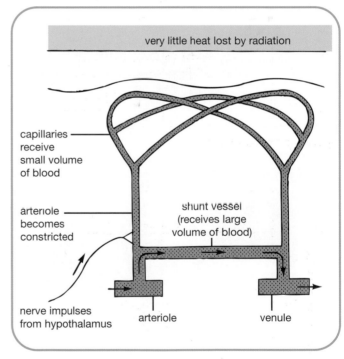

Figure 33.13 Vasoconstriction in skin

Decreased rate of sweating

Since sweating is reduced to a minimum, heat is conserved.

Contraction of erector muscles

This process (see Figure 33.14) is more effective in furry animals (and birds) than in human beings. It results in hairs being raised (or feathers being fluffed out) from the skin surface. A wide layer of air which is a poor conductor of heat, is trapped between the animal's body and the external environment. This layer of **insulation** reduces heat loss.

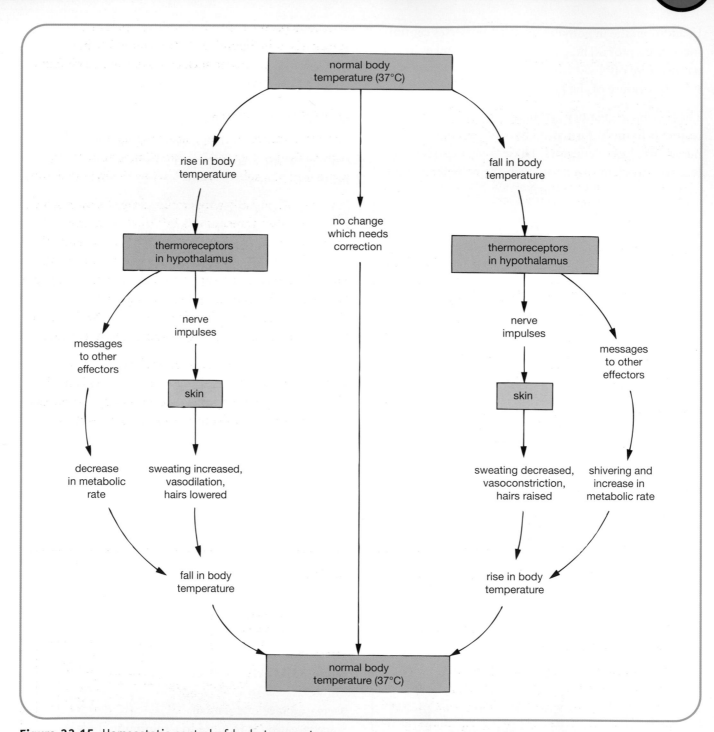

Figure 33.15 Homeostatic control of body temperature

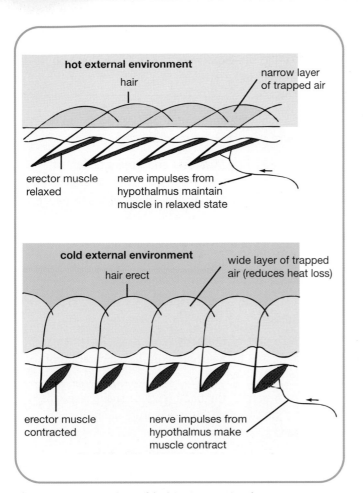

Figure 33.14 Action of hair erector muscles

Within the figure:

hot external environment
hair
narrow layer of trapped air
erector muscle relaxed
nerve impulses from hypothalmus maintain muscle in relaxed state

cold external environment
hair erect
wide layer of trapped air (reduces heat loss)
erector muscle contracted
nerve impulses from hypothalmus make muscle contract

The homeostatic control of body temperature is summarised in Figure 33.15. It includes further corrective mechanisms such as **shivering** and **changes in metabolic rate**.

Voluntary responses

The mechanisms of temperature regulation summarised in Figure 33.15 are all **involuntary** and controlled at a subconscious level by the hypothalamus.

However when body temperature drops below normal, nerve impulses transmit this information to the cerebrum (thinking part of the brain). This makes the person 'feel cold' and become aware of the problem. They then take an appropriate course of action (e.g. put on extra clothing, turn up the heating, exercise vigorously, consume a hot drink etc.) and by doing so help to return their body to its normal temperature.

When the body temperature rises above tolerable limits, a reverse set of behavioural responses is made. This ability to make appropriate **voluntary responses** is an important part of control of body temperature.

Practical Activity and Report

Investigating human body responses to sudden heat loss

Information

- A **thermistor** is a device which responds to tiny changes in temperature.

- In this investigation you are going to measure any changes that occur in the surface temperature of one part of the body (the left hand) when another part of the body (the right hand) undergoes sudden heat loss.

- At the same time you are going to measure any changes that occur in the temperature of the armpit.

- The left hand represents the body shell; the armpit represents the body core.

You need

1 bucket of crushed ice in water
2 thermistors

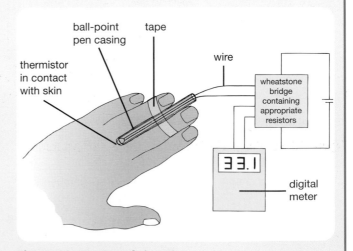

Within the figure:
ball-point pen casing
tape
thermistor in contact with skin
wire
wheatstone bridge containing appropriate resistors
33.1
digital meter

Figure 33.16 Use of thermistor

315

2   **digital multimeters**
2   **set of leads and wheatstone bridges containing appropriate resistors (see Figure 33.16)**
2   **6 volt power packs**
1   **stopclock**

(Alternatively the thermistors can be connected up to interface with a computer)

What to do

1   Read all of the instructions in this section and prepare your results table before carrying out the experiment.

2   With the aid of your partner, secure thermistor 1 between two fingers of your left hand as shown in Figure 33.16.

3   Position thermistor 2 in your left armpit.

4   Switch on the multimeters and allow 2 minutes for the thermistors to equilibrate with their surroundings.

5   Note the starting temperature of thermistor 1 (left hand – body shell) and that of thermistor 2 (armpit – body core).

6   Start the clock and then plunge your right hand into the bucket of icy water.

7   Record the temperature of the thermistors every 30 seconds for as long as you can bear it up to a maximum of 5 minutes.

8   If other students have carried out the same investigation, pool your results.

Reporting

Write up your report by doing the following:

1   Rewrite the title given at the start of this activity.

2   Put the subheading '**Aim**' and state the aim of your experiment.

   a) Put the subheading '**Method**'.

   b) Draw a simple labelled diagram of a thermistor in use.

   c) Using the impersonal passive voice, briefly describe the experimental procedure that you followed and state how you obtained your results.

4   Put the subheading '**Results**' and draw a final version of your table of results.

5   Put a subheading '**Analysis and Presentation of Results**'. Present your results as two line graphs with shared axes on the same sheet of graph paper.

6   Put a subheading '**Conclusions**' and write a short paragraph to state what you have found out from a study of your results. This should include answers, in sentences, to the following questions:

   a) What overall trend is shown by each set of results?

   b) Which part of the body (core or shell) remains unaffected by a sudden drop in temperature of the right hand?

   c) Which part of the body (core or shell) responds to a sudden drop in temperature of the right hand?

   d) Why is such a compensatory reduction in temperature (which is brought about by constriction of blood vessels) of survival value?

7   Put a final subheading '**Evaluation of Experimental Procedure**'. Give an evaluation of your experiment (keeping in mind that you may comment on any stage of the experiment that you wish).

Try to incorporate answers to the following questions in your evaluation. Make sure that at least one of your answers includes a supporting argument.

   a) Why is it important that the results are obtained from two parts of the *same* person's body?

   b) Why are the thermistors allowed to equilibrate with their surroundings for 2 minutes?

   c) Why is it impossible sometimes to obtain a set of results for a continuous period of 5 minutes?

   d) What is the purpose of pooling the class results?

Advantage of homeostasis

The evolution of systems that respond to homeostatic control is of survival value because these maintain the body's internal environment at a relatively steady optimum state. This allows the body to function efficiently despite wide fluctuations in the external environment (many of which would be unfavourable).

Extreme conditions

Homeostasis only works within certain limits. If a person is exposed to extremely adverse conditions in the external environment (e.g. freezing temperature, total lack of drinking water etc.) for a prolonged period, the body responds by exerting negative feedback control as described above. However a homeostatic system eventually breaks down when its corrective mechanisms can no longer return the body to its steady state. In extreme cases death results.

Testing Your Knowledge

1 Is a human being an *ectotherm* or an *endotherm*? Explain your answer. (2)

2 **a)** Which part of the brain contains the centre responsible for regulation of body temperature? (1)

   b) By what means does this structure obtain information about the body's surface temperature? (1)

3 **a)** Name TWO effectors to which the hypothalamus sends nerve impulses when the body temperature decreases to below normal level. (2)

   b) Describe the responses made by these effectors and explain how they return the body temperature to normal. (4)

4 With reference to homeostasis, explain why a person suffering from prolonged overcooling may eventually die of hypothermia. (2)

Applying Your Knowledge

1 Figure 33.17 shows a simplified version of the homeostatic control of blood water concentration.

   a) Redraw the diagram and complete the blanks. (2)

   b) Name the hormone represented by the letters ADH and state the effect that it has on kidney tubules. (2)

   c) What relationship exists between concentration of ADH in the bloodstream and volume and concentration of urine produced by the kidneys? Explain your answers. (3)

   d) In addition to their role in the above control system, the osmoreceptors send nerve messages to the cerebrum producing the sensation of thirst. Explain how this could act as a successful corrective mechanism. (2)

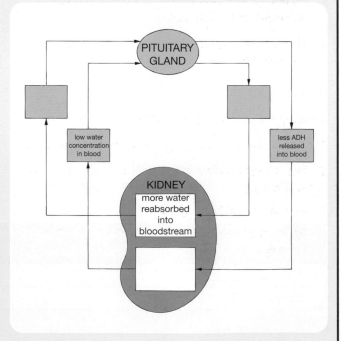

Figure 33.17

2 Figure 33.18 shows the effect of consuming 50 g of glucose (after a period of fasting) on the concentrations of fatty acids, glucose and insulin in the bloodstream.

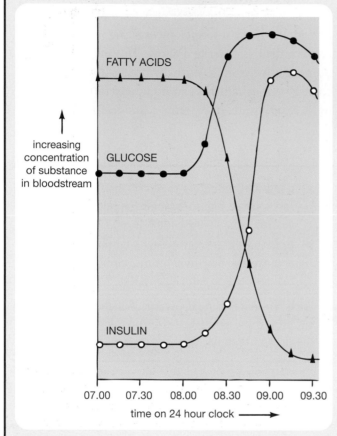

Figure 33.18

a) i) During which period of time was the person's blood sugar concentration at a steady level?

ii) By what means is this steady level maintained? (2)

b) i) At what time was the glucose consumed?

ii) What initial effect did the intake of glucose have on blood sugar level and concentration of insulin in the blood?

iii) Why was there a time lag between these two effects? (3)

c) What evidence is there from the graph that insulin suppressed the breakdown of stored fat? (1)

3 The data shown in Table 33.2 were obtained during an investigation into the functions of the human kidney.

a) With reference to the homeostatic control of osmoregulation, explain why the concentration of urea is higher in urine than in glomerular filtrate. (1)

b) By how many times is the concentration of **i)** sodium ions, **ii)** urea, more concentrated in urine than in glomerular filtrate? (2)

c) i) State ONE way in which the data in the table would differ if they referred to a newly diagnosed sufferer of *diabetes mellitus* prior to treatment.

ii) What treatment would be given? (2)

4 Sodium accounts for about 90% of the positive ions in the human body's extracellular fluid. Figure 33.19

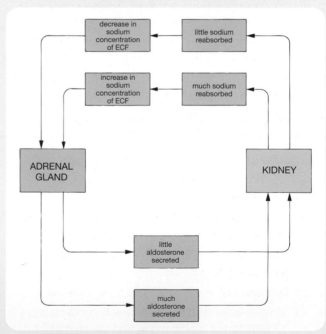

Figure 33.19

	concentration of substance present in glomerular filtrate (g/100 cm³)	concentration of substance present in urine (g/100 cm³)
sodium ions	0.30	0.60
glucose	0.10	0.00
urea	0.03	2.10

Table 33.2

shows how the hormone aldosterone helps to maintain the sodium concentration of the extracellular fluid.

a) Give an example of an extracellular fluid present in the human body. (1)

b) Identify the **i)** receptor; **ii)** effector in the example of homeostasis shown in Figure 33.15. (2)

c) What type of message passes from the receptor to the effector in this case? (1)

d) Name the corrective mechanism which restores the sodium concentration to its set point when salt intake is too high. (1)

e) Redraw the diagram in the form illustrated in Figure 33.5. (2)

f) **i)** A reduction in the volume of the body's extracellular fluid, caused by diarrhoea or a haemorrhage, is followed by an increase in aldosterone production. Suggest why.

   ii) This increase in aldosterone production brings about a decrease in the blood's water concentration which, in turn, triggers increased production of a second hormone. Identify the latter. (2)

g) Predict with reasons the ultimate fate of a person fed a sodium-free diet for a prolonged period. (2)

5 Figure 33.20 represents a section through human skin.

a) Name the part of the brain to which heat and cold receptors relay information about the external environment. (1)

b) By what means does the thermoregulatory centre of the brain communicate information to structures X and Y in order to effect control of body temperature? (1)

c) **i)** In what way should structure X respond following a drop in body temperature?

   ii) Explain how this response would help to conserve heat. (2)

d) **i)** In what way would structure Y respond to an increase in body temperature?

   ii) Explain how this response would help to promote heat loss. (2)

6 The data in Table 33.3 refer to the body temperature of a student who exercised vigorously and then, after a rest, plunged into a cold bath. Body temperature was measured every 2 minutes by inserting a sterilised thermistor under the student's tongue.

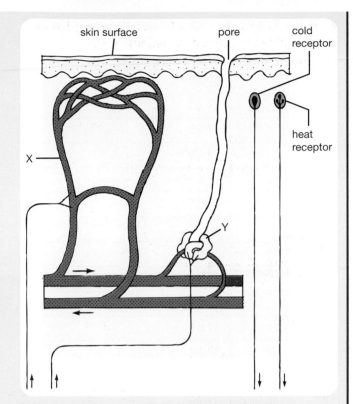

Figure 33.20

a) Plot a line graph of the data. (3)

b) Using FOUR arrows, indicate on your graph that:
   i) exercise was begun at minute 2; **ii)** exercise was stopped at minute 22; **iii)** immersion in the cold bath occurred at minute 34; **iv)** exit from the cold bath took place at minute 40. (4)

c) In general what trend in body temperature occurs during:
   i) the period of vigorous exercise?
   ii) the time in the cold bath?
   iii) Why is there a slight delay before each of these trends begins? (3)

d) The student's skin was flushed from minute 18 onwards.
   i) Suggest why.
   ii) What is the benefit to the body of flushed skin? (2)

e) **i)** At which ONE of the following times in minutes was the student found to be shivering?
   A 2   **B** 22   **C** 32   **D** 42
   ii) What is the survival value of shivering? (2)

time (min)	body temperature (°C)
0	37.00
2	37.00
4	37.00
6	37.05
8	37.10
10	37.10
12	37.15
14	37.20
16	37.30
18	37.40
20	37.50
22	37.60
24	37.60
26	37.50
28	37.45
30	37.40
32	37.40
34	37.35
36	37.35
38	36.90
40	36.65
42	36.70
44	36.80
46	36.85
48	37.00
50	37.00

Table 33.3

7 Write an essay on negative feedback control in the human body with reference to control of body temperature and blood sugar level. (10)

34 Regulation of populations

Factors affecting population size

A **population** is a group of individuals of the same species which makes up part of an ecosystem. A population's size is comparable to the volume of water in a bath. It is affected by several factors as shown in Figure 34.1

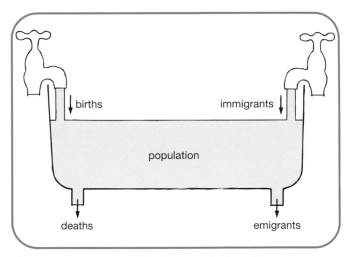

Figure 34.1 Factors affecting population size

The birth rate of a population is a measure of the number of new individuals produced by a population during a certain interval of time. The **death rate** is a measure of the number of individuals within a population that died during the same interval of time.

Population dynamics

Population dynamics is the study of population changes (growth, maintenance and decline) and the factors which cause these changes.

Population density

The number of individuals of the same type present per unit area (or volume) of a habitat is called the population density.

Stability

When a population colonises a new environment, it grows in number until it reaches a certain size which the available environmental resources can just maintain. This limit is called the **carrying capacity** of the environment.

The population is now in a state of **dynamic equilibrium** and remains relatively stable despite short-term oscillations in number from generation to generation. Birth rate now equals death rate and the population neither increases nor decreases in size.

Sheep

Figure 34.2 shows the growth curve of the population of sheep following their introduction to the Australian island of Tasmania in the 19th century. Although the numbers varied from one census to another, the

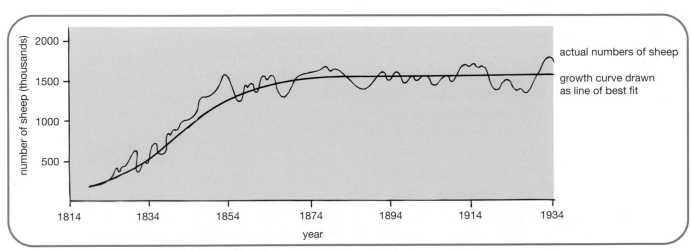

Figure 34.2 Population growth and stability

population was found to remain fairly stable once it had reached the carrying capacity of the environment in the 1850s.

Factors influencing population change

Environmental resistance

Each species has an enormous **reproductive potential** (see Table 34.1). However under natural conditions in an ecosystem, a population is prevented from increasing in size indefinitely by **environmental resistance**. This consists of factors such as: availability of food, water, oxygen, light, space and shelter; predation; disease; and climate. These factors which affect the size of a population fall into the following two categories.

animal	average number of offspring per breeding pair per year
fox	5
red grouse	8
rabbit	24
mouse	30
trout	800
cod	4 000 000
oyster	16 000 000

Table 34.1 Reproductive potential

Density-independent factors

A **density-independent** factor is one which affects the growth of a population regardless of the population's density. A forest fire, for example, will wipe out most of a population whether it is densely packed or sparsely distributed in an ecosystem.

Similarly, extremes of weather such as excessive rainfall, floods, drought, storms and spells of unusually high or low temperature all act as density-independent factors. They tend to cause a sudden drastic reduction in population number. However, given sufficient time, the population normally makes a full recovery.

Density-dependent factors

A **density-dependent** factor is one which only affects the population once it has grown to a certain size (and

density). In the absence of environmental resistance, a small population continues to grow until it reaches a certain density. Its growth rate is then affected by one (or more) of the following density-dependent factors, each of which has a regulatory effect.

Competition for food

When the number of animals present in a population outstrips the available food supply, **competition** occurs between individuals (see page 208). Some fail to breed and may even starve to death.

Experiments using flour beetles show that when the mass of food per beetle drops below a certain level, frequency of mating declines, egg production decreases and cannibalism sets in. As a result population growth is affected (see Figure 34.3).

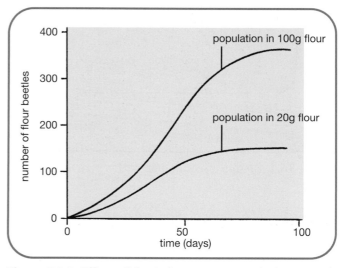

Figure 34.3 Effect of food shortage on population growth

Parasitism and disease

A **parasite** is an organism which benefits at the expense of another organism (the host) by feeding on it. The denser the population of the host, the greater the chance of the parasite obtaining food and producing offspring which will in turn find new hosts.

Compared with a sparse population, a dense host population is more prone to attack and damage by parasites. Some parasites cause **disease** directly whereas others transmit disease-causing micro-organisms from one host to another. Either way, the subsequent disease reduces the host population in number.

A dense population of a domesticated organism (e.g. a

crop of cereal plants of the same genotype) grown as a vast monoculture is especially vulnerable to parasitic attack.

Predation

A dense population of prey organisms is more prone to attack by their **predators**. Figure 34.4 summarises the results from a survey carried out on the nests of the great tit. Weasels smash open the bird's eggs and eat the chicks.

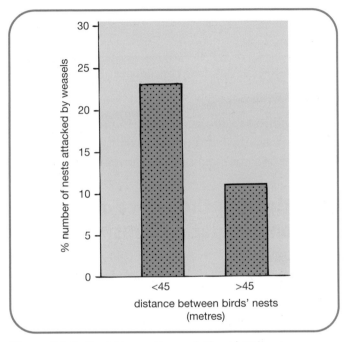

Figure 34.4 Predation and population density

From the graph it can be seen that a higher number of nests are attacked when they are close together than when they are spaced out. This shows that predation by weasels is a density-dependent factor.

Predator-prey interactions

A delicate balance exists between populations of predators and their prey. An increase in number of prey (perhaps due to climatic conditions favouring growth of their plant food) leads to an increase in predation.

As the size of the prey population decreases, competition between predators for the remaining prey becomes more and more intense until eventually the number of predators drops. This in turn allows the prey population to build up again which leads to a corresponding increase in the predator population and so on.

This self-regulating series of **interdependent fluctuations** is summarised in Figure 34.5. The predator curve takes the same shape as that of the prey but lags behind it since time is required for each change in the sequence of events to take effect.

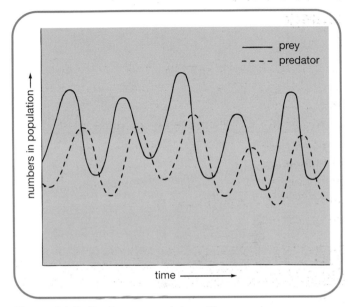

Figure 34.5 Predator–prey interactions

Two examples of animals showing this relationship are *Hydra* which feeds on waterfleas (see Figure 34.6) and lynx which preys upon the snowshoe hare (see Figure 34.7 and Question 3 on page 327).

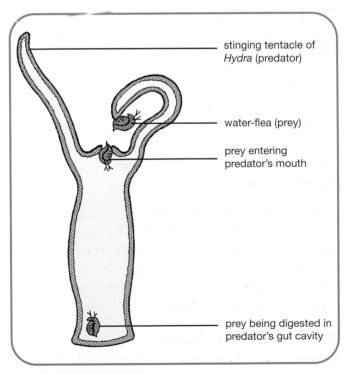

Figure 34.6 *Hydra* and water flea

Figure 34.7 Lynx and snowshoe hare

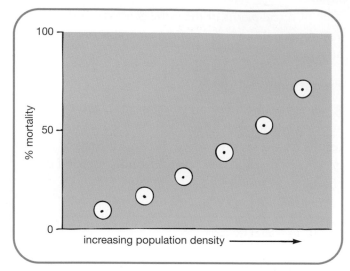

Figure 34.8 Effect of density-dependent factor

In a natural environment, the predator–prey relationship is not always as straightforward as suggested by Figure 34.5. A prey organism is often hunted by several predators. In addition, predators will turn to alternative prey if their favourite food is in short supply.

Effect of two types of factor on percentage mortality

When the effect of a density-dependent factor (e.g. shortage of food) on a population is graphed as rate of mortality against increasing population density, a direct relationship is found to exist as shown in Figure 34.8.

When the effect of a density-independent factor (e.g. extremely low temperature) on a population is graphed as rate of mortality against increasing population density, a direct relationship is not found to exist. The effect is either constant (100% mortality) or random at different population densities as shown in Figure 34.9.

Investigating the effect of soil type on population density of springtails

Springtails are tiny soil insects (see Figure 34.10). They live in air spaces in top soil and possess a springing organ which enables them to jump an enormous distance relative to their body size. Population density of springtails can be measured as number of animals/cm³ soil.

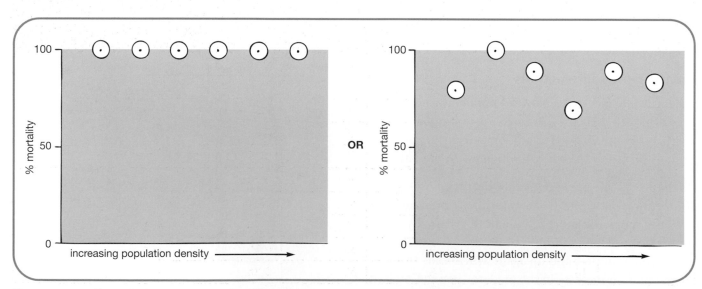

OR

Figure 34.9 Effect of density-independent factor

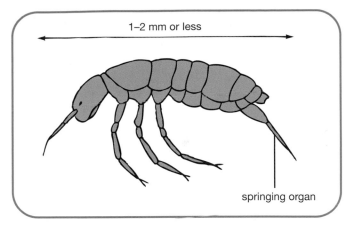

1–2 mm or less

springing organ

Figure 34.10 Springtail

In this investigation the two soil types are: 'cultivated' soil in an empty flower bed and nearby 'uncultivated' soil under dense bushes.

Samples are taken at random from both soils using a corkscrew soil **auger**. The auger is screwed in a clockwise direction into the soil to a depth of 100 mm and then pulled up carefully. The soil clinging to it is transferred to a plastic measuring cylinder. This container is gently tapped to compact the sample until it resembles the soil in the sample site.

The volume of each soil sample is recorded and the soil kept for further use in a plastic bag. The moisture content of the soil at each sample site is recorded using a **moisture meter** (with a scale of 1–8 where 1 = dry and 8 = wet).

Back in the laboratory, each soil sample is transferred to a **Tullgren funnel** (see Figure 34.11). Springtails (and other tiny animals) present in the sample move downwards away from the light source and fall through the holes in the gauze platform into the water.

After a standard length of time (e.g. 24 hours), the number of springtails extracted from each sample is

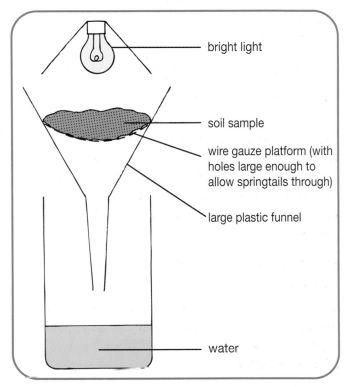

bright light

soil sample

wire gauze platform (with holes large enough to allow springtails through)

large plastic funnel

water

Figure 34.11 Tullgren funnel

soil type	sample site	moisture meter reading	number of springtails	volume of soil sample (cm³)	population density (springtails/cm³)	average population density (springtails/cm³)
'cultivated' soil in empty flower bed	1	4	2	45	0.04	0.03
	2	3	3	50	0.06	
	3	4	1	50	0.02	
	4	3	0	45	0.00	
	5	4	3	55	0.05	
'uncultivated' soil under bushes	6	6	10	50	0.20	0.19
	7	7	8	45	0.18	
	8	7	11	45	0.24	
	9	6	9	50	0.18	
	10	6	8	50	0.16	

Table 34.2 Springtail population density results

recorded and the animals returned to their natural habitat. Table 34.2 gives a typical set of results.

Discussion of results

In this investigation, the springtails were found to be more densely populated in the moister soil under the bushes. Since these animals lose water through their permeable skins and are killed by severe drought (acting as a density-independent factor), it is possible that their population density is affected by the **moisture content** of the soil.

However it must be remembered that some other difference between the two soils could also affect the population density of the animals. Springtails feed on dead **organic matter** which is less plentiful in the flower bed (and may therefore be acting as a density-dependent factor).

In addition, many springtails are probably killed by the repeated disturbance of the flower bed during cultivation (acting as a density-independent factor). To find out exactly which factor (or combination of factors) determines the population density of spring-tails would require a complex series of experiments each involving one altered variable at a time.

Homeostatic control

Regulation is the tendency of a population to decrease in size when it rises above a certain level and increase in size when it falls below that level.

In a natural ecosystem, such control is brought about by one or more density-dependent factors acting on the population. This is a further example of **homeostasis** (negative feedback control) as shown in Figure 34.12. It is the means by which a population is maintained at a relatively stable equilibrium, enabling it to make maximum use of available resources.

Testing Your Knowledge

1 **a)** What is meant by the *carrying capacity* of the environment in relation to a population of animals? (1)

   b) What relationship exists between a population's birth rate and death rate when it reaches the carrying capacity of the environment? (1)

2 Explain what is meant by the term *environmental resistance*. (1)

3 **a)** Define the terms density-dependent factor and density-independent factor. (2)

   b) Name THREE density-dependent factors that could exert a regulatory effect on a population of herbivorous animals. (3)

   c) What general name is given to this process of negative feedback control? (1)

   d) Give TWO examples of a density-independent factor. (2)

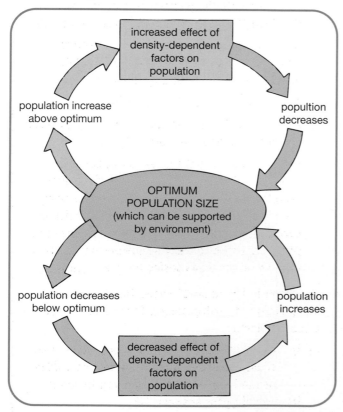

Figure 34.12 Homeostatic control of population size

Applying Your Knowledge

1 The graphs in Figure 34.13 show the results of population censuses carried out over a period of twenty years. Graph A refers to a species of bird living in a wood. Graph B refers to a species of flowering plant living on a moist meadow.

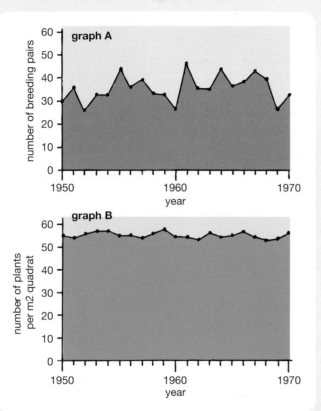

Figure 34.13

a) State the population density of the plant species in 1960. (1)

b) Estimate the environment's carrying capacity of the bird species. (1)

c) Suggest why the censuses were continued for such a long period of time. (1)

d) Rewrite the following generalisation made from the graphs by completing the blanks using the words given in brackets.

Once the _____ size of a species of _____ or animal has reached the _____ capacity of the _____, it remains relatively _____ despite short-term _____ in number.

(carrying, environment, oscillations, plant, population, stable) (1)

2 Imagine a swarm of one million locusts. This population could be decreased in number by increased predation, a thunderstorm, intense drought or shortage of food. For each of these factors, state whether it operates in a density-dependent or density-independent manner. (4)

3 The graph in Figure 34.14 shows the fluctuations in population numbers of the snowshoe hare and its main predator, the lynx.

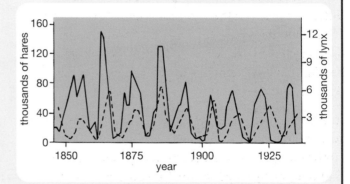

Figure 34.14

a) i) From Figure 34.14, identify which line represents the predator and which the prey.
   ii) Give TWO reasons to justify your choice. (3)

b) Which of the following periods (in years) is closest to the average length of a complete population cycle of the prey animal?

   A 5   B 10   C 25   D 50 (l)

c) Account for the regular fluctuations in population number of predator and prey. (1)

d) It is now known that the regular cycles in prey numbers still occur in the absence of predators. The plants eaten by the hares respond to heavy grazing by producing unpalatable shoots rich in toxins for 2–3 years. Suggest therefore why some scientists prefer to describe the predator cycles as *tracking* rather than *causing* the prey cycles. (1)

4 The graph in Figure 34.15 shows the fluctuations in a population of phytoplankton (microscopic algae) in a Scottish loch.

a) i) Name TWO density-independent factors that could have been responsible for the sudden rise in numbers of the plankton in spring.
   ii) Explain your answers. (4)

➔

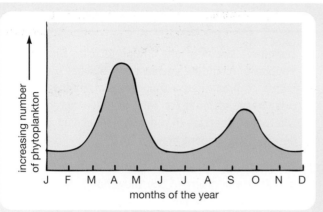

Figure 34.15

b) These plants are eaten by water fleas. Make a copy of the graph and then draw in a second curve to show the likely fluctuations in water flea population over the course of one year. (2)

5 The graph in Figure 34.16 shows the results from a study of the populations of two organisms, a predator and its well-fed prey, over a period of several weeks.

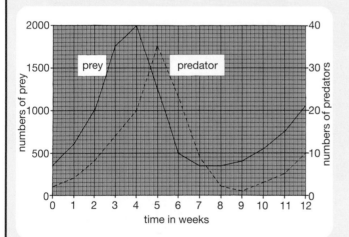

Figure 34.16

a) i) How many days did the predator population take to reach its maximum size?

ii) By how many individuals did it increase in number during this period of time? (2)

b) By how many times did the prey outnumber the predators at week 4? (1)

c) Account for: **i)** the decrease in number of predators between weeks 7 and 8; **ii)** the increase in number of prey between weeks 8 and 9. (2)

6 The diagram in Figure 34.17 shows the results of removing starfish (the dominant predator) from a marine food web.

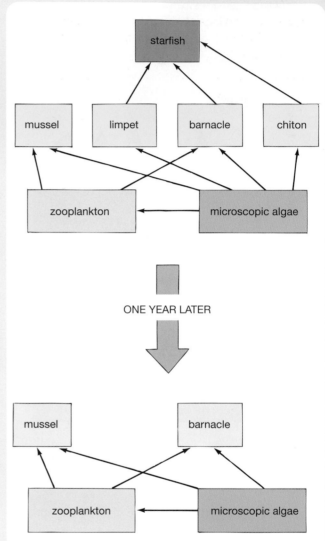

Figure 34.17

a) i) Name TWO prey animals that disappeared from the food web after one year.

ii) Suggest why they disappeared. (2)

b) Why, in this case, is the presence of the predator beneficial to the two prey organisms you gave as your answer to **a)**? (1)

7 In 1954/55, 90% of Britain's rabbit population died of a viral disease called *myxomatosis*. The virus was passed from animal to animal in underground burrows by the rabbit flea. The survivors were those rabbits that lived a relatively solitary existence above ground and rarely caught one another's fleas.

a) i) Is *myxomatosis* a density-dependent or density-independent factor governing the size of a rabbit population?

ii) Explain your answer. (2)

b) i) Construct a food web to include the following organisms: fox, grass, rabbit, sheep, tree seedling.

ii) *Myxomatosis* disturbed this food web for many years. Suggest ONE advantage and ONE disadvantage to the farmer that resulted during this time. (4)

8 Give an account of the factors that influence population change. (10)

35 Monitoring populations

Need to monitor wild populations

Many species of wild plants and wild animals are kept under close surveillance by scientists in order to obtain information about their **population numbers** and the factors which affect them. These data are needed for a variety of purposes including:

- the management of species which provide humans with **food** or **raw materials**;
- the control of **pest** species;
- the assessment of levels of pollution using **indicator** species;
- the protection and conservation of **endangered** species.

Food species

If the death rate of a species exceeds its birth rate then the size of the population decreases. In the long term, this leads to extinction of the species.

Fish

Populations of **edible fish** species are monitored to estimate their remaining stock since it is essential that fish are not removed at a rate which exceeds their maximum rate of reproduction.

There are several well-documented cases of fish stocks that have already crashed due to **over-fishing**. These

include herring, haddock and cod (see Figure 35.1) in the seas around Britain. Despite reduced fishing levels in recent years, fish stocks have failed to make a full recovery. It is for this reason that EU regulations insist that catches must stay within certain **fixed quotas**.

Red deer

Scotland has a large population of **red deer** (see Figure 35.2) living in the wild. Although the very young members of the population are vulnerable to attack by foxes and golden eagles, the adult deer have had no natural predator (apart from humans) since the wolf became extinct in Scotland over 250 years ago.

Figure 35.2 Herd of red deer

In an attempt to prevent uncontrolled population growth occurring amongst red deer, humans kill several thousand animals every year for their meat (venison). This planned harvesting of a wildlife species is called **culling**.

In the early 1960s, the number of Scottish red deer stood at around 150 000 which was regarded by many experts as an optimum level. However by 2005, despite a cull of around 3000 animals every year, the population had increased to around 400 000 (see Figure 35.3).

Many experts, including those representing WWF (World Wildlife Fund), believe, therefore, that the

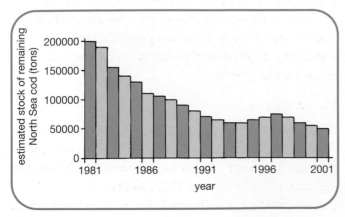

Figure 35.1 Effect of over-fishing on cod stock

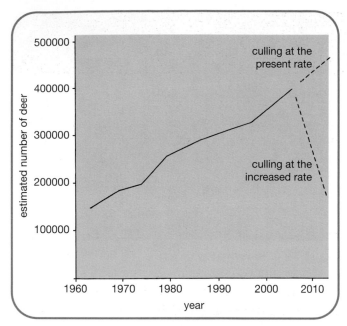

Figure 35.3 Potential effects of culling red deer

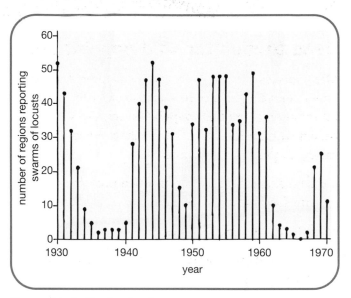

be tackled by special planes equipped with tanks of insecticide spray.

Figure 35.4 Monitoring locust swarms

annual cull should be increased to keep the population stable at 150 000 while at the same time boosting the supply of a valuable food resource.

Pest species

Species that pose a threat to mankind's health or economy are called **pests**. These include insects, fungi and vermin that spread diseases or ruin crops. Monitoring populations of pest species provides information needed in our attempts to control them. Such control is often carried out by **chemical** means (e.g. application of pesticide) and occasionally by **biological** means (e.g. introduction of ladybirds to control greenfly).

Locust

Insect pests thrive on crop monocultures and consume about one third of the world's harvest. Some pest species undergo sudden irregular **population explosions**. Such population explosions in the **desert locust** have been shown to coincide with years of higher rainfall. The extra moisture provides ideal conditions for the eggs to hatch in the desert sand.

By monitoring climatic factors and the emerging swarms of locusts (see Figure 35.4), scientists can warn farmers in advance to prepare for an invasion. Swarms, containing millions of insects capable of eating hundreds of thousands of tons of crops in **one day**, can

Pathogenic fungi

Scientists study **epidemics** caused by **pathogenic fungi** such as powdery mildew and black stem rust that attack cereal crops (see Figure 35.5). They construct an epidemic model which includes the host species (susceptible, infected and recovered individuals) and the pest species. From the results of the investigations, they are able to estimate the pest's rate of transmission. This is found to vary from one epidemic to another depending on factors such as climate and density of host plants in the monoculture.

Armed with essential data about an advancing epidemic, scientists can warn farmers to adopt **preventative measures** and protect their crops before the air-borne fungal spores arrive. This allows the most effective use to be made of the pesticide while preventing needless application at times when the crops are not under threat.

Brown rats

Some mammalian pests such as the **brown rat** (see Figure 35.6) carry micro-organisms that cause human diseases. *Leptospira* is a bacterium carried by brown rats that causes Weil's disease. This condition is characterised by jaundice and kidney failure and can be fatal to humans. For many years it was thought that up

331

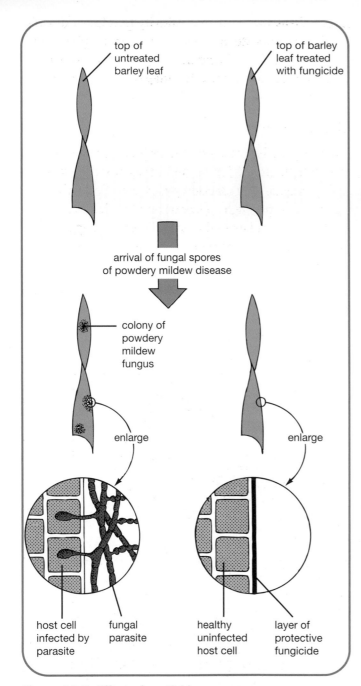

Figure 35.5 Effect of pesticide

to 90% of brown rats in Britain carried the pathogenic strain of this bacterium. Studies of rat populations have shown however that only about 7% of individuals carry the bacterium. More alarming was the discovery that rat numbers are now at an all time high in Britain and that these pests carry at least five other disease-causing micro-organisms previously unknown in British rats.

Figure 35.6 Brown rat with young

Testing Your Knowledge

1 Give FOUR types of wild population whose numbers are monitored by scientists. (4)

2 a) Suggest why it is important that every effort is made during commercial fishing to net only the larger members of a shoal and leave the smaller ones in the sea. (2)

 b) What name is given to the planned harvesting of a wildlife species? (1)

3 a) Name TWO types of pest species. (2)

 b) 'The successful control of pests is directly related to the understanding of population dynamics.' Justify this statement with reference to TWO name examples. (4)

Indicator species

Certain wildlife species often serve as **indicators** of the state of the environment's health by their abundance or scarcity.

Freshwater invertebrates

The presence of large populations of **mayfly** and **stonefly** nymphs in a river ecosystem shows that the water is clean and rich in dissolved oxygen. On the other hand, an abundance of **rat-tailed maggots** and **sludgeworms** in the water indicates that it has a low oxygen content and is badly polluted with organic waste.

Lichens

Lichens are simple plants which grow on the trunks and branches of trees. Three examples are shown in Figure 35.7.

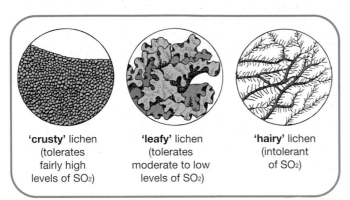

'**crusty**' lichen
(tolerates
fairly high
levels of SO₂)

'**leafy**' lichen
(tolerates
moderate to low
levels of SO₂)

'**hairy**' lichen
(intolerant
of SO₂)

Figure 35.7 Lichens

Since lichens vary in their sensitivity to **sulphur dioxide** (SO_2), they indicate the level of atmospheric pollution by this harmful gas. Using the data obtained from a lichen survey, an **air pollution map** of an area can be constructed as shown in Figure 35.8.

Birds of prey

Predatory birds at the top of their food pyramid (see Figure 35.9) are the first to suffer the consequences of environmental degradation by over-use of **pesticide** sprays on crops. High concentrations of chemical residues accumulate in the birds' tissues and egg shells and indicate the state of the local environment's health.

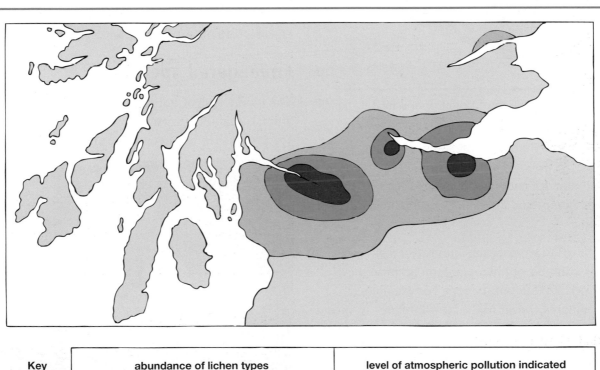

Key	abundance of lichen types	level of atmospheric pollution indicated
	no lichens or few crusty only	high
	many crusty, few leafy	moderate
	some crusty, many leafy	low
	some crusty, many leafy, some hairy	zero

Figure 35.8 Air pollution map based on lichen survey

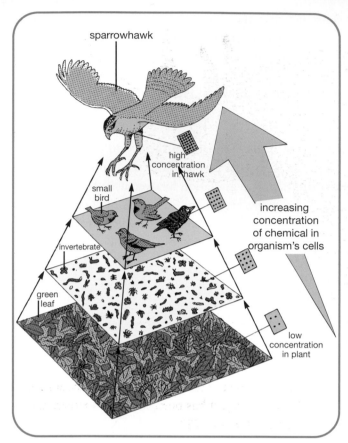

Figure 35.9 Accumulation of chemical in a food pyramid

Phytoplankton

Phytoplankton are microscopic plants which inhabit the world's oceans and act as the first link in marine food chains.

Scanners fitted to weather satellites reveal that enormous numbers of phytoplankton occur in the oceans at certain times of the year. For example, explosive blooms appear in the North Atlantic in spring when warmth and increasing daylengths coincide with rich supplies of nutrients stirred up from the deep by winter storms (see Figure 35.10).

Since phytoplankton absorb vast amounts of carbon dioxide during photosynthesis, it is possible that they are helping to protect the world against the greenhouse effect.

It is important to monitor the populations of these tiny producers because they indicate the state of the ocean ecosystem. The future survival of the Earth may even depend on their activities if tropical forests continue to be cleared at the present rate.

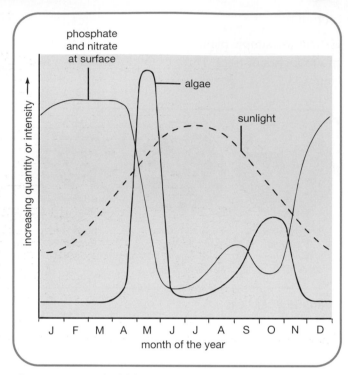

Figure 35.10 Algal bloom in North Atlantic Ocean

Endangered species

The monitoring of populations of wild plants and animals enables humans to recognise and protect rare species for their aesthetic value and genetic diversity.

Figure 35.11 *Diapensia lapponica*

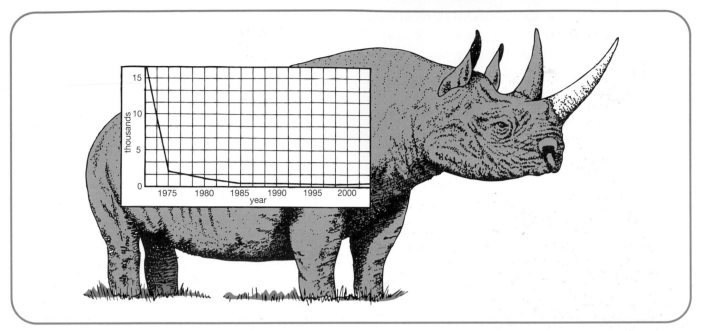

Figure 35.12 Population decline of black rhinoceros in Kenya

Wild plants

In Britain, 21 native species of wild plants are now known to be very rare and in need of rigorous protection. One of these is *Diapensia lapponica* (see Figure 35.11) which is found only at one site on a Scottish mountainside.

To prevent such endangered species from becoming extinct, the areas that they live in are often designated as nature reserves. In addition, an act of parliament makes it illegal to remove any part of these plants since picking even just one flower or leaf could reduce the plant's chance of survival.

Black rhinoceros

Figure 35.12 shows the dramatic decline in number of the black rhinoceros due to over-hunting. At last steps

are being taken to save this animal from extinction. A World Wildlife project has been set up to increase the population to 1000 by 2020.

Whale

Although estimates of whale populations tend to be subject to a large margin of error, it is known that their numbers have dropped dramatically since the start of commercial whaling (see Table 35.1).

This downward spiral will only be reversed in the future if the current moratorium (suspension of activity) on commercial whaling is respected by all countries of the world.

More is at stake than the survival of the world's largest mammal. Like the panda and the elephant, the whale has become an emblem of conservation. Experts fear

species of whale	estimated number before commercial whaling	estimated number today	percentage remaining
Blue	200 000	2000	1
Humpback	100 000	9000	9
Fin	450 000	70 000	15.6
Sei	200 000	28 000	14
Right	50 000	3000	6

Table 35.1 Decline in whale numbers

that if these 'high-profile' species cannot be saved then there is little hope of preventing a multitude of other species of 'lower profile' (but of equal importance) from slipping into extinction (also see chapter 20).

Case study of a population

In recent years, millions of tons of poisonous wastes have been pumped or dumped into the North Sea. These have included sewage sludge, fertilisers, incinerator fall-out, heavy metals, organo-chlorines, dioxins, radioactive waste and oil.

In 1988 populations of the **common seal** were found to be suffering from a **virus** (unique to seals but similar to canine distemper) which attacked their immune system. This led to an epidemic which wiped out an estimated 60% of the seal population.

Many scientists were cautious about linking this disease with pollution since, strictly speaking, the seals died of a natural cause. In 1989 blood samples were taken from common seals and about half of the animals tested were found to possess antibodies to the disease. This meant that the others (mostly young pups) were susceptible to the virus. This proved to be true later that year when the disease struck again, especially amongst young animals.

Poisoned food chain

Organo-chlorines are toxic chemicals used as pesticides. Since they are **non-biodegradable**, they persist and accumulate in food chains with the final consumer being the most severely affected. This is shown in Figure 35.13.

Although certain organo-chlorines such as DDT have now been banned from use by most countries, they still persist in almost all ecosystems. Many of the North Sea common seals that died of the virus were found to contain high concentrations of organo-chlorines and other pollutants in their tissues and blubber.

Some scientists are of the opinion therefore that an indirect link existed between the viral disease and pollution. They argue that animals living in an ecosystem under such constant stress by pollution must be more susceptible to disease. Some experts suggest that the pollutants may, in some way, have impaired the seals' immune system.

Testing Your Knowledge

1 a) In general, what is meant by the term *indicator species*? (1)

   b) What is indicated by the presence of a dense population of each of the following in an environment?
      i)   rat-tailed maggots in a river;
      ii) 'hairy' lichens on walls and fences;
      iii) stonefly nymphs in river water;
      iv) algal blooms in the Atlantic Ocean. (4)

2 a) Give TWO reasons for protecting and conserving endangered species. (2)

   b) Why does the law in Britain make it illegal to remove any part of an endangered plant species? (2)

3 a) What is meant by the expression '*moratorium on commercial whaling*'? (1)

   b) Some countries in the world still refuse to heed the moratorium. Why is it essential that they are persuaded to do so in the future? (1)

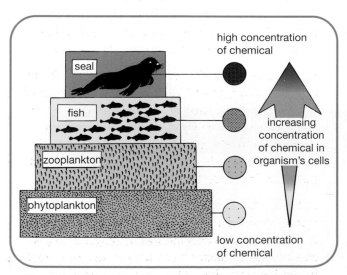

Figure 35.13 Accumulation of chemical in a food chain

Applying Your Knowledge

age of fish (years)	total number of plaice caught				
	year 1	year 2	year 3	year 4	year 5
2	89	78	83	421	573
3	547	625	602	2127	2268
4	2485	2331	2491	2683	2594
5	2137	2027	2076	1527	1332
6	891	766	750	363	257
7	417	389	428	215	209
8	235	201	217	92	76
9	67	94	86	41	38
10	50	81	63	20	15
11	43	57	38	17	12
12+	77	94	69	32	28

Table 35.2

1 The data in Table 35.2 refer to the catch of plaice in the North Sea by a fleet of trawlers over a 5-year period.

   a) Which age of plaice is the most common in each year's catch? (1)

   b) i) What trend is shown by the data when read vertically downwards from 5-year-old fish?
      ii) Explain why. (2)

   c) Suggest why the '12+ years' entry is always greater than the '11 years' entry. (1)

   d) Make a generalisation about the way in which the catch data for years 4 and 5 differ from those for years 1–3. (2)

   e) The fishermen were pleased with their catch in years 4 and 5. Suggest why they should be concerned about the future (if the trend shown by the data continues). (2)

   f) What is the percentage decline in number of 10-year-old fish in the 5-year study? (1)

2 Sheep ticks will cling to any woolly surface that brushes against them in spring. A method of estimating the abundance of sheep ticks on a Scottish farm is to drag a woollen blanket over randomly chosen areas of the field under investigation.

   a) Why are several areas sampled? (1)

   b) Why are the sample areas chosen at random? (1)

   c) How could an estimate of the abundance of sheep ticks for the whole field be calculated? (2)

   d) Suggest TWO courses of action open to the farmer if sheep ticks are found to be abundant in a field intended for sheep grazing. (2)

3 Read the following passage and answer the questions that follow it.

The cottony cushion scale insect first appeared in the Californian orange groves in 1868 following its accidental introduction from Australia. In the absence of natural enemies, it quickly multiplied and within 20 years was causing major damage to the citrus trees. The fruit growers were close to ruin.

Scientists studying the pest's natural enemies found that a certain type of Australian beetle (a species of ladybird) could be used to clear the orange trees of the scale insect (see Figure 35.14). Following the introduction of the predators, the pest was brought under control within 2 years (see Figure 35.15) and the citrus fruit industry was saved.

Scientists monitoring populations found that the cottony cushion scale insect was continuing to survive in low numbers but was no longer posing a major threat to orange trees since it was being kept in check by ladybirds. When, at a later date, the ➜

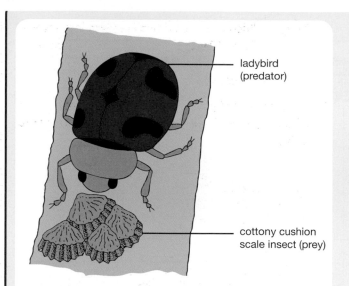

ladybird
(predator)

cottony cushion
scale insect (prey)

Figure 35.14

orange groves were treated with pesticide, most of
the scale insects died. However a few were resistant
to the chemical and survived. All of the ladybirds
died. This resulted in a new epidemic of scale insects
which was only curbed by the reintroduction of
ladybirds.

a) Construct a food chain to show the relationship
between the organisms mentioned in the
passage. (1)

b) Study the graph in Figure 35.15 and identify
which line is: **i)** the predator; **ii)** the prey. **iii)**
Explain your choice in each case. (3)

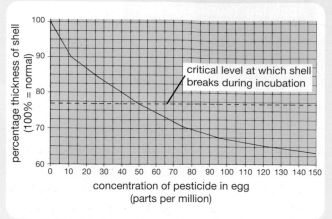

Figure 35.16

c) How many years did it take for the pest
population to reach a level where it was inflicting
major economic damage? (1)

d) During which year were the predators
introduced? (1)

e) Why did the number of predators drop after
1895? (1)

f) Describe the relationship between predator and
prey during the years 1900 to 1940. (2)

g) Initially fruit growers were very pleased with the
outcome of the pesticide treatment carried out in
the 1940s.
  i) Suggest why.
  ii) Why was their delight short-lived?

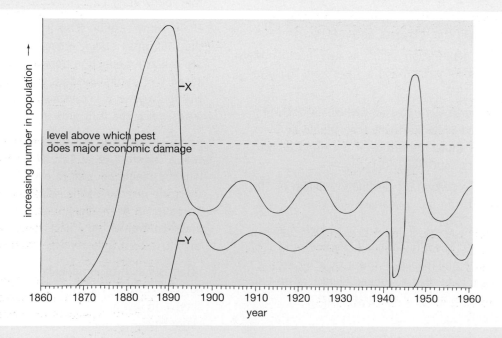

Figure 35.15

iii) By what means was the problem overcome? (3)

h) Predict the form that the lines in the graph took after 1960. (1)

i) Why is it important that scientists continue to monitor populations of cottony scale insects? (1)

4 The graph in Figure 35.16 shows the relationship between concentration of pesticide in the eggs of a predatory bird and thickness of egg shell.

a) Describe the relationship shown by the graph. (1)

b) i) If this species of bird lays eggs containing 25 ppm of pesticide residue, by what percentage will the shell's normal thickness be reduced?

ii) How many more parts per million of pesticide would need to be present to reach the shell's critical level?

iii) Predict the bird's reproductive success when 100 ppm of pesticide are present in its eggs. Explain your answer. (4)

5 The graph in Figure 35.17 shows the results of an analysis of the breast muscles of several species of birds for an organo-chlorine pesticide.

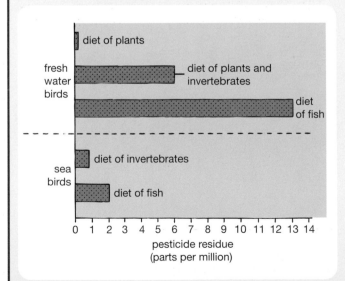

Figure 35.17

a) i) From the graph, state the general relationship that exists between a bird's diet and the concentration of pesticide residue in its muscle tissues.

ii) Account for this relationship. (3)

b) i) Which of the environments was less severely affected by the pesticide, as indicated by these data?

ii) Suggest why. (2)

6 The graphs in Figure 35.18 show some of the effects of adding excessive amounts of untreated sewage to a river.

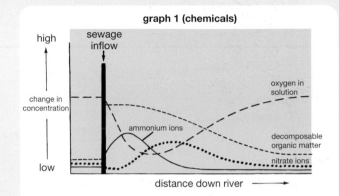

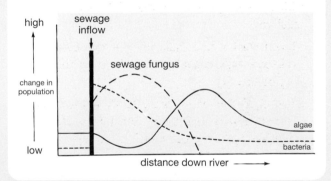

Figure 35.18

a) Broadly speaking, what relationship exists between concentration of dissolved oxygen and numbers of bacteria present in the river? Explain why. (1)

b) i) What evidence is there from graph 2 that sewage fungus is not a native member of the river community?

ii) Account for the initial increase in population size of sewage fungus.

iii) Name a factor that could have been responsible for the subsequent decrease in sewage fungus.

iv) State whether the factor you gave as your answer to **iii)** would operate in a density-dependent or density-independent manner. (4)

c) i) What factor could account for the initial decrease in number of algae following the sewage inflow?

ii) Is this factor density-dependent or density-independent? Explain your answer. (2)

d) With reference to the nitrogen cycle, explain why the maximum concentration of nitrate ions occurs in the water after the maximum concentration of ammonium ions. (1)

e) Suggest which factor was responsible for the algal population boom. Why should this be so? (1)

f) Name a natural outside factor that could hasten the river's return to a normal healthy state. (1)

7 Give an account of the need by scientists to monitor populations of wild animals, plants and micro-organisms. (10)

36 Succession in plant communities

Plant succession

Simplified example

Imagine a bare infertile field left abandoned to nature (see Figure 36.1). Although many types of seeds land on it from neighbouring ecosystems, at first the only successful plants are annual (and a few perennial) weeds. If, for example, acorns from a neighbouring oak forest arrive, the seedlings fail. This is because the soil is poor and too exposed (hot and dry in summer; early to freeze in winter).

After a short number of years, the **pioneer community** of weeds is gradually replaced by sturdy grasses and taller shrubby perennials which become dominant by choking out the smaller plants and depriving them of light.

Amongst this community, oak seedlings are eventually able to survive once the soil becomes richer and moister and the shrubs become bushy enough to provide shelter. After many years the oaks develop into trees whose dense canopy of leaves prevent much light reaching the ground. This causes many species of grasses and shrubs to die out but still leaves a diverse community of shade-tolerant plants (e.g. bluebell and wood-sorrel – see Figure 36.2) on the woodland floor.

Figure 36.2 Community of shade-tolerant plants

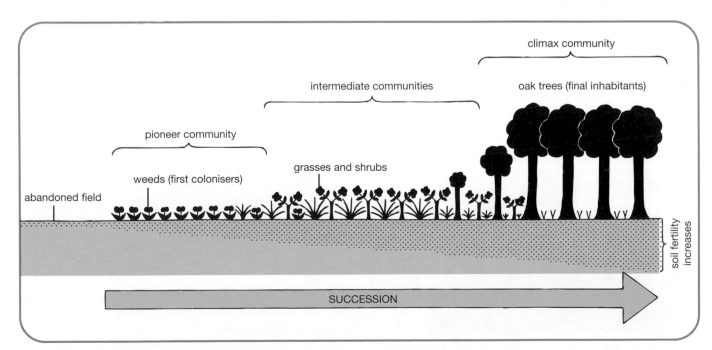

Figure 36.1 Succession on an abandoned field

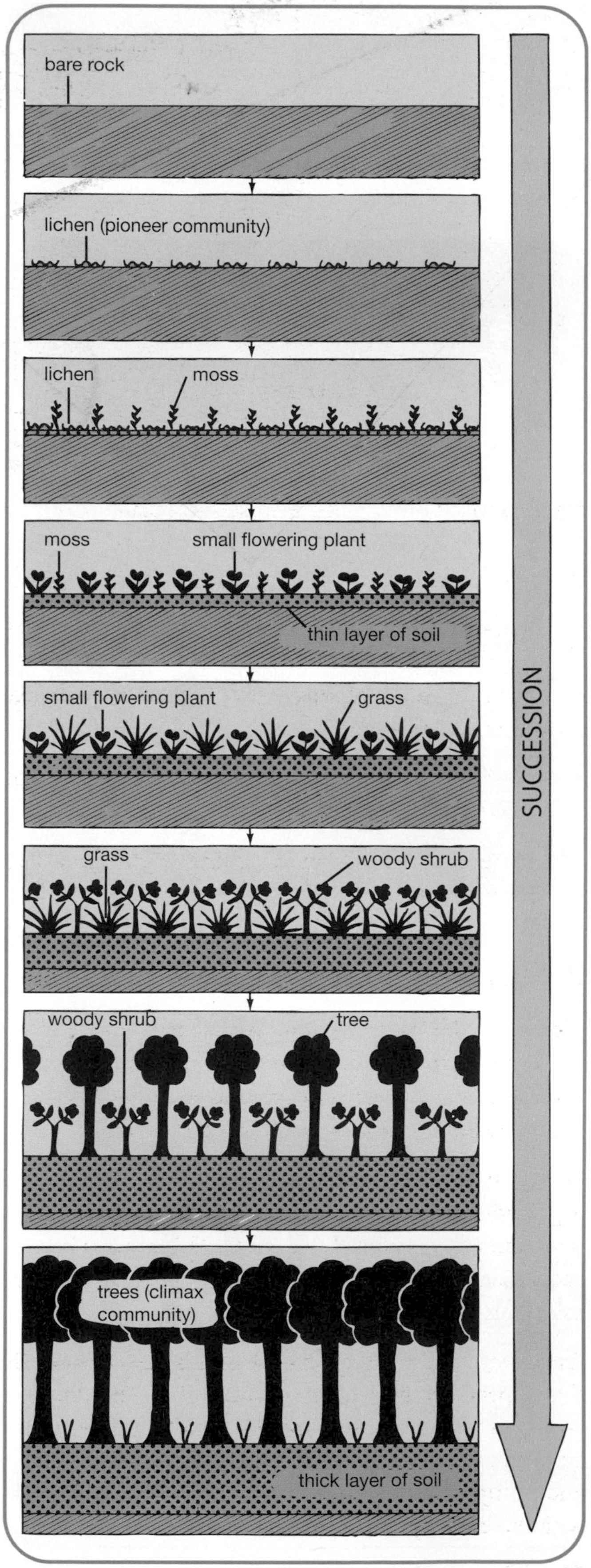
bare rock
lichen (pioneer community)
lichen
moss
moss
small flowering plant
thin layer of soil
small flowering plant
grass
grass
woody shrub
woody shrub
tree
trees (climax community)
thick layer of soil
SUCCESSION

Figure 36.3 Primary succession

Thus many years after the original field was abandoned, an oak forest stands in its place. The oak forest is called the **climax community**. This sequence of events (summarised in Figure 36.1) involves a regular progression from a pioneer community of plants to a climax community. It is **unidirectional** and it is called **succession**.

Primary and secondary succession

Primary succession (see Figure 36.3) occurs during the colonisation of a barren area which has not previously been inhabited e.g. bare rock. In this case the pioneer community consists of lichens (see Figure 36.4) which are able to withstand drying out. They also make acids which break down the rock. Mosses are the next stage in the succession. As the rock disintegration process continues and dead plants accumulate, a layer of proper soil gradually develops.

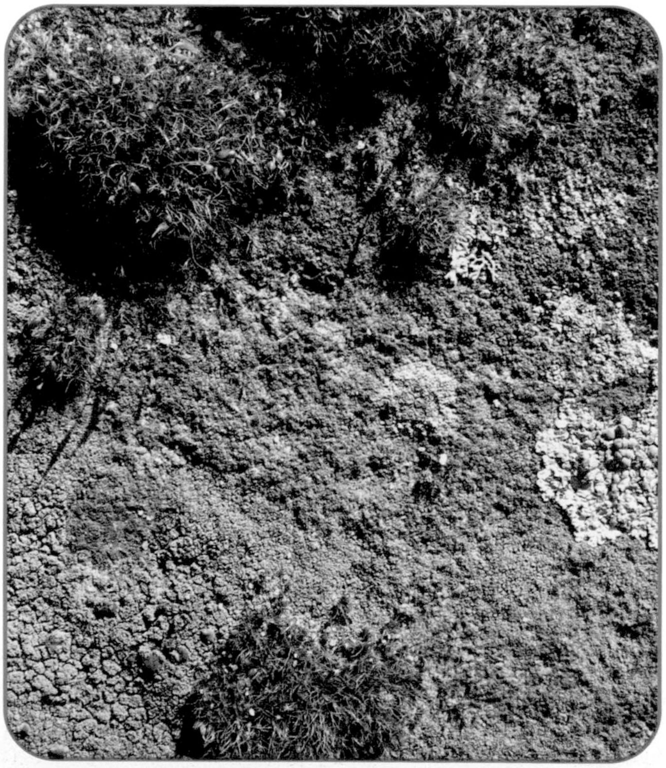

Figure 36.4 Early stage in primary succession

Eventually the mosses are succeeded by small flowering plants and then shrubs and trees as before. Such primary succession takes a considerable length of time (e.g. a thousand years).

Secondary succession occurs during the colonisation of

an area which has previously been occupied by a well-developed community but has become barren. An example would be a forest cleared for agricultural use and later abandoned by the farmer or a forest that has been destroyed by fire.

Secondary succession takes a shorter time (e.g. one hundred years) because soil containing nutrients is already present and the conditions are normally more favourable to colonisation.

Cause of succession

The main driving force behind succession is the effect that the plants have on the habitat. Succession occurs because each community acts on and modifies the habitat. After a period of relatively short-term stability, a community ends up making the habitat less favourable for itself and more favourable for a different community which therefore succeeds it.

For example, during the early stage of the succession shown in Figure 36.1, the members of the pioneer community add humus (dead organic matter) to the soil. This takes the form of dead leaves and other plant parts. These are acted upon by decomposers (bacteria and fungi – see Figure 36.5) which break them down into small fragments and release nutrients from them. This process improves the soil's texture and mineral content. It gradually provides growing conditions suitable for the larger grasses and shrubs which eventually choke and shade out the earlier colonisers before in turn being supplanted by the climax community (see Figure 36.6).

Figure 36.5 Decomposers

Figure 36.6 "Nothing succeeds like succession!"

Characteristics of a climax community

A climax community is:

- the final product of long-term unidirectional change within a community;
- self-perpetuating and, under natural conditions, not replaced by another community;
- a mature community in dynamic equilibrium with its environment.

Biomass and species diversity

Biomass is the total weight of organic matter that makes up a group of organisms. It is normally expressed as dry weight because the water content of living organisms varies over short periods of time.

During succession, the height and biomass of the vegetation in the community increases reaching a maximum level at the climax stage. This process is accompanied by the soil becoming deeper and richer in organic matter and minerals.

As the vegetation and the soil change, so do the animals that depend on them for food and shelter. The climax community is therefore normally found to support the greatest diversity of animal species (especially insects) and to support the most complex food webs compared with the earlier plant communities.

The differences between a pioneer community and a

climax community of forest are summarised in Figure 36.7.

characteristic	pioneer community	climax community
growth rate	rapid	slow
height of vegetation	low	high
life span	short	long
relative number of seeds	many	few
relative size of seeds	small	large
relative distance seeds are dispersed	long	short
food productivity	low	high
biomass	small	large
species diversity	low	high
nutrient supply in soil	low	high
food chains and webs	few and simple	many and complex

Figure 36.7 Comparison of pioneer and climax communities

Types of climax community

The types of plant that can survive and the direction taken by succession in a region are dependent upon the climatic and soil factors present. The different natural climax communities that occur in Western Europe (see Figure 36.8) are therefore found to correlate closely with the different climatic regions (Figure 36.9).

Human intervention

Only a few hundred years ago, much of Britain was covered with a climax community of deciduous forest. However most of this has now been cleared to provide land for **agriculture, conifer forestry** and **human settlement**. Since this land is rarely left abandoned to nature for extensive periods, succession leading to the climax community of deciduous forest is normally prevented.

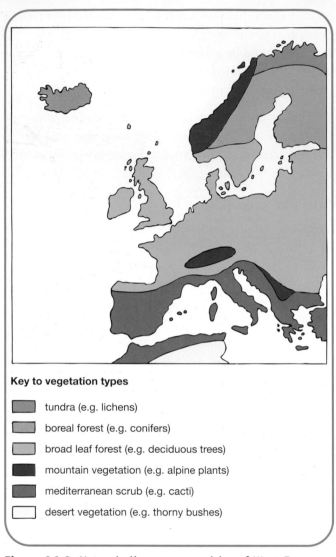

Key to vegetation types

- tundra (e.g. lichens)
- boreal forest (e.g. conifers)
- broad leaf forest (e.g. deciduous trees)
- mountain vegetation (e.g. alpine plants)
- mediterranean scrub (e.g. cacti)
- desert vegetation (e.g. thorny bushes)

Figure 36.8 Natural climax communities of West Europe

Intensive grazing

The process of succession towards a forest climax community is interrupted by animals such as sheep grazing intensively on grassland that was once forest. Any tree saplings that appear are rapidly consumed before they can develop into trees.

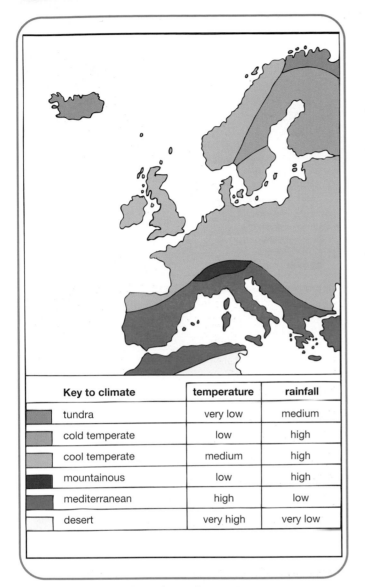

Key to climate	temperature	rainfall
tundra	very low	medium
cold temperate	low	high
cool temperate	medium	high
mountainous	low	high
mediterranean	high	low
desert	very high	very low

Figure 36.9 Climatic regions of West Europe

Testing Your Knowledge

1 a) With reference to plant communities, explain what is meant by the term *succession*. (2)

   b) Give TWO differences between primary and secondary succession. (2)

2 a) State TWO characteristics of a climax community. (2)

   b) Compare a pioneer community and a climax community with reference to species diversity, biomass and complexity of food webs. (3)

3 a) Name the TWO environmental factors which determine the climax community typical of a geographical region. (2)

   b) Give THREE examples of climax communities that occur in Europe. (3)

   c) Scotland's climax community of deciduous forest is far less extensive now than it was 500 years ago. Explain why. (2)

Applying Your Knowledge

1 a) What name is given to the series of events represented by Figure 36.10? (1)

b) i) Match each of the following processes with one of the three types of lettered arrow: *self-perpetuation, habitat modification* and *extinction*.

ii) Justify your choice in each case. (6)

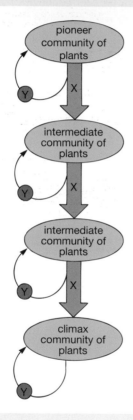

Figure 36.10

2 The graph in Figure 36.11 represents the process of succession on an abandoned field. The thickness of a line corresponds to a species' relative importance at a given time. A broken line indicates the presence of seeds which failed to become established.

a) In which year was the field abandoned? (1)

b) i) How many different species of plant were established in the field in year 20?

ii) Identify by their letters those grass species which landed on the field as seeds in year 8.

iii) Of these grass species, how many were represented by seeds that failed in year 8?

iv) Suggest why these failed. (4)

c) Explain why trees succeeded shrubs. (1)

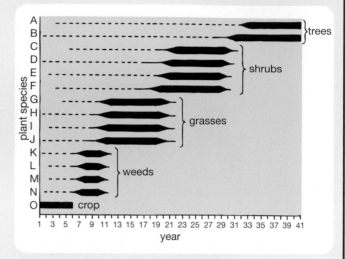

Figure 36.11

3 The five boxed statements in Figure 36.12 refer to the process of plant succession in a fresh water lake. Arrange them into the correct sequence. (1)

(a) Alder trees thrive in the marshy soil aided by N-fixing bacteria in their roots and replace the sedges.

(b) Reeds grow in the silt at the lake's edge and accumulate more sediment making the lake shallower still.

(c) Further silt gathers in the marshy soil inhabited by alders and eventually becomes rich and dry enough to suppoort oak trees.

(d) A river flows into a lake carrying silt which is deposited on the bed of the lake making it shallower.

(e) Sedges form mounds of marshy soil in shallower water at the lake's edges and replace the reeds.

Figure 36.12

4 a) Copy the axes shown in Figure 36.13 and draw a
      line graph to show how biomass varies as
      succession proceeds. (1)

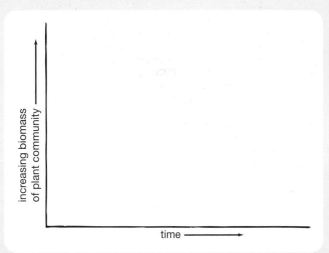

Figure 36.13

  b) Explain your answer. (1)

5 Figure 36.14 shows a simplified version of succession
   on a barren stretch of sand blown up from a sea
   shore.

   a) Does the diagram illustrate an example of primary
      or secondary succession? Explain your choice of
      answer. (1)

   b) i)   What effect does colonisation by couch grass
           have on the habitat?
      ii) Which community is eventually able to
           succeed couch grass?
      iii) What TWO effects does this next community
           have on the habitat?
      iv) Which community is eventually able to
           succeed it? Why? (6)

   c) Which lettered region on the diagram can support
      the British climax community? (1)

6 Give an account of primary succession involving
   communities of land plants. (10)

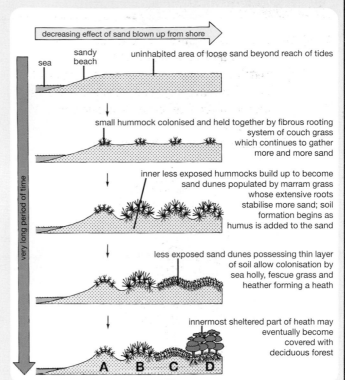

Figure 36.14

What You Should Know

(See Table 36.1 for Word bank)

biomass	feedback	pest
carrying	food	pioneer
climatic	homeostasis	population
climax	independent	receptors
conserve	indicator	replaced
density-dependent	internal	resistance
dynamics	limits	set point
effectors	modifies	stable
external	monitored	succession
favourable	negative	webs

Table 36.1 Word bank for chapters 33–36

1 To function efficiently, many aspects of the human body's _____ environment must be maintained within tolerable _____.

2 Physiological _____ is the name given to this maintenance of the internal environment despite changes in the _____ environment.

3 Homeostasis operates on the principle of _____ feedback control. By this means a change in the internal environment is detected by _____ which send messages to _____. These trigger responses which negate the deviation from the norm and return the internal environment to its _____.

4 Population _____ is the study of population changes and the factors which cause them.

5 A population increases until it reaches the _____ capacity of the environment and then it remains relatively _____.

6 A _____ is prevented from increasing in size indefinitely by several factors known collectively as environmental _____.

7 Some of these factors affect the growth of the population in a manner _____ of population density. Other factors operate in a _____ manner.

8 In a natural ecosystem, a population is kept relatively stable by density-dependent factors effecting negative _____ control (homeostasis).

9 The population numbers of many wild plants and animals are _____ by humans.

10 This provides data which help humans to manage species used for _____, to control _____ species, to assess pollution levels using _____ species and to protect and _____ endangered species.

11 The change which involves a regular progression from a _____ community of plants to a climax community is called _____.

12 During succession each community enjoys a period of short-term stability during which time it _____ the habitat. In doing so it makes the habitat less _____ to itself and more favourable to its successor.

13 The _____ community is the final product of succession whose nature is determined by soil and _____ factors. It enjoys stability and is not _____ by another community.

14 Compared with a pioneer community, a climax community possesses a more diverse range of species, a higher _____ and a more complex set of food_____.

Appendix 1

Microscopes

Magnification and resolution

Magnification is the apparent enlargement of an object. By magnifying material, a microscope allows structures which are invisible to the naked eye to be examined.

Resolution (resolving power) is the ability of a microscope to produce a separate image of each of two structures located closely together.

When two objects lie more closely together than the limit of resolution of the microscope, they appear as a single fused image. Further magnification simply increases the size of the fused image.

Figure Ap1.1 compares two types of microscope. A **light microscope** employs a beam of light. Glass lenses are used to magnify the image. An **electron microscope** employs a beam of electrons. Electromagnets are used instead of glass lenses to magnify the image.

The maximum useful magnification possible using a light microscope is approximately 1500 times. An electron microscope has a much higher resolving power and can achieve a useful magnification of over 500 000 times. Also see Figure Ap1.2.

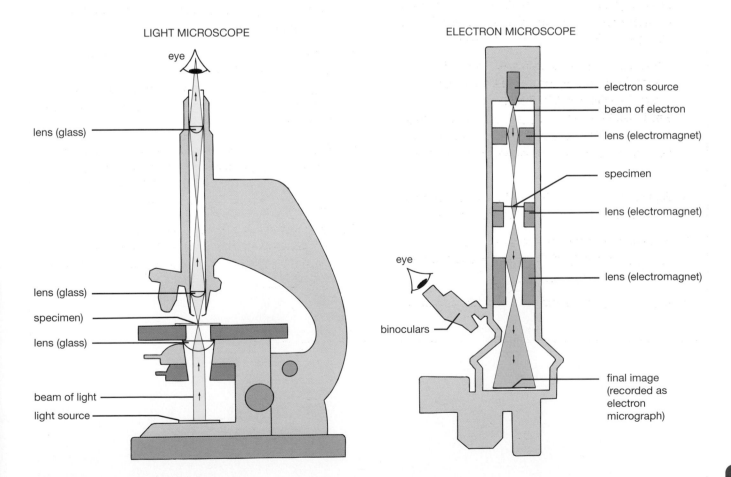

Figure Ap1.1 Two types of microscope

Figure Ap1.2 Electron microscope specimens

Ultrastructure of a Cell

Electron micrographs

An image of material viewed under an electron microscope can be recorded as a black and white photograph called an **electron micrograph**. Figures Ap2.1 and Ap2.2 are diagrams based on several electron micrographs. They show how the electron microscope reveals the presence in a cell of many tiny structures which cannot be seen using a light microscope. Many of these specialised structures are called **organelles** and they make up the cell's **ultrastructure**.

Membrane systems within a cell

The **membranes** possessed by the various organelles present in a cell share the same basic structure as the plasma membrane (see page 10 for the fluid-mosaic model). However they vary from one another in function and arrangement.

The nucleus, for example, has a double membrane perforated by pores which allow the exit of mRNA for protein synthesis. The inner membrane of a mitochondrion is convoluted giving it a large surface area upon which the biochemical processes of aerobic respiration can take place. The membranes of the endoplasmic reticulum and Golgi apparatus become pinched off releasing vesicles containing protein. The membrane round a lysosome isolates digestive enzymes until they are required for phagocytosis.

Thus the various membrane systems operate as **selective barriers** each contributing in its own way to the integrated working of the cell.

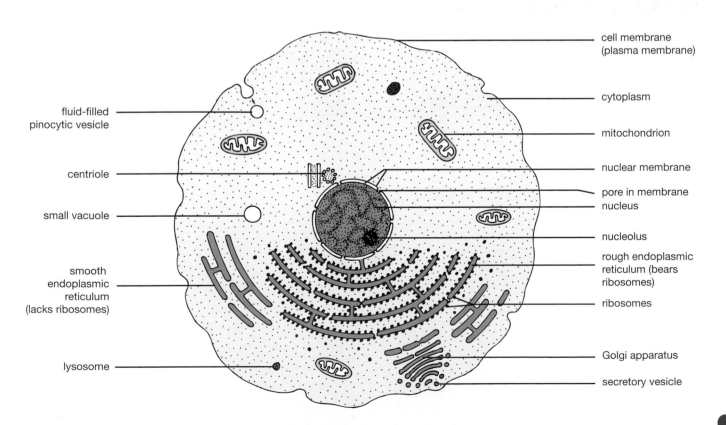

Figure Ap2.1 Ultrastructure of generalised animal cell

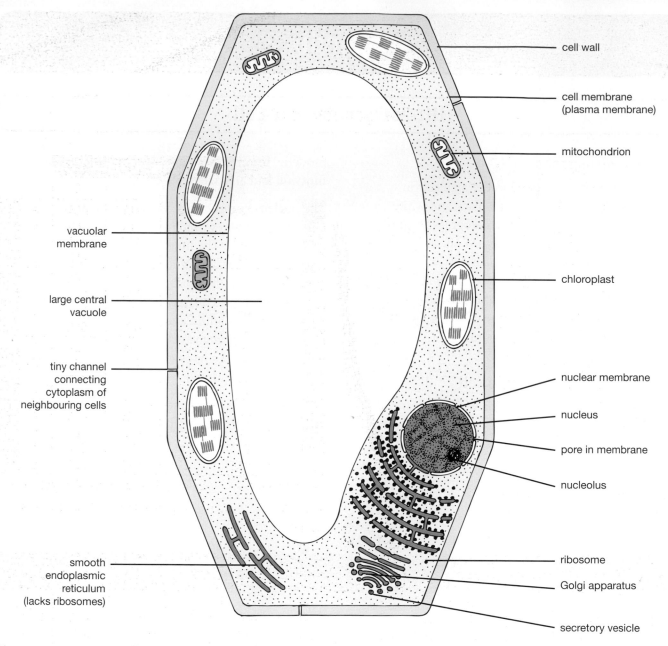

Figure Ap2.2 Ultrastructure of generalised plant cell

cell wall

cell membrane
(plasma membrane)

mitochondrion

chloroplast

nuclear membrane

nucleus

pore in membrane

nucleolus

ribosome

Golgi apparatus

secretory vesicle

vacuolar
membrane

large central
vacuole

tiny channel
connecting
cytoplasm of
neighbouring cells

smooth
endoplasmic
reticulum
(lacks ribosomes)

The genetic code

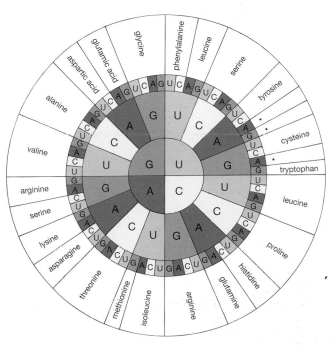

Figure Ap3.1 Alternative presentation of mRNA's 64 codons and amino acids coded

	second letter of triplet				
	A	G	T	C	
A	AAA	AGA	ATA	ACA	A
	AAG	AGG	ATG	ACG	G
	AAT	AGT	ATT	ACT	T
	AAC	AGC	ATC	ACC	C
G	GAA	GGA	GTA	GCA	A
	GAG	GGG	GTG	GCG	G
	GAT	GGT	GTT	GCT	T
	GAC	GGC	GTC	GCC	C
T	TAA	TGA	TTA	TCA	A
	TAG	TGG	TTG	TCG	G
	TAT	TGT	TTT	TCT	T
	TAC	TGC	TTC	TCC	C
C	CAA	CGA	CTA	CCA	A
	CAG	CGG	CTG	CCG	G
	CAT	CGT	CTT	CCT	T
	CAC	CGC	CTC	CCC	C

(left axis: first letter of triplet; right axis: third letter of triplet)

(A = adenine, G = guanine, T = thymine, C = cytosine)

Table Ap3.1 DNA's bases grouped into 64 (4 × 4 × 4) triplets

codon	anti-codon	amino acid	codon	anti-codon	amino acid	codon	anti-codon	amino acid	codon	anti-codon	amino acid
UUU	AAA	} phe	UCU	AGA	} ser	UAU	AUA	} tyr	UGU	ACA	} cys
UUC	AAG		UCC	AGG		UAC	AUG		UGC	ACG	
UUA	AAU	} leu	UCA	AGU		UAA	AUU	} •	UGA	ACU	•
UUG	AAC		UCG	AGC		UAG	AUC		UGG	ACC	tryp
CUU	GAA	} leu	CCU	GGA	} pro	CAU	GUA	} his	CGU	GCA	} arg
CUC	GAG		CCC	GGG		CAC	GUG		CGC	GCG	
CUA	GAU		CCA	GGU		CAA	GUU	} glun	CGA	GCU	
CUG	GAC		CCG	GGC		CAG	GUC		CGG	GCC	
AUU	UAA	} ileu	ACU	UGA	} thr	AAU	UUA	} aspn	AGU	UCA	} ser
AUC	UAG		ACC	UGG		AAC	UUG		AGC	UCG	
AUA	UAU		ACA	UGU		AAA	UUU	} lys	AGA	UCU	} arg
AUG	UAC	met	ACG	UGC		AAG	UUC		AGG	UCC	
GUU	CAA	} val	GCU	CGA	} ala	GAU	CUA	} asp	GGU	CCA	} gly
GUC	CAG		GCC	CGG		GAC	CUG		GGC	CCG	
GUA	CAU		GCA	CGU		GAA	CUU	} glu	GGA	CCU	
GUG	CAC		GCG	CGC		GAG	CUC		GGG	CCC	

(U = uracil)

Table Ap3.2 mRNA's codons, tRNA's anticodons and amino acids coded

abbreviation	amino acid
ala	alanine
arg	arginine
asp	aspartic acid
aspn	asparagine
cys	cysteine
glu	glutamic acid
glun	glutamine
gly	glycine
his	histidine
ileu	isoleucine
leu	leucine
lys	lysine
met	methionine
phe	phenylalanine
pro	proline
ser	serine
thr	threonine
tryp	tryptophan
tyr	tyrosine
val	valine
•	chain terminator

Table Ap3.3 Key to amino acids

Answers

1 Cell variety in relation to function

1 A red blood cell's small size and biconcave shape allow it to present a relatively large surface area of cell to the surrounding environment. It also contains haemoglobin which has an affinity for oxygen. These features suit it to its function of uptake and transport of oxygen. (2)

2 **a)** X is a root hair and it absorbs water and mineral salts. (2)

   b) Y consists of sieve tubes and companion cells. Each sieve tube has sieve plates but no nucleus whereas each companion cell lacks sieve plates but has a nucleus. A sieve tube's function is to transport sugars; a companion cell's function is to control a sieve tube. (4)

   c) Yes. A root is arranged into different tissues with each type specialized to perform one (or more) particular function(s). (1)

3 **(i)** They both possess columnar epithelial cells and goblet cells.

   (ii) Unlike the windpipe lining, the epithelium from the intestine has enzyme-producing cells and lacks cilia. (4)

4 **a)** Sperm. (1)

   b) It has a tail for swimming towards an egg. Its rich supply of mitochondria provides plenty of energy for active movement. (2)

5 **a)** By photosynthesis. (1)

   b) **(i)** By using its photoreceptor and eye spot.

   (ii) By using its flagellum. (2)

   c) It lacks the wall typical of plant cells yet it has chloroplasts which animal cells do not possess. (2)

6 In order to survive, a living organism's body must be able to carry out the many functions essential for the maintenance of life. A unicellular organism's body consists of only one cell; a multicellular organism's body consists of several cells.

   The single cell of a unicellular organism (e.g. *Paramecium*) has to be able to carry out all the functions necessary for life. It is versatile and carries out each job in a simple way. For example locomotion is brought about by the beating of its hair-like cilia.

Osmoregulation is effected by contractile vacuoles which remove excess water. The one cell making up *Paramecium's* body is not specialized and is unable to concentrate on one specific job only.

A multicellular organism (e.g. human being) has many cells with which to perform the functions essential for life. These cells are arranged into tissues. A tissue is a group of cells specialised to perform a particular function. Instead of each cell performing all the jobs, the cells in a tissue perform one job in an efficient and advanced way.

This is made possible by the fact that the cells in a tissue are structurally adapted to suit the function that they carry out. For example motor neurons possess motor fibres enabling them to transmit nerve impulses efficiently; red blood cells contain haemoglobin and have a biconcave shape which presents a large absorbing surface to the surrounding environment for the uptake of oxygen.

Other specialised tissues in the human body are responsible for digestion, excretion, protection, movement etc. Such a division of labour results in each tissue performing its own particular function very efficiently. This enables a multicellular organism to operate at a much more advanced level than a unicellular one. (10)

2 Absorption and secretion of materials

1 **a)** Fluid mosaic model. (1)

   b) A = phospholipid; B = protein. (1)

   c) **(i)** C and D.

   (ii) It provides the means by which small molecules can pass through the membrane. (2)

2 **a)** T.   **b)** F. protein.   **c)** T.

   d) F. endocytosis.   **e)** F. against. (5)

3 **a)** **(i)** Oxygen.

   (ii) Carbon dioxide. (2)

   b) Protein (or starch). (1)

4 **a)** **(i)** X = diffusion; Y = active transport.

   (ii) Y.   **(iii)** X. (4)

   b) **(i)** It will decrease.

    (ii) It requires energy from respiration but respiratory enzymes do not work well at low temperatures. (2)

5 a) **(i)** Hypotonic.   **(ii)** Since the ratio of final mass/initial mass for potato cells in 0.2 M sucrose is more than 1, their final mass must have been greater than their initial mass showing that water passed into the cells from the sucrose solution by osmosis. (2)

b) **(i)** Hypertonic.   **(ii)** Since the ratio of final mass/initial mass for potato cells in 0.4 M sucrose is less than 1, their final mass must have been less than their initial mass showing that water passed out of the cells into the sucrose solution by osmosis. (2)

c) 0.26 M. (1)

6 a) **(i)** Solution 5.   **(ii)** Solution 1. (2)

b) 1 = Y, 2 = X, 3 = Y, 4 = Z, 5 = X, 6 = Y. (3)

c) Solution 5. (1)

7 A, E, C, B, F, D. (1)

8 a) Ion concentration is greater in the cell sap than in the pond water. (1)

b) Potassium = 1200: 1 and sodium = 71:1. (1)

c) The data support the selective theory since they show that the plant accumulates different concentrations of different ions (rather than accumulating equal concentrations of them all). (1)

d) **(i)** It disputes the suggestion.   **(ii)** In every case given in this example, ion uptake is from low to high concentration which is the reverse of diffusion. (2)

9 a) **(i)** It results in an increase in rate of ion uptake.

    (ii) Some other factor (e.g. mass of sugar available for energy release) has become limiting.

    (iii) It is an inverse relationship. Since sugar provides the energy for ion uptake, the number of units of sugar present in the cell sap decreases as the number of units of ion absorbed increases. (4)

b) **(i)** Approximately 37°C.   **(ii)** Respiratory enzymes are denatured at higher temperatures. (2)

c) Following drainage, air entered the spaces between the soil particles making oxygen available to the root hairs for respiration and energy release. This allowed increased active uptake of essential ions to occur which led to improved growth of the crop plants. (1)

10 The phospholipid molecules in a plasma membrane arrange themselves into a bilayer with their water-soluble heads to the two outside surfaces and their water-insoluble tails to the inside. This arrangement creates a fluid yet stable boundary around the cell. It regulates the entry and exit of materials by allowing only tiny molecules such as water, oxygen and carbon dioxide to pass through it by diffusion.

    The protein molecules in the cell membrane form a mosaic and vary in size. They perform several functions. Some provide the membrane with structural support. Some contain channels which allow additional small molecules to pass through the membrane; others act as carriers which actively transport molecules and ions across the membrane and require energy to do so.

    Some proteins in the cell membrane are enzymes which promote biochemical reactions. Some proteins act as receptors for hormones arriving at the outer surface of the cell; others act as antigenic markers which identify the cell's tissue type. (10)

11 See pages 11–12. Use should be made of information in Table 2.1. (10)

3 ATP and energy release

1 a) See Figure A3.1 (2)

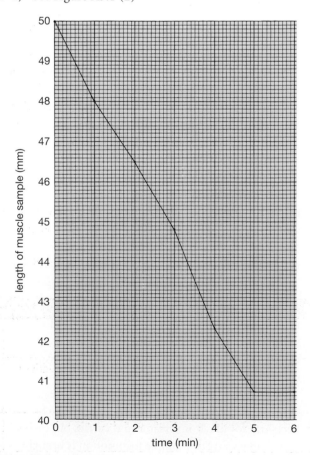

Figure A3.1

b) (i) Between minutes 3 and 4.

(ii) Between minutes 5 and 6. (2)

c) ATP→ADP + Pi + energy

d) Run a control without ATP. (2)

2 a) and **b)** See Figure A3.2. (4)

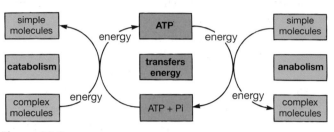

Figure A3.2

3 a) Production of ATP. (1) **b)** ADP. (1)

c) (i) It increased. **(ii)** The process of respiration (which releases CO_2) was no longer limited by shortage of inorganic phosphate to make ATP. (2)

d) Ethanol (formed during fermentation) began to poison some of the yeast cells. (1)

4 a) 56%. (1) **b)** 1267.2 kJ. (1)

c) Synthesis of protein, transmission of nerve impulses. (2)

5 A molecule of adenosine triphosphate (ATP) is produced when a molecule of adenosine diphosphate (ADP) combines with an inorganic phosphate group (Pi). This process is called phosphorylation and it requires energy. It is summarised in the following equation:

ADP + Pi + energy → ATP

When the process of respiration occurs in a living cell, glucose is broken down by a series of enzyme-controlled steps and energy is released. It is this chemical energy that is used to synthesise ATP.

Many molecules of ATP are present in every living cell. ATP can rapidly revert to ADP and Pi and release its energy. ATP is therefore able to make energy available for cellular processes which need energy to drive them, such as synthesis of nucleic acids, muscular contraction and transmission of nerve impulses.

ATP's role in the cell is to act as the means by which energy is transferred from energy-releasing reactions (e.g. respiration) to energy-consuming reactions (e.g. protein synthesis).

A cell's ATP supply does not run out because a constant turnover of molecules occurs within the cell, with some ATP molecules undergoing breakdown and releasing energy while others are being regenerated using energy released during respiration.

In green plant cells, ATP is also regenerated from ADP and Pi and used during photosynthesis. (10)

4 Chemistry of respiration

1 a) Its inner membrane is folded into cristae which present a large surface area upon which chemical processes can take place. (1)

b) A nerve cell is more active and requires more energy (released by aerobic respiration in mitochondria) to function. (1)

2 a) See Table A4.1. (5) **b)** Krebs' cycle. (1)

c) Krebs' cycle and oxidative phosphorylation. (2)

3 The concentrations of lactic acid and carbon dioxide increase. Since both of these are acidic, they depress the pH. (2)

4 a) (i) Oxygen. **(ii)** Anaerobic. (2)

b) (i) Cytochrome system. **(ii)** On cristae in mitochondria. (2)

c) (i) X = Adenosine triphosphate, Y = Adenosine diphosphate, Z = Inorganic phosphate.

(ii) X. (4)

process	site where process occurs in cell	reaction(s) involved in process	products
A glycolysis	[cytoplasm]	splitting of [glucose] into pyruvic acid	ATP and NADH$_2$ and [pyruvic acid]
B [Krebs'] cycle	central matrix of [mitochondrion]	removal of [hydrogen] from carbon compounds by oxidation; release of carbon dioxide	[carbon dioxide] and NADH$_2$
C hydrogen transfer	[cristae] of mitochondrion	release of energy from hydrogen	[ATP] and [water]

Table A4.1

5 a) The clip that is open should be closed and a mass of sodium hydroxide pellets equal to that in tube Y should be put into the bottom of tube X. (2)

b) By using a thermostatically controlled water bath. (1)

c) (i) A will move down; B will move up.

(ii) Carbon dioxide given out by the worm is absorbed by the sodium hydroxide. Oxygen taken in by the worm causes a decrease in volume of the enclosed gas. Level B rises to fill this space causing level A to drop. (2)

d) (i) Their movement will be much slower.

(ii) Respiration will proceed at a reduced rate since respiratory enzymes will be affected by the low temperature. (Unlike a human, a worm cannot maintain a high body temperature in cold conditions.) (2)

e) To keep the experiment fair and prevent the introduction of a further variable factor. (1)

f) To invalidate the argument that the results were simply due to some space inside the respirometer being occupied and not due to the worm respiring. (1)

g) Before beginning the experiment, a syringe would be fitted to tube Y with its needle through the stopper. After the experiment had been running for one hour, the syringe would be used to find out the volume of air needed to return B to its original level. This would give a reading of the worm's oxygen consumption/per hour. (2)

6 a) Q = 3, R = 3, S = 6, T = 1, U = 4, V = 4. (3)

b) Q = pyruvic acid, R = lactic acid, S = citric acid, T = carbon dioxide, W = water. (5)

c) (i) H. **(ii)** A. **(iii)** A.

(iv) B, C, D, E, F and G. (4)

7 See pages 29–31. Use should be made of information in Figure 4.1. (10)

5 Role of photosynthetic pigments

1 a) 0.45. (1)

b) See Figure A5.1. (1)

c) Y = xanthophyll. Rf = 0.7. (1)

d) Using a centrifuge. (1)

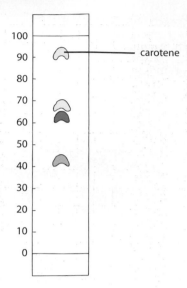

Figure A5.1

e) To produce a concentrated spot of pigments. (1)

f) To prevent the pigments from dissolving in the main bulk of the solvent at the bottom of the tube. (1)

g) Propanone. (1)

2 ◆ Grind up equal masses of both leaf types in sand and propanone.

◆ Filter to give two pigment extracts.

◆ Repeatedly spot and dry each extract onto its own strip of chromatography paper.

◆ Run the two chromatograms in boiling tubes allowing the solvent of petroleum ether and propanone to ascend the papers.

◆ Compare the chromatograms and find out whether the arrangement and number of pigments differ or not. (5)

3 a) Colour/wavelength of light. (1)

b) They are absorbed by the filter. (1)

c) To make the results more reliable. (1)

d) To allow the plant to return to equilibrium. (1)

e) See Figure A5.2. (2)

f) Photosynthetic rate was greatest in blue light. (1)

g) See Table A5.1. (1)

4 a) Chlorophyll a and b. (2)

b) (i) Carotene and xanthophyll.

(ii) Yes. In this case it accounts for the difference

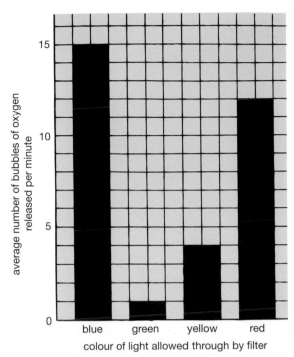

Figure A5.2

unit of length	abbreviation	fraction of one metre
metre	m	1
millimetre	mm	10^{-3}
micrometre	μm	10^{-6}
nanometre	nm	10^{-9}

Table A5.1

between the action spectrum (rate of photosynthesis at different light wavelengths) and absorption spectrum (percentage of light absorbed by chlorophyll at different light wavelengths). (4)

5 a) Blue and red. (2)

b) Most photosynthesis occurs at these regions on the strand of alga. These sites therefore release most oxygen which in turn attracts many aerobic bacteria. (2)

6 A chloroplast is a large discus-shaped organelle present in the cytoplasm of a green plant cell. It is bounded by a double membrane and contains structures called grana and lamellae.

Grana are stacks of flattened sacs containing photosynthetic pigments, normally chlorophyll a and b, carotene and xanthophyll. Grana are green in colour.

Lamellae are tubular extensions for... between grana. They do not contain chlorop... not green.

The colourless background material in a chloroplast is called stroma. It lacks chlorophyll but contains enzymes required to promote the carbon fixation stage of photosynthesis. The stroma may contain starch grains.

Chlorophyll a and b are the principal photosynthetic pigments. When they are extracted and tested using a spectrometer, they are found to absorb light mainly in the red and blue regions of the spectrum of white light. They absorb very little of the green light and give most of it back out again making them appear green to the eye.

The variation in degree of absorption by chlorophyll depending on the wavelength of light can be displayed as a graph called the absorption spectrum of chlorophyll. Its main peaks are in the red and blue regions.

An action spectrum is a graph displaying the results obtained by investigating the effectiveness of each wavelength of light at bringing about the process of photosynthesis in a living green plant.

A close correlation is found to exist between an action spectrum and the absorption spectrum of chlorophyll. This provides strong evidence that chlorophyll traps certain wavelengths of light for photosynthesis. The light energy is then converted into chemical energy to make carbohydrate. (10)

7 a) P = chlorophyll a. A = chlorophyll c, xanthophyll and carotene. (1)

b) Fungus. (1)

c) See Figure A5.3. (3)

d) See x-axis of graph in Figure A5.3. (1)

e) (i) Pigment 2 = phycocyanin.

(ii) Pigment 1 = phycoerythrin. (2)

f) See Figure A5.4. (2)

g) Since phycoerythrin absorbs green light, this extends the range of wavelengths that can be absorbed (beyond those taken in by chlorophyll a) and helps the plant to survive in a dimly lit habitat. (1)

6 Chemistry of photosynthesis

1 a) Green chlorophyll. (2)

b) (i) X = oxygen, Y = $NADPH_2$, Z = ATP.

(ii) Oxygen **(iii)** $NADPH_2$ and ATP. (6)

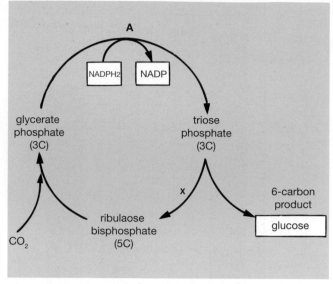

Figure A6.1

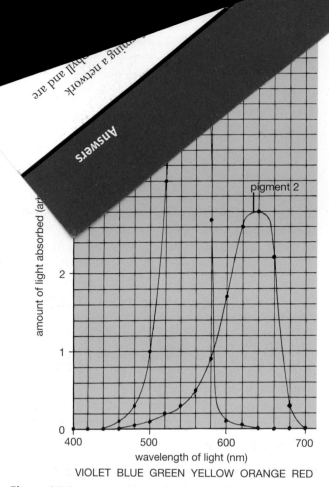

wavelength of light (nm)

VIOLET BLUE GREEN YELLOW ORANGE RED

Figure A5.3

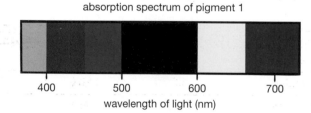

absorption spectrum of pigment 1

wavelength of light (nm)

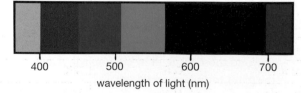

absorption spectrum of pigment 2

wavelength of light (nm)

Figure A5.4

2 ADP + Pi + light energy → ATP. (2)

3 **a)** (2) and **b)** (2) See Figure A6.1.

  c) **(i)** Ribulose bisphosphate. **(ii)** See Figure A6.1. (2)

  d) **(i)** Adenosine triphosphate. **(ii)** See Figure A6.1.

  (iii) To provide the energy needed to drive the reaction. (3)

  e) **(i)** Glycerate phosphate. **(ii)** Because ATP and NADPH$_2$ from the light-dependent reaction would not be available to convert it to triose phosphate. (2)

  f) **(i)** Ribulose bisphosphate.

  (ii) Because no CO_2 would be available for it to combine with and change into glycerate phosphate. (2)

4 **a)** Glycerate phosphate and ribulose bisphosphate. (2)

  b) GP becomes converted to RuBP in light. (2)

  c) In light, both substances remain at a steady level but in darkness the quantity of GP rises and that of RuBP falls. This provides evidence that the conversion occurs in light (with a steady state being maintained as the cycle turns) but that the conversion stops in darkness (with GP accumulating and RuBP disappearing as the cycle becomes disrupted). (2)

5 **a)** Light intensity. (1)

  b) Carbon dioxide concentration. (1)

  c) 20. (1) **d)** Temperature. (1)

6 During the light-dependent stage of photosynthesis in green plant cells, light energy (mainly red and blue) is absorbed by green chlorophyll. This light energy, which becomes converted to chemical energy, is used to split molecules of water into hydrogen and oxygen. The process is called photolysis of water. The oxygen is

released as a by-product and diffuses out of the cell. The hydrogen combines with a hydrogen acceptor called NADP to form $NADPH_2$. Some energy is used to regenerate ATP from ADP + Pi by a process called photophosphorylation.

The light-dependent stage occurs in a chloroplast's grana. It is followed by the second stage of photosynthesis called carbon fixation in the stroma of the chloroplasts. This process takes the form of a cyclical series of enzyme-controlled reactions known as the Calvin cycle.

One of the chemicals in the cycle is ribulose bisphosphate (RuBP), the carbon dioxide acceptor. A molecule of RuBP combines with a molecule of CO_2 to form 3-carbon glycerate phosphate (GP). GP then becomes converted to 3-carbon triose phosphate (TP) if a supply of $NADPH_2$ (to provide hydrogen) and ATP (to provide energy) are available from the light-dependent stage.

As the cycle turns, some molecules of TP are converted back to RuBP (using energy from ATP) ready to repeat the cycle. Other molecules of TP are released from the cycle and combine together to form molecules of glucose which may, in turn, become converted to starch or cellulose. (10)

7 DNA and its replication

1 a) Deoxyribose sugar and phosphate. (2)

  b) See Figure A7.1. (1)

Figure A7.1

2 30%. (1)

3 c, a, d, b, f, e. (1)

4 a) A human red blood cell lacks a nucleus. (1)

  b) That of a chicken does have a nucleus but that of a cow does not. (1)

  c) For each type of animal, the DNA content of its sperm is half that of its kidney cells. (1)

  d) DNA content is at its highest following DNA replication during interphase. It returns to normal when the growing cell undergoes mitosis and its nucleus divides into two. (2)

5 a) 1 = chromosome, 2 = DNA, 3 = base. (3)

  b) 1. (1)   c) 3. (1)

  d) It is too short. (1)

6 a) Unwinding of double helix. (1)

  b) It must open up by weak hydrogen bonds between base pairs breaking. (1)

  c) (i) 7, 11, 28, 31.   (ii) 4, 20.   (iii) Four. (7)

  d) D. (1)   e) E. (1)

7 See pages 54–55. Use should be made of information in Figure 7.6. (10)

8 RNA and protein synthesis

1 a) E, C, A, G, B, D, F. (1)   b) See Figure A8.1. (2)

Figure A8.1

2 a) 1 = C, 2 = T, 3 = T, 4 = A, 5 = U, 6 − A, 7 = G, 8 = C, 9 = G. (2)

  b) P = transcription and release of mRNA. Q = translation of mRNA into protein. (1)

  c) See Table A8.1. (2)   d) CAA. (1)

amino acid	codon	anticodon
alanine	GCG	CGC
arginine	CGC	GCG
cysteine	UGU	ACA
glutamic acid	GAA	CUU
glutamine	CAA	GUU
glycine	GGC	CCG
isoleucine	AUA	UAU
leucine	CUU	GAA
proline	CCG	GGC
threonine	ACA	UGU
tyrosine	UAU	AUA
valine	GUU	CAA

Table A8.1

  e) U = proline, V = glutamine, W = glutamic acid, X = cysteine, Y = arginine, Z = isoleucine. (2)

f) **(i)** ACACUUGCGGGC.

(ii) TGTGAACGCCCG. (2)

3 See Table A8.2. (3)

stage of synthesis	site
formation of mRNA	nucleus
collection of amino acid by tRNA	cytoplasm
formation of codon-anticodon links	ribosome (in cytoplasm)

Table A8.2

4 a) The increase in relative numbers of ribosomes in cells during day 1 to 5 indicates that this is the period of most rapid protein synthesis and growth of the new leaf. After day 5 growth slows down and eventually comes to a halt at day 11 when the leaf has reached its full size. (2)

b) A basic number of ribosomes will always be needed by a cell of a fully grown leaf to make proteins such as enzymes essential for biochemical pathways (e.g. photosynthesis). (1)

5 a) A = cell membrane, B = vesicle,
C = Golgi apparatus, D = pore,
E = nuclear membrane, F = chromosome,
G = ribosome, H = endoplasmic reticulum. (4)

b) Protein. (1)   **c)** G. (1)

6 This is because protein structure is determined by the base sequence present in DNA which varies from person to person. (2)

7 Transcription
The sequence of bases along part of a DNA strand (one gene) contains the instructions for determining the sequence of amino acids in a particular protein. This information is carried from the nucleus to ribosomes in the cytoplasm by messenger RNA (mRNA).

Transcription is the process by which mRNA is formed from appropriate part of a DNA strand. The DNA molecule uncoils and part of it opens up by weak bonds between its bases breaking.

Free RNA nucleotides become attached to their complementary DNA nucleotides on one of the open strands by their bases. Adenine on DNA, for example, pairs with uracil on RNA; guanine pairs with cytosine.

In the presence of the enzyme RNA polymerase and ATP (for energy), strong chemical bonds form between adjacent nucleotides giving the single-stranded mRNA its sugar-phosphate 'backbone'. The weak bonds between DNA and mRNA break and mRNA is released.

Translation
Molecules of transfer RNA (tRNA) are present in the cytoplasm. Many types exist and each has a single triplet of bases exposed. This is called an anticodon and it corresponds to an amino acid. Each type of amino acid molecule becomes attached to its tRNA and is transported to a ribosome.

The ribosome becomes associated with one end of the mRNA molecule and two molecules of tRNA enter the ribosome and become attached by their anticodons to the complementary codons (triplets of bases) on the mRNA. A strong peptide bond forms between the two adjacent amino acid molecules.

As the ribosome moves along the mRNA, the process continues with further tRNA molecules entering in turn and their amino acids being added to the growing polypeptide chain. By this means mRNA's genetic code of triplets (codons) is translated into protein. The tRNA and mRNA molecules are normally reused several times. (10)

9 Functional variety of proteins

1 See Table A9.1. (5)   **2** See Figure A9.1. (3)

name of protein	type of protein (G or C)	role of this protein
antibody	G	to defend the body against antigens
catalase	G	to catalyse the breakdown of hydrogen peroxide to water and oxygen
cytochrome	C	to transfer hydrogen during aerobic respiration allowing the formation of ATP
haemoglobin	C	to transport oxygen round the body
insulin	G	to promote the conversion of glucose to glycogen

Table A9.1

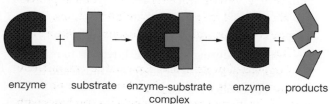

enzyme substrate enzyme-substrate complex enzyme products

Figure A9.1

3 a) (i) C. **(ii)** A. **(iii)** B. (3)

  b) (i) B.

    (ii) Collagen consists of strong fibres as shown by B.

    (iii) Its fibres are strong and inelastic which allows them to function perfectly as part of a bone or tendon. (3)

  c) (i) C. **(ii)** It has an active site allowing it to fit its substrate like a key fits a lock. (2)

4 a) Casein contains them all. Group 2 rats gained weight throughout the experiment. Zein lacks two essential amino acids. Group 1 rats lost weight throughout the experiment. (4)

  b) (i) Zein. **(ii)** Their diet could have been changed to casein or to zein supplemented with the two essential amino acids that it lacks. (3)

  c) 35 g. (1) **d)** 20%. (1)

5 Each molecule of protein consists of many amino acids linked together by strong peptide bonds into one or more polypeptide chains. Further weak bonds between certain amino acids in a polypeptide chain make it coil into a spiral.

When several polypeptide chains become linked together by cross-bridges, this results in the formation of a fibrous protein. If several polypeptides become folded together and held by bonds in a roughly spherical shape, the protein formed is described as globular. When a globular protein combines permanently with a non-protein chemical, the molecule formed is called a conjugated protein.

Many types of fibrous protein exist and each possesses a structure suited to its function. Collagen, for example, is strong and inelastic to serve as a component of bones.

Globular proteins play a variety of roles. Some are hormones (chemical messengers). In animals these are transported in the bloodstream to target tissues where they exert an effect. For example, in humans, insulin made in the pancreas promotes the conversion of excess glucose to glycogen in the liver.

Some globular proteins support the cell membrane; others act as antibodies and defend the body against antigens(e.g. viruses). All enzymes are globular proteins and are required to speed up chemical reactions. Each enzyme combines with its specific substrate like a lock and key.

Conjugated proteins also carry out several different functions. Mucus, for example, lubricates and protects parts of the human body. Haemoglobin combines with oxygen in the lungs and delivers it to respiring cells.

The functioning of a healthy living organism is therefore dependent upon many different types of protein each carrying out its particular function. (10)

10 Viruses

1 a) They are able to multiply. (1)

  b) It lacks sub-cellular structures such as ribosomes, nucleus, mitochondria and cell membrane. (1)

2 C, F, B, G, H, D, A, E. (1)

3 a) See Table A10.1. (4)

life form	nm	µm	mm	m
human red blood cell	10 000	10	1×10^{-2}	1×10^{-5}
Escherichia coli	3000	3	3×10^{-3}	3×10^{-6}
tobacco mosaic virus	250	0.25	2.5×10^{-4}	2.5×10^{-7}
bacteriophage	225	0.225	2.25×10^{-4}	2.25×10^{-7}
polio virus	30	0.03	3×10^{-5}	3×10^{-8}

Table A10.1

  b) (i) 100 times. **(ii)** 40 times. (2)

4 a) No. Retroviruses contain RNA not DNA. (1)

  b) X = viral DNA. Y = viral RNA. Z = protein. (3)

  c) (i) Budding. **(ii)** Death. (2)

  d) Figure 10.9 shows the production of only one viral particle. In reality, many copies of the virus would be produced. (1)

5 a) DNA, helical, enveloped. (1)

  b) (i) X = human immunodeficiency virus (HIV), Y = tobacco mosaic virus.

    (ii) Reverse transcriptase. (3)

  c) Herpes. (1)

  d)

1	RNA	… go to 2
	DNA	… go to 5
2	helical	… go to 3
	polyhedral	… go to 4
3	naked	**tobacco mosaic virus**
	enveloped	**influenza virus**
4	naked	**bushy stunt virus**
	enveloped	**human immunodeficiency virus**

5   helical   . . . go to 6
      polyhedral   . . . go to 7
6   naked   **coliphage virus**
      enveloped   **smallpox virus**
7   naked   **bacteriophage virus**
      enveloped   **herpes** (4)

6 See pages 73–74. Use should be made of information in Figure 10.5. (10)

11 Cellular defence mechanisms

1 **a)** A = lymphocyte, B = phagocyte. (1)

   b) 1 = nuclear membrane, 2 = ribosome,
3 = endoplasmic reticulum, 4 = Golgi apparatus,
5 = bacterium, 6 = vacuole, 7 = lysosome,
8 = vesicle. (4)

   c) 7 contains digestive enzymes, 8 contains antibodies. (2)

   d) The enzymes are kept enclosed inside lysosomes and vacuoles. (1)

2 **a)** W = 2, X = 1, Y = 4, Z = 3. (2)

   b) **(i)** Method 4.

      (ii) The body responds by making antibodies and remembers how to do so in the future. Method 3 leaves the body unprotected when the horse's antibodies disappear. (2)

3 **a)** In response to the antigen (even in a damaged state) lymphocytes make antibodies which give immunity. In addition some of these lymphocytes act as memory cells allowing a rapid response to the antigen if invasion occurs in the future. (2)

   b) Artificially. (1)

4 **a)** **(i)** IgG.   **(ii)** The baby's own immune system makes antibodies after a few months. (2)

   b) **(i)** IgM.   **(ii)** IgE. (2)   **c)** 1.4–4.0 mg/ml. (1)

   d) The concentration of IgM increases before that of IgG in both the primary and the secondary response. (1)

   e) During the primary response, the concentration of IgG takes a longer time to begin rising than during the secondary response. The highest concentration of IgG reached during the primary response is less than that reached during the secondary response. The concentration of IgG decreases after reaching its maximum level in the primary response but remains level at its highest concentration in the secondary response. (3)

   f) Since the response to the second injection was faster than that to the first, this suggests that certain lymphocytes already 'knew what to do' from the previous time. (1)

5 See pages 78–79. Use should be made of information in Figure 11.3. (10)

6 **a)** See Figure A11.1. (2)

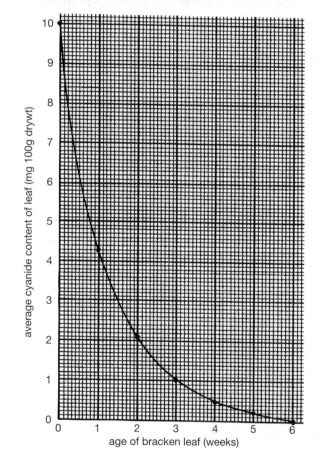

Figure A11.1

   b) The younger the leaf, the higher its cyanide content. (1)

   c) The presence of a higher concentration of cyanide in young leaves protects them against hungry animals at a time when the leaves are delicate and especially vulnerable to attack. (1)

7 **a)** A disease-causing organism. (1)

   b) **(i)** Fungus.

      (ii) By secreting enzymes which digest away the cell walls of the host. (2)

c) To secrete resin. (1)

d) It has probably been halted in its tracks and isolated by sticky acidic resin. (1)

e) It protects young leaves inside from attack by insects and pathogenic micro-organisms. (1)

8 a) Willow tree. (1) **b)** Larva. (1)

c) It is unlikely to suffer further damage to its tissues by the invading insect. (1)

9 See pages 83–86. Answer should take the form of a summary of the main points. (10)

12 Meiosis

1 a) 4. **b)** 46. **c)** 33. **d)** 16. **e)** 14. **f)** 38. (6)

2 B, D, A, C. (1)

3 a) 12. **b)** 6. **c)** 6. **d)** 3. **e)** 3. (5)

   f) **(i)** See Figure A12.1.

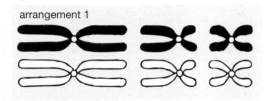
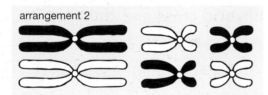

Figure A12.1

      (ii) See Figure A12.2. (4)

4 a) 2. (1)

   b) See Figure A12.3. (2)

5 See Figure A12.4. (2)

6 See Table A12.1. (7)

13 Monohybrid cross

1 a) **cross**   ss × Ss

      gametes   all s ↓ S and s

      F_1 ss and Ss (2)

gametes from arrangement 1

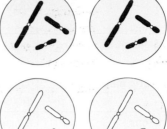

gametes from arrangement 2

Figure A12.2

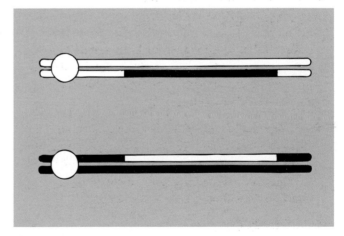

Figure A12.3

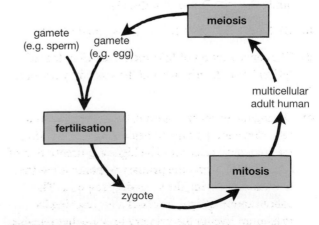

Figure A12.4

	mitosis	meiosis
site of division	occurs all over body of growing animal	occurs only in gamete mother cells in sex organs
pairing and movement of chromosomes	homologous chromosomes do not form pairs; chromosomes line up singly on equator	homologous chromosomes form pairs; chromosomes line up in pairs on equator
exchange of genetic material	chiasmata not formed and no crossing over occurs	chiasmata may be formed and crossing over may occur
number of divisions	single division of nucleus	double division of nucleus
number and type of cells produced	following cell division, two genetically identical daughter cells formed	following cell division, four genetically different gametes formed
effect on chromosome number	chromosome number unaltered	chromosome number halved
effect on variation	does not increase variation within a population	increases variation by providing opportunity for independent assortment and crossing over to occur

Table A12.1

b) 50% (1)

2 a) 6. (1)   b) W. (1)   c) B 1 in 2. (1)

3 The rabbit of unknown genotype would be crossed with a rabbit possessing uniform coat. If the unknown rabbit were homozygous, the following results would be obtained:

cross   SS × ss

gametes   all S ↓ all s

F$_1$ all Ss (all spotted)

If the unknown rabbit were heterozygous, the following results would be obtained:

cross   Ss × ss

gametes   S and s ↓ all s

F$_1$ Ss and ss (1 spotted : 1 uniform) (4)

4 b) Pink × pink.   c) Red × red.   d) Ivory × ivory.

e) Red × pink.   f) Ivory × pink. (5)

5 a) and b)

(i)   cross   RR × RR

gametes   all R ↓ all R

F$_1$ all RR (all red, no ratio)

(ii)   cross   RW × RW

gametes   R and W ↓ R and W

F$_1$ RR, RW, RW, WW (1 red : 2 roan : 1 white)

(iii) cross   RW × WW

gametes   R and W ↓ all W

F$_1$ RW and WW (1 roan : 1 white)

(iv)   cross   RR × WW

gametes   all R ↓ all W

F$_1$ all RW (all roan, no ratio) (12)

14 Dihybrid cross and linkage

1 a) and b) See Figure A14.1. (8)

2 a) ggll. (1)   b) gl. (1)   c) GL, gl, Gl and gl. (4)

d) Second parent = GgLl, B = Ggll, C = GGLl and D = GgLl. (4)

3 a) 1:1. (1)   b) BbYy. (1)

c) Totals = 45, 15, 15 and 5. Ratio = 9:3:3:1. (1)

d) (i) It would have indicated that they were located on the same chromosome.

(ii) They are located on different chromosomes. (2)

4 i) Genes R, S and T which have their loci on the same chromosome, are said to be linked genes.

ii) Crossing over, the process which can separate the alleles of such genes, occurs at points called chiasmata.

iii) There is a greater chance of crossing over occurring between genes S and T than between genes R and S.

Let round = R and pear shape = r
Let round = C and yellow = c

cross 1 RRCC x rrcc

gametes all RC x all rc

F₁ all RrCc

cross 2 RRCC x rrcc

gametes RC,Rc,rC,rc RC,Rc,rC,rc

male gametes

	RC	Rc	rC	rc
RC	<u>RRCC</u>	RRCc	RrCC	RrCc
Rc	RRcC	<u>RRcc</u>	RrcC	Rrcc
rC	rRCC	rRCc	<u>rrCC</u>	rrCc
rc	rRcC	rRcc	rrcC	<u>rrcc</u>

(female gametes)

F₂ phenotypic ratio = 9 round red:
                        3 pear-shaped red:
                        3 round yellow:
                        1 pear-shaped yellow

Figure A14.1

iv) If a cross-over occurred between loci S and T then the recombinant gametes formed would be <u>rSt</u> and <u>RsT</u>, and the parental gametes would be <u>rST</u> and <u>Rst</u>. (8)

5 a) (i) BbMm, bbmm. (ii) 1:1.

(iii) Brown coat, normal movement and albino coat, waltzer movement. (iv) 1:1. (4)

b) Bbmm and bbMm. (2)

6 a) and b)

cross 1 wwff × WWFF

gametes all wf ↓ all WF

F₁ all WwFf

cross 2 WwFf × wwff

gametes WF, Wf, wF, wf ↓ all wf

F₂

WwFf,	Wwff,	wwFf,	wwff
81	14	16	89
white,	white,	brown,	brown,
full	shrunk	full	shrunk
	(R)	(R)	

(R = recombinant) (5)

c) (i) 15%. (ii) Because the two genes are on the same chromosome and show linkage. (2)

d) 15 units. (1)

7 a) See Figure A14.2. (4) b) 38%. (1)

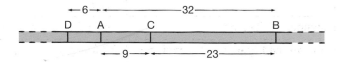

Figure A14.2

8 ◆ Obtain male grey-bodied, straight-winged and virgin female yellow-bodied, curved-winged fruit flies.

◆ Using ether, anaesthetise both strains of flies and put a few males and a few females into several culture tubes to allow for non-recovery from the ether and to increase the reliability of the results.

◆ Incubate the tubes at a suitable warm temperature and remove the parents after the eggs have been laid so that they will not be confused with the next generation.

◆ Once the F₁ adults emerge, set up a testcross between several F₁ males and virgin female yellow-bodied, curved-winged flies using the above procedure.

◆ When the F₂ generation emerge, anaesthetise them and classify them according to their phenotypes.

◆ Pool the results with those of other groups or repeat the experiment several times to further increase the reliability of the results.

◆ Calculate the percentage number of recombinants in the F₂ generation. If it is around 50% then the two genes are on different chromosomes. If it is significantly less than 50% then the two genes are linked (i.e. located on the same chromosome). (10)

15 Sex linkage

1 a) **original cross** $X^rY \times X^RX^R$

gametes X^r and Y ↓ all X^R

F₁ genotypes X^RX^r and X^RY

F₁ phenotypes 1 red-eyed : 1 red-eyed

and sex ratio female male (4)

b) second cross $X^R X^r \times X^R Y$

   gametes X^R and $X^r \downarrow X^R$ and Y

   F_2 genotypes $X^R X^R, X^R X^r$ $X^R Y,$ $X^r Y$

   F_2 phenotypes   2   :   1   :   1

   and sex ratio   red-eyed   red-eyed   white-eyed

                   females   male   male (4)

c) The phenotypic ratio would have been 3 red-eyed: 1 white-eyed. Within each eye colour group, 50% would have been male and 50% would have been female. (1)

2 See Figure A15.1. (4)

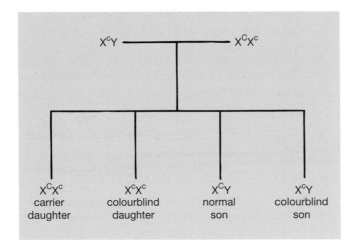

$X^c Y$ ——————— $X^C X^c$

$X^C X^c$ $X^c X^c$ $X^C Y$ $X^c Y$
carrier   colourblind   normal   colourblind
daughter   daughter   son   son

Figure A15.1

3 a) $A = X^H Y, B = X^H X^H, C = X^h Y, D = X^H X^h$. (2)

b) Male parent $= X^H Y$ and is normal. Female parent is $X^H Y^h$ and is normal (but a carrier). (2)

c) None. (1)

4 a) F. **b)** T. **c)** T. **d)** F. **e)** T. (5)

5 a) It is impossible for a male to be tortoiseshell. Since the Y chromosome carries no allele for coat colour, the heterozygous genotype cannot arise in a male. (1)

b) possible cross 1 $X^B X^B$ $X^G Y$

                               $\downarrow$

   offspring $X^B X^G$ (and $X^B Y$)

   possible cross 2 $X^G X^G$ $X^B Y$

                               $\downarrow$

   offspring $X^G X^B$ (and $X^G Y$) (4)

6 a) All the F_1 offspring would be white. A cross involving a non sex-linked gene would give the same results. (2)

b) cross $X^w X^w \times X^W Y$

   gametes   all $X^w \downarrow X^W$ and Y

   F_1 $X^W X^w$ and $X^w Y$

                     1   :   1

                 white male   red female

A ratio of 1 white male: 1 red female in the F_1 is different from the all white F_1 that would result from a cross involving a non sex-linked gene. Such a 1:1 ratio for both sex and phenotype would therefore verify that the gene for plumage colour is sex-lined. (3)

7 See page 117. Use should be made of information in Figure 15.3. (10)

16 Mutation

1 a) Deletion. **b)** Substitution. **c)** Inversion.
   d) Insertion. (4)

2 a) Cell 1 = deletion. Cell 2 = duplication. (2)

b) Deletion. (1)

3 a) **(i)** 0.5 in 1000. **(ii)** 1 in 1000. **(iii)** 11 in 1000. (3)

b) Egg mother cells of older women seem to be more prone to non-disjunction of chromosomes during meiosis. (1)

4 a) 22,   22 + XX,   22 + Y,   22 + Y. (2)

b) 44 + X,   44 + XXX,   44 + XY,   44 + XY. (2)

c) 25%. (1)

5 See pages 121–126. Answer should take the form of a summary of the main points. Use should be made of information in Figures 16.2, 16.5 and 16.9–16.12. (10)

6 500 per million. (1)

7 a) aa $\times$ aa. (1) **b)** 1 in 71429 chance. (1)

8 A gene mutation involves a change in one or more nucleotides in a DNA strand. There are four main types of gene mutation.

During substitution, one nucleotide drops out of the DNA chain and is replaced by another possessing a different base. This may result in the production of a triplet of bases (a codon) which corresponds to a different amino acid from before. Following transcription and translation a protein molecule will be formed with one 'wrong' amino acid in its chain.

During inversion, two or more nucleotides become disconnected from the chain and then join up again but in reverse order. This may result in the production of one (or two) different codons from normal and ultimately a protein with one (or two) 'wrong' amino acids in its chain.

Substitution and inversion are called point mutations since they bring about only minor changes (e.g. a change in one amino acid in the chain). Often the organism is hardly affected.

During deletion, a nucleotide is lost permanently from the DNA chain. During insertion, a nucleotide is added to the DNA chain.

These two changes are called frameshift mutations. Each leads to a major change since it causes a large portion of DNA to be misread (from where the mutation occurs, all the way along to the end of the gene). This involves many different codons and an equally large number of 'wrong' amino acids.

The protein formed following deletion or insertion is so different from the normal protein that it is non-functional. Since most proteins are essential to an organism, a mutant lacking a functional version of one of its proteins often fails to survive and the mutation is described as being lethal. (10)

17 Natural selection

1 Despite <u>over-production</u> of offspring, a population explosion does not occur because the environment does not provide enough food. <u>Competition</u> for this (and other) limited resources occurs. Since <u>variation</u> exists amongst the members of the population, <u>natural selection</u> will favour those that are better adapted in some way (e.g. faster, stronger, etc.). They will survive and the weaker animals will lose out in the struggle and die. (4)

2 a) The black one. (1)

   b) See Figure A17.1. The ratio = 1:1. (2)

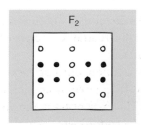

Figure A17.1

c) See Figure A17.2. **(i)** F3. **(ii)** F5. (2)

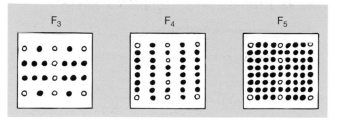

Figure A17.2

3 a) An inverse relationship. The percentage number of melanic moths decreases as the number of lichen species increases. (1)

   b) Melanic (dark) moths lose their selective advantage in clean places outside the city where they are no longer camouflaged and therefore easily spotted by predators. (2)

   c) The line graph for pale moths would have shown a trend similar to that of the lichens. (1)

   d) Polluted areas are becoming cleaner in response to the Clean Air Acts. As a result the melanic moths are losing their selective advantage. (1)

4 a) D. (1)   b) Resistance to the antibiotic. (1)

   c) It developed its resistance before coming in contact with the antibiotic. (1)

5 a) Bacteria resistant to erythromycin enjoyed a selective advantage and became more common, so fewer and fewer cases responded to treatment as the years went by. (1)

   b) Very few or no cases would have been successfully treated by erythromycin. (1)

   c) **(i)** They prescribed a different antibiotic.

      (ii) Some mutant bacteria were already resistant to this second antibiotic. (2)

6 a) They were resistant. (1)   b) When the <u>pesticide</u> was first used it killed the <u>susceptible majority</u> of insect pests. Any insects that survived by <u>natural selection</u> belonged to a <u>resistant strain</u>. In the absence of <u>competition</u>, these resistant insects increased in number, unaffected by DDT. (5)

7 a) It kills or prevents growth of many bacteria in the mouth thereby reducing tooth decay. (1)

   b) Bacteria resistant to the antibiotic will be selected. They will multiply, fail to respond to the antibiotic and cause tooth decay. (1)

8 a) (i) Early dog violet. **(ii)** Heath dog violet. (1)

b) (i) Heath dog violet. **(ii)** Neither one. (2)

c) (i) Damp, neutral.

(ii) Highly competitive sturdy grasses keep them out. (3)

9 See pages 134–135. Use should be made of information in Figures 17.3 and 17.4. (10)

10 a) Consumption of warfarin interferes with the blood-clotting mechanism and the rat bleeds to death if it becomes cut or injured. (1)

b) It thins their blood and prevents the formation of unwanted internal clots. (1)

c) Mutation. (1)

d) Strictly speaking it is incompletely dominant. Heterozygote W^sW^r, which is resistant yet only needs some extra vitamin K in its diet, is phenotypically intermediate to the two homozygous forms. (2)

e) It has been favoured by natural selection since members of the sensitive strain normally die after eating the warfarin. (1)

f) Extinction. (1)

g) Very few W^rW^r will be fortunate enough to find a continuous supply of food containing the large amount of vitamin K that they need. (1)

h) (i) $W^sW^r \times W^sW^r$

$\downarrow$

$W^sW^s, W^sW^r, W^sW^r, W^rW^r$

(ii) Yes, Since W^sW^r has the selective advantage (not needing unrealistic amounts of vitamin K yet still being resistant to warfarin), W^s will still keep occurring and surviving in the heterozygotes. (2)

18 Speciation

1 No because offspring resulting from the cross are not fertile. (2)

2 D, A, C, E, B. (1)

3 a) One. They share the same gene pool. (2)

b) Reproductive. (1)

c) Two. The large dogs would be unable to successfully breed with the small dogs. (2)

4 a) Some of the mutations unique to its isolated gene pool have been selected making it different from the mainland species. (2)

b) Continued isolation for a very long time. (1)

5 a) They had evolved elsewhere and reached Australia by boat. (2)

b) They had evolved elsewhere and sailed or flew to New Zealand. (2)

6 See pages 143–144. Use should be made of information in Figure 18.1. (10)

7 a) Quantitative = body length (or tail length).

Qualitative = ventral colour (or presence or absence of pectoral spot). (2)

b) They are able to interbreed and produce fertile offspring. (1)

c) (i) The genetic distance between this type of mouse and the one in Norway is much less than that between this mouse and the one on the Scottish mainland suggesting that the ancestor was Norse.

(ii) It lends support to Berry. After the Ice Age, when *Apodemus* returned to the Scottish mainland (probably from southern parts of UK), a barrier of water separated the mainland from the Shetlands. So the route from Norway seems more likely. (3)

d) A common ancestor could have spread throughout the islands along with the human inhabitants in olden times when conditions were primitive. Different groups of mice have been kept isolated by water barriers ever since and have started to take different courses of evolution in isolation (but not enough time has elapsed yet for complete speciation to have occurred). (2)

19 Adaptive radiation

1 a) It shows the evolution of a group of related organisms along several different lines by becoming adapted to make use of a wide variety of foodstuffs. It therefore illustrates an example of adaptive radiation. (2)

b) The variety of foods on the islands was unavailable on the mainland. (1)

c) Water. (1)

d) (i) Narrow. **(ii)** Wide. (2)

e) **(i)** Two. **(ii)** Three. **(iii)** Two. (1)

f) It is able to make optimum use of a resource in the absence of competition. (1)

2 a) (i) 18 cubic units.

(ii) Desert dweller = 78 square units. Eskimo = 72 square units.

(iii) The desert dweller's long slim body shape presents a larger surface area from which excess heat can be lost thus keeping her/him cool in a hot climate. The Eskimo's short dumpy body shape presents a smaller surface area from which heat can be lost thus helping to conserve heat and keep her/him warm in a cold climate. (4)

b) (i) Adaptive radiation means the evolution of a group of related organisms along different lines by adapting to different environments. In the human example each group has taken its own course of evolution from a common ancestor. Natural selection has tended to favour the tallest slimmest people in the hot climate and the shortest most dumpy in the cold one.

(ii) They are no longer separated by effective barriers. (3)

3 a) See Table A19.1. (2)

animal	description of canine tooth	function
walrus	tusk	digging up shellfish
tiger	fang-like	stabbing prey
human	pointed crown	biting and tearing

Table A19.1

b) Divergent. They have the same underlying structure and have evolved from a common ancestor but have become adapted to suit different functions. (2)

4 They do not share a closely related common ancestor. They have arisen independently and then, by coincidence, have taken similar courses of evolution by becoming adapted to suit equivalent ecological niches on two separate continents. (2)

5 Adaptive radiation is the evolution, over a very long period of time, of a group of related organisms along several different lines. During this process the members of each line become adapted to life in one of a variety of environments. This results in the different species that evolve each being able to make optimum use of the resources available in a particular environment in the absence of competition from other related species.

Darwin's finches on the Galapagos Islands provide a good example of adaptive radiation. It is thought that they evolved from a common ancestor on the mainland of South a America. This one species of finch occupies an ecological niche as a ground-dwelling seed-eater. On the mainland other ecological niches (e.g. tree-dwelling insect-eater) are already occupied by other birds.

It is thought that a group of mainland finches were carried out to the islands by very high winds many years ago. These birds found themselves in a variety of environments with ecological niches unoccupied by other birds.

Some finches landed on islands where the dominant vegetation was cactus, others where it was forest and so on. Speciation took place with the water between the islands acting as one of the isolating mechanisms. Mutations occurred within each gene pool but were not shared by those on other islands. Any mutant alleles that made the birds' beaks better suited to dealing with the locally available food enjoyed a selective advantage.

Natural selection occurred and the beneficial characteristics were passed on from generation to generation. Over a very long period of time, this resulted in the evolution along different lines of distinct groups, some with short, wide beaks perfectly suited to breaking open hard seeds, other with long, pointed beaks suited to eating insects and so on. By this means adaptive radiation led to the evolution of many different species of finch. (10)

6 a) 3. (1)

b) Isabella (or Española). (1)

c) Santa Cruz. (1)

d) (i) The percentage number of species only found on a particular island increases as the island's location becomes more distant from the centre of the group.

(ii) Isabella and Santa Fe have a similar percentage number of finches peculiar to them yet they differ enormously in size.

(iii) The more isolated islands have had less exchange with other gene pools and so a relatively higher number of their finches have taken their own separate course of evolution. (4)

e) X = large seed-eating from Pinta. Y = large cactus-eating from Pinta. Z = medium seed-eating from Culpepper. (1)

f) (i) Type X (large seed-eating).

(ii) One type might become extinct (or the two species might survive as small populations living in fierce competition or …?). (2)

20 Extinction and conservation

1 a) (i) C (ii) D. (iii) B. (iv) E. (v) A. (1)

b) D. (1) c) C. (1) d) B. (1)

2 a) Proterozoic. (1)

b) (i) Algae. (ii) No. (2)

c) (i) Devonian.

(ii) One which possesses transport tissue made of vessels (e.g. xylem). (2)

d) (i) Seed ferns and primitive conifers.

(ii) Seed ferns. (3)

e) (i) Tree species.

(ii) The clearing of tropical rain forest is having a serious effect. (2)

3 a) (i) Trilobites. (ii) Fish. (2)

b) (i) Corals. (ii) Fish. (iii) Trilobites. (3)

c) (i) 230 million years ago.

(ii) 220 million years ago. (2)

d) (i) 65 million years ago.

(ii) Number of species of mammals increased.

(iii) The evolution of milk to feed their young and improved level of parental care have increased their chance of survival and contributed to the rise in their numbers. (3)

4 It is possible that as the Earth became colder, all eggs were incubated at a lower temperature which produced one but not both sexes. This made reproduction impossible. (2)

5 The artist is trying to convey the message that the consumption of hamburgers can lead to the clearing and loss of rainforest because in some countries the forest is cleared to create grassland to feed beef cattle. (2)

6 a) 4000 times. (1)

b) (i) 100 000.

(ii) Overhunting and destruction of habitat by humans. (2)

7 See pages 157–158. Use should be made of information in Figure 20.1. (10)

8 a) 1 = B, 2 = E, 3 = D, 4 = A, 5 = C. (1)

b) This is a matter of opinion. (2)

9 See pages 158–160. Use should be made of information in Figure 20.4. (10)

21 Artificial selection

1 No. Eventually all the alleles for the desirable features will have been accumulated and the strain will not improve any further. (2)

2 If it possesses several desirable characteristics, a crop of uniform plants is of great use of humans. However the crop could eventually suffer inbreeding depression making it economically inferior. (2)

3 The F_1 generation shows hybrid vigour. (2)

4 a) F_1. (1)

b) F_1 is taller, has longer mean ear length and has higher mean yield than either parent. (3)

c) P_1. (1)

d) Inbreeding depression means the decline of a certain characteristic as a result of continuous selfing as shown by plant height in this example. (1)

e) (i) The yield has increased dramatically and less land is needed.

(ii) By doing so they are guaranteed a bumper F_1 crop whereas grains kept back by themselves would give poorer and poorer plants. (3)

5 a) Dominant. The heterozygote has bar eyes. (1)

b) Duplication. (1)

c) 16. (1)

6 a) 4, 7, 2, 5, 1, 6, 3. (1)

b) (i) Endonuclease.

(ii) So that they sticky ends will be complementary allowing the plant gene to be inserted into the bacterial plasmid. (2)

c) Ligase. (1)

d) They hope to transfer these genes for desirable characteristics to rape and cabbage crop plants. (2)

7 a) See Figure A21.1 (2)

b) X = genetically modified, Y = control. (1)

c) (i) 7. (ii) 70. (1)

d) 90%. (1) e) The tomato would fail to ripen. (1)

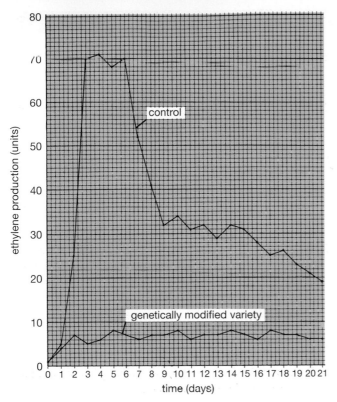

Figure A21.1

f) Genetically modified tomatoes have a longer shelf life. (1)

8 ◆ Obtain DNA from a normal human and cut the DNA containing the required gene into fragments using endonuclease.

◆ Extract plasmids carrying the gene for resistance to an antibiotic (say X) from a species of bacterium.

◆ Cleave open the plasmids using endonuclease.

◆ Seal DNA fragments into the plasmids using ligase.

◆ Attempt to insert the plasmids into bacterial host cells.

◆ Grow the bacteria on medium containing antibiotic X to select those bacterial cells that have taken up the plasmid.

◆ Test samples of DNA from the surviving colonies with a gene probe from the somatotrophin gene.

◆ Identify which bacterial colony has cells containing the somatotrophin gene.

◆ Use this strain as a biochemical factory to make somatotrophin.

◆ Give somatotrophin to children who are potential sufferers of this type of dwarfism. (10)

9 a) Somatic fusion. (1)

b) So that the protoplasts can meet and fuse. (1)

c) (i) So that the protoplasts neither gain nor lose water from their surroundings by osmosis.

(ii) They would absorb water by osmosis and burst. (2)

10 Genotypic and phenotypic variation exist amongst the members of a species. For thousands of years, people have selected from each generation of useful plants and animals, those individuals possessing desirable characteristics. These individuals are then used as the parents of the next generation and this process of artificial selection is repeated time and time again through the generations. The organisms lacking the useful characteristics are prevented from breeding.

Selective breeding over many generations has produced crop plants (e.g. wheat) with higher yields and better resistance to disease than their ancestors. Amongst animals such as cattle, pigs and sheep, meat production has been greatly increased. In addition some strains of cattle give an enormous yield of milk. Many breeds of domestic dog have also been produced, each selected to perform a service useful to mankind (e.g. sheepdog with excellent herding abilities).

The members of a desirable strain are often interbred to produce true-breeding offspring with the same characteristics as their parents. The disadvantage of this practice is that it increases the chance of individuals arising that are homozygous for harmful recessive alleles (e.g. low resistance to disease). These poorer individuals are said to show inbreeding depression. To avoid this and introduce new alleles, farmers often employ hybridization.

A hybrid is the offspring resulting from two genetically dissimilar parents. If these parents possess different desirable features, there is a chance that some of their offspring will inherit the good traits from both parents and be better than either one. Such heterozygous offspring are said to show hybrid vigour. Amongst farm animals, hybridisation often leads to higher average birth weight and improved meat yield. Unfortunately the heterozygotes are not true-breeding. (10)

22 Maintaining a water balance – animals

1 a) The left bar in the graph shows that the freshwater fish is hypertonic to the surrounding water (which has a much lower concentration of salt). As a result the fish tends to gain unwanted water by osmosis.

The right bar in the graph shows that the saltwater fish is hypotonic to the surrounding water (which has a much higher concentration of salt). As a result the fish tends to lose water by osmosis. (2)

b) See Table A22.1. (6)

Table A22.1

	freshwater bony fish	saltwater bony fish
relative size of glomeruli	large	small
relative number of glomeruli	many	few
relative filtration rate of blood	high	low
relative volume of urine	large	small
relative concentration of urine	dilute	concentrated
direction in which salts are actively transported by chloride secretory cells	into body	out of body

Table A22.1

c) Diffusion is the movement of molecules down a concentration gradient from high to low whereas active transport is the movement of molecules against a concentration gradient from low to high. Diffusion does not require energy; active transport does require energy. (2)

2 a) Four times. (1) **b)** 30 : 1. (1)

c) **(i)** 4.8. **(ii)** 1.2. (2)

d) **(i)** 108 cm³. **(ii)** 300%. (2)

3 a) **(i)** Shaded. **(ii)** Unshaded. (1)

b) On moving from fresh water to salt water, the filtration rate of the eel's kidney decreases, a reduced volume of urine is made and the chloride secretary cells actively transport salt out of the body. Each of these is reversed when the fish moves from salt water to fresh water. (3)

c) The adult eel migrates to salt water to spawn whereas the adult salmon migrates to fresh water to spawn. (1)

d) Fish able to adapt and migrate are able to exploit a wider range of habitats. (1)

4 a) B. (1)

b) **(i)** A. **(ii)** 0.5. (2)

c) **(i)** A. **(ii)** Oxygen is used during aerobic respiration to generate the energy needed to actively transport molecules against a concentration gradient. (2)

d) See Table A22.2. (2)

habitat	species of crab	reason
river estuary	A	It is able to exert some degree of osmoregulation. This suits it to life in a habitat where the salt concentration of the water varies with the tides and the river flow.
deep sea water	B	Although it is unable to osmoregulate, it is well suited to its habitat because the salt concentration of the sea water remains constant at 3.5%.

Table A22.2

5 a) A camel cannot retreat into a cool underground burrow during the day. (1)

b) Sweating. (1)

c) **(i)** No. The liquid in the stomach is not pure water. It contains fermented food.

(ii) Yes. The liquid is isotonic to blood so it would help to relieve the dehydration. (2)

d) **(i)** No. Any water gained from fat metabolism is lost during the increased breathing needed to absorb extra oxygen.

(ii) As an energy store. (2)

e) Yes. It insulates the camel against the desert heat keeping it cool and reducing its need to sweat. (2)

f) The water is lost from tissues other than the blood so its heart continues to pump the blood (which is unaffected) round the body as efficiently as normal. (1)

6 See pages 184–185. (10)

23 Maintaining a water balance – plants

1 a) To prevent air entering the xylem vessels and forming air locks. (1)

b) The rubber tubing must be a tight fit to prevent entry of air and leakage of liquid contents. (1)

c) Cohesion–tension theory. The fact that the ascending water molecules are able to pull up a column of mercury (a very dense liquid) without the system breaking down shows that the water molecules must be cohering strongly together. (2)

2 a) It decreases during the day and increases at night. (2)

b) During periods of maximum transpiration, the columns of water molecules in the xylem vessels are pulled thin by the tension from above. At the same time they adhere to the insides of the xylem vessel walls pulling them inwards. This has an overall effect of reducing the tree's diameter. (2)

3 a) See Table A23.1. (2)

	apparatus X	apparatus Y
initial mass (g)	283.80	455.72
mass after 2 days (g)	259.32	399.72
mass of water lost (g)	24.48	56.00
volume of water lost (cm³)	24.48	56.00
volume of water required to restore initial level (cm³)	25.00	56.00

Table A23.1

b) 0.51 g. (1)

c) More water is taken is by the plant than given back out again because the plant retains a little for photosynthesis and maintenance of turgor. (1)

d) (i) Rate of water loss will decrease in both X and Y.

(ii) Rate of water loss will come to a halt (or very nearly) in X but it will continue as normal in Y. In X the stomata will close in darkness and prevent transpiration. In Y loss of water by evaporation will not be affected by darkness. (4)

e) (i) Weight atmometer.

(ii) It acts as a type of control by indicating when X is acting as a free evaporator and when it is affected by physiological factors such as stomatal closure. (2)

4 a) See Figure A23.1. (4)

b) B. None of its stomata were blocked. (1)   **c)** A. (1)

d) The fact that the bubble moved a short distance in A. (1)

e) Rate of water uptake by the plant. (1)

f) Lower. A comparison of C and D shows that D (lower surface free of Vaseline) lost more water than C (upper surface free of vaseline). (2)

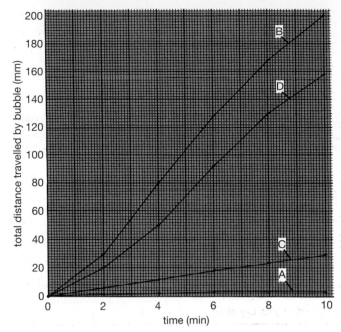

Figure A23.1

5 a) Each leaf is reduced to a needle which has a thick epidermis and sunken stomata. (3)

b) The possession of needles reduces its overall transpiring surface. Therefore, in winter, it will not lose excessive quantities of water which might be impossible to replace during freezing weather. (2)

6 a) (i) Increase in temperature reduces time taken.

(ii) Increase in wind speed reduces time taken.

(iii) Darkness increases time taken. (3)

b) There are two variable factors involved at the same time. (1)

c) 9. (1)   **d)** 4 and 8. (1)   **e)** 5 and 6. (1)

f) (i) Air humidity.

(ii) As air humidity increases, transpiration rate decreases.

(iii) During transpiration, water vapour molecules diffuse along a concentration gradient from high concentration inside the leaf's moist air spaces to low concentration in the air outside. As the humidity of the air outside increases, the concentration of water vapour molecules increases and the concentration gradient becomes less steep resulting in decreased rate of water loss by transpiration. (5)

7 a) Xerophytes. (1)

b) (i) Superficial rooting system and succulent tissues.

(ii) The roots grow parallel to the soil surface enabling them to absorb maximum water when rain does fall. Succulent tissues store the water helping the plant to resist drought. (4)

c) There are fewer stomata and they are sunken in pits. (2)

8 a) See Table A23.2. (6)

plant type	letter in diagram	two structural features of stem that led to choice of answer
hydrophyte	B	narrow diameter, large air spaces
young mesophyte	C	separate vascular bundles, lack of woody rings
old mesophyte	A	wide diameter, woody rings present

Table A23.2

b) They keep it floating at the water surface for light. They store oxygen for respiration. (2)

9 The transpiration stream is the name given to the continuous passage of water and nutrient ions flowing up through a plant from its roots to its transpiring leaf surfaces.

Water enters the root hairs by osmosis and passes along a water concentration gradient from high to low across the root cells to the xylem vessels. This osmotic flow exerts a force called root pressure which helps to make water rise up the plant.

The upward flow of water in the transpiration stream is promoted to a small extent by capillarity, the force of attraction between the water molecules and the sides of the xylem vessels (which are very narrow tubes).

Transpiration is the loss of water by evaporation from the plant's leaves through tiny pores called stomata. As water is lost from moist air spaces in the leaf, water moves by osmosis from xylem vessels along a concentration gradient from high to low across the leaf cells to replace the losses. By this means, a transpiration pull is exerted by the leaves. It acts as the main force bringing about the ascent of water in the plant's transpiration stream. It is made possible by the fact that water molecules adhere closely to one another and can therefore be pulled up through the plant as a continuous column.

Plants need water as a raw material for photosynthesis and for the maintenance of turgor to ensure the support of soft tissues. Plants need mineral ions for healthy growth and development (e.g. magnesium is needed to make chlorophyll). The flow of water and mineral ions through the transpiration stream ensures that the plant's cells receive a continuous supply of these essentials. In addition, evaporation of water by transpiration from leaf surfaces has a cooling effect in hot weather. (10)

10 See pages 197–199. Use should be made of information in Figures 23.12–23.18. (10)

24 Obtaining food – animals

1 See Table A24.1. (5)

feature of experimental design	reason
same number of animals of same species used and identical environmental condition maintained in each dish	to ensure that only one variable factor is under investigation
several animals used in each dish	to ensure that the results are not based on only one animal whose behaviour might not be typical of the species in general
hungry animals used	to ensure that the animals are motivated and ready to search for food
glass bead similar in size to the piece of liver used in the control	to ensure that the animals are not simply attracted to any physical object
experiment repeated many times	to ensure that the results are typical and reliable

Table A24.1

2 a) (i) A repeat of the experiment with a chemical-free trail.

(ii) Otherwise it could be argued that the ants were simply following the pattern of disturbed soil and not the chemical. (2)

b) Ants locate food by following a trail containing chemical from a previous ant's body. (1)

c) A repeat of the experiment without food at the end of the trail. (1)

d) As long as food is present, the ants passing along the trail keep adding more chemical which is picked up by the sense organs of later ants and so on. (1)

e) No chemical is added to the trail when the food runs out so the trail soon lacks scent and fails to attract any ants. (1)

3 a) Number of bats; species of bat; size of environmental chamber; temperature of environmental chamber. (4)

b) Sense of hearing (one group lack it). (1)

c) Bats with their ears plugged will fail to catch prey and will tend to collide with the wires because they will be unable to pick up echoes from objects in their path. (2)

4 a) 20–120 m². (1)   **b)** Insufficient food. (1)

c) Energy expended covering the large distances is greater than the energy gained from the food obtained. (1)

d) 70 m². This gives the greatest net energy gain (benefits–costs). (2)

5 a) **(i)** 5 joules/second.   **(ii)** 6 joules/second.

(iii) 5 joules/second. (3)

b) **(ii)**. (1)

c) The predator may have to expend too much energy subduing the prey. (1)

6 a) A = dry and cold, B = wet and hot. (2)

b) Wet and cold. Since this set of conditions provides each species with one (but not both) of the conditions that it favours, the two species begin roughly equal and fierce competition follows. (2)

c) It is supported in that under both sets of extreme conditions, one species ousts the other. (1)

d) Graph 1 shows that when narrow tubes are present, the number of species C increases; graph 2 shows that when wide tubes are present, the number of species C drops to zero. It is possible that A eats C but that C can escape and multiply by inhabiting glass tubes which are too narrow for predator A to enter. (2)

e) Interspecific. (1)

7 a) See Table A24.2. (1)

b) Q. It was never an overall winner in any set of 20 contests. (2)

winner	net number of contests won
T	14
R	14
P	16
R	20
S	4
R	6
P	8
T	10
R	18
P	12

Table A24.2

c) R. It beat all the other birds. (2)

d) R, P, T, S, Q. (1)

8 See pages 205–208. Use should be made of information in Figure 24.6 and 24.7. (10)

9 Animals compete for food whenever it is in short supply. Interspecific competition involves individuals of different species. Sometimes this becomes so intense that one species ousts the other. The loser is forced to migrate or face extinction.

When two species of *Paramecium* are cultured together in a medium containing a limited supply of food, both species initially increase is number. However, after a few days, the larger species starts to decrease in number and is eventually wiped out by the smaller one which can probably make more efficient use of the food supply.

A similar situation exists amongst the squirrels in Britain. The introduction of the American grey squirrel has resulted in the decline, almost to extinction, of the native red squirrel. The grey squirrel competes aggressively for food and is able to use a wide variety of foodstuffs. The red squirrel is losing out because it is more timid and its digestive system is less versatile.

Sometimes interspecific competition is reduced by the two rival species eating different foods. Cormorants are birds that feed on fish and other marine animals. When common and green cormorants share the same ecological niche, one species of bird tends to concentrate on flat fish and sea bed animals while the other feeds on eels and fish at the surface of the sea.

Intraspecific competition means competition between members of the same species. This form of competition is even fiercer than interspecific competition because the rivals all have exactly the same requirements. They are unable to reach a compromise like the two species of cormorant.

Intraspecific competition often takes the form of territoriality (competition for territories). A territory is an area which contains enough food for a solitary male animal (e.g. a robin) and his mate and, later, their young. The male defends his territory fiercely against rivals.

Red grouse birds establish territories on heather moorland. When food is plentiful, a male defends a small territory since this supplies his needs. When food is scarce he has to defend a larger territory. When times are hard, intraspecific competition is most intense, there is not enough land to go round and weaker birds fail to establish a territory. They do not breed and may even die. This is an example of natural selection in action, weeding out the weaker members of a species. (10)

25 Obtaining food – plants

1 a) See Table A25.1. (4)

feature of experimental design	reason
equal mass of cotton wool and equal volume of water added to each carton	to ensure that the cartons differ only by the one variable factor under investigation
yoghurt cartons are encased in black paper or painted black	to create competitive conditions where light is available from above only
only recently purchased cress seeds are used	to ensure that seeds are viable and capable of growth
experiment is repeated many times and results are pooled	to improve the reliability of the results

Table A25.1

b) Add transparent lids to the cartons. (1)

2 a) Different numbers of seeds were planted. (1)

b) Size of pots, mass of cotton wool, volume of water added. (3)

c) (i) See Table A25.2.

(ii) To standardize the results making them directly comparable. (3)

d) See Figure A25.1. (3)

e) As competition increases, the percentage number of healthy seedlings in the population decreases. (1)

number of seeds planted	number of healthy seedlings with green leaves	% number of healthy seedlings with green leaves
100	87	87
200	178	89
300	204	68
400	228	57
500	225	45

Table A25.2

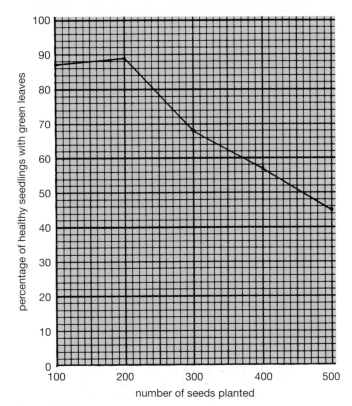

Figure A25.1

f) Light. (1)

g) See Figure A25.2. (2)

3 a) See Figure A25.3. (2)

b) (i) As the number of grains increases so does the yield.

(ii) Eventually the grains will be so numerous on the unit of soil that they will be competing with one another for any resources that are in limited supply. (2)

c) A parasite could spread more easily from plant to plant. (1)

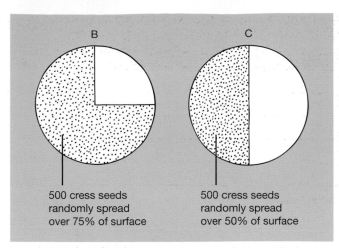

Figure A25.2

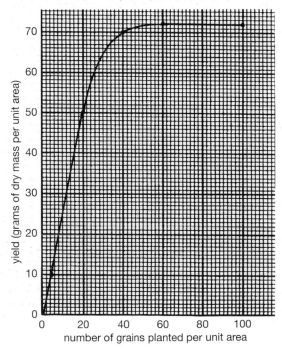

Figure A25.3

d) A (since there would be plenty of room for weeds). (1)

e) 60. (1)

4 a) B. (1)

b) The leaf (or flower). Perhaps the inhibitor is washed off the leaf (or flower) surfaces as water flows over them from above. A comparison of pots B and C suggests this to be the case. (2)

c) To show that the physical effect of water landing on flax plants from above is not the factor that inhibits their growth. (1)

d) More plants should be used in each pot. The experiment should be repeated several times to see if the results can be successfully repeated. (2)

5 a) The average number of algal species remains low. This situation occurs because the dominant algae survive the low grazing pressure and 'choke out' the delicate algae. (2)

b) The average number of algal species reaches its highest level. This situation occurs because the dominant algae are held in check by grazing snails and the more delicate algae are able to prosper in the absence of intense competition. (2)

c) The average number of algal species reaches its lowest level. This situation occurs because all types of algae are being overgrazed, many to the point of extinction. (2)

6 X had 20; Y had 11. Cropping maintains species diversity by keeping the sturdier plants in check and preventing them from choking out the more delicate ones. (3)

7 a) True. The plant must get beyond this point regularly so that photosynthesis exceeds respiration and a store of food can be made for use during the night and on dull days. (2)

b) i) 1 hour 40 minutes approximately.

ii) 6am and 10pm.

iii) 16 hours. **(iv)** It would have been less. (4)

8 a) See Table A25.3. (5)

feature of experimental design	reason
<u>discs immersed in sodium hydrogen carbonate solution instead of water</u>	to ensure that CO_2 supply is not a limiting factor
lamp placed at an equal distance from the syringes in each treatment	<u>to ensure that an additional variable factor is not introduced</u>
<u>discs of equal size used from sun and shade plant</u>	to ensure that an additional variable factor is not introduced
experiment repeated and results pooled	<u>to improve the reliability of the results</u>
<u>green cellophane filter used</u>	to absorb all colours in spectrum of white light except green

Table A25.3

b) The experiment could be carried out in a darkened room where the only source of light was the lamp completely surrounded by a green filter. (2)

9 a) (i) 2.

(ii) The extract was rich in the two yellow pigments. (2)

b) A rich supply of yellow pigments enables shade plants to make use during photosynthesis of the green and blue-green light that has passed down through the canopy of deciduous leaves above them. (2)

26 Coping with dangers

1 Habituation. It prevents the animal from wasting energy on many needless repeats of its escape response. (3)

2 a) To make the results reliable and ensure that they are not based on the behaviour of only a few animals which might not be typical of the whole species. (1)

b) By tapping the side of the tank. (1)

c) It withdraws its head into its tube. (1)

d) (i) In both cases, after several trials, the number of worms showing the escape response had dropped to zero.

(ii) At trial 20 when use of the second type of stimulus (moving shadow) began, the percentage number of worms showing the escape response was as high or higher than it had been at trial 1 when use of the first stimulus (mechanical shock) began. However all of the worms had become habituated to the first type of stimulus by trial 20, so the second type of stimulus must have been acting independently of the first. (2)

e) (i) 25. (ii) Since it is short-lived, the animal soon begins to show its escape response again which saves its from real danger. (2)

3 a) Learning curve. (1)

b) He was still learning how to do the job. (1)

c) Between 3 and 4. He had learned how to do the job and had remembered what to do from one day to the next thereby improving his performance. (2)

d) He had reached his best level of performance. (1)

4 a) Counter shading. (1)

b) Viewed from above, the bird's dark colour tends to blend with the dark land mass below. Viewed from

underneath, the bird's light colour tends to blend with the bright sky above. (2)

c) Individual. (1)

5 Camouflage relies on the caterpillars being widely scattered. If they cluster together they become conspicuous and are easily picked off by predators. Bright warning colours are more effective when the caterpillars are clustered together because the message becomes even more obvious to potential predators. (3)

6 a) Successful camouflage works best when the organism does not move. (1)

b) Ability to sting; possession of poison. (1)

c) The mimic. The predator is deceived into avoiding palatable prey. The model's security is threatened if the predator takes a chance after finding a mimic palatable. (3)

d) By imitating the sound made by the stinging/poisonous model, the harmless mimic scares off the predator. (1)

e) They share any losses suffered while the predator is learning the warning signs. (1)

f) Batesian. (1)

g) Mullerian. Since the warning colours are ineffective in darkness, an odour warns potential predators instead. (2)

7 See pages 233–236. Answers should take the form of a summary of the main points. (10)

8 See pages 237–239. Answer should take the form of a summary of the main points. Use should be made of information in Figures 26.17–26.23. (10)

27 Plant growth

1 a) Cell elongation and vacuolation. (2)

b) (i) See Figure A27.1.

(ii) Cell elongation occurs between marks 1 and 4. Cells behind mark 4 have reached their full length. (4)

2 a) Q, S, P, T, R. (1)   b) P, Q, S and T. (1)

c) (i) R.   (ii) It is hollow (with its end wall digested away). It bears lignin on the inside of its cell walls.

(iii) Its hollow structure allows it to transport water up the plant. Its lignified walls give the plant support. (5)

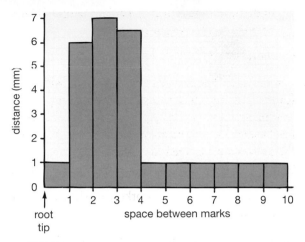

Figure A27.1

3 a) 2, 3, 1. (1) **b)** Cambium. (1)

c) **(i)** Between stages 2 and 3. **(ii)** Between stages 3 and 1. (2)

d) Secondary xylem. (1)

e) The cells in X are still capable of cell division. The cells in Y have become specialised to do a job and are no longer able to divide. (2)

4 a) 6 years. (1) **b)** Year 4. (1)

c) **(i)** Year 5. **(ii)** The fact that the ring is very narrow indicates a year of poor growth. This could have been due to decreased food production by damaged leaves. (2)

5 a) A = autumn wood. B = spring wood. (1)

b) Cambium. (1)

c) They are formed in spring when the cambium is most active. (1)

d) Grain of timber. (1)

e) The climate hardly changes from one season to the next so growth continues at a fairly constant rate throughout the year. (1)

6 a) An apical meristem is found at the root tip and shoot tip of a plant. Each apical meristem is a region of undifferentiated cells capable of dividing repeatedly and bringing about primary growth of the plant.

Within an apical meristem such as a shoot apex, the cells at the tip undergo mitosis and cell division. The undifferentiated cells formed are small and cubical in shape and possess dense cytoplasm. Immediately behind this region, the cells, which are slightly older, are undergoing the next stage in the process of growth. They are absorbing water by osmosis and becoming elongated and vacuolated.

The cells behind the elongated cells are found to be undergoing differentiation. This means that they are becoming specialised to perform a particular function as part of a permanent stem tissue. For example, in those cells destined to become xylem vessels, rings or spirals of lignin are deposited on the insides of the cell walls. The nucleus, cytoplasm and end walls disintegrate leaving strong, hollow tubes suitable for water transport and support.

Groups of meristematic cells round the sides of the shoot tip develop into side buds and leaves. The pattern of activity in a shoot tip is repeated in a root tip. The meristematic cells in a root tip are protected by a cap since they are being constantly pushed down through the soil by elongation of the cells behind them. (5)

b) A perennial plant continues to grow for many years and needs additional xylem vessels for support and water transport. These are formed by secondary thickening in the stem. The secondary xylem is produced by a lateral meristem called cambium.

Cambium is present in a young stem's vascular bundles between the xylem and the phloem. Radial division of cambium cells produces a complete ring. Tangential division of the cells in the ring of cambium produces an inner ring of unspecialised cells. These differentiate into xylem vessels by the process described above.

In temperate climates secondary thickening is most active in spring when large xylem vessels called spring wood are formed. It tails off in autumn forming smaller vessels and stops in winter. Such intermittent activity produces annual rings of wood which vary in overall thickness depending on the climatic conditions that were present during their development. (5)

7 a) See Figure A27.2. (2)

b) Light (intensity) and temperature. (1)

c) Workforce wear sterile clothing. Sterile nutrient agar is used to grow the plants. (2)

d) 'Fruit that all ripen at the same time.' (1)

e) Micropropagation. (1)

f) A new disease-causing organism. The members of the crop population are genetically identical so if one is susceptible to the disease, they will all be susceptible. (2)

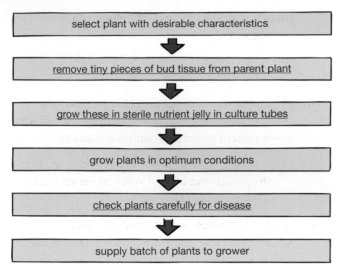

Figure A27.2

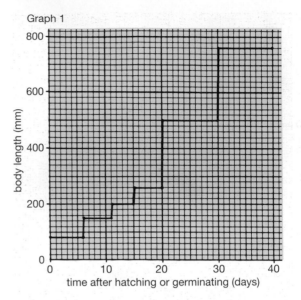

Graph 1

y-axis: body length (mm)
x-axis: time after hatching or germinating (days)

28 Growth patterns

1 a) (i) Fresh weight is a less reliable indicator of growth because it is affected by temporary fluctuations in the organism's water content.

(ii) Measuring dry mass involves removing all of the organism's water. Since this results in the death of the organism, it is not always the most practical method of measuring growth. (3)

b) At regular intervals of time, measure

(i) total cell number **(ii)** dry mass **(iii)** height

(iv) fresh mass **(v)** body length. (5)

2 a) See Figure A28.1. (4)

b) (i) Graph 2. **(ii)** Graph 1. (1)

c) Graph 2's sigmoid form with a dip at the start is typical of an annual plant's growth with food reserves being used up at the start during seed germination. Graph 1's step-like form is typical of an insect's growth curve with abrupt increases in body length following moults. (2)

3 a) See Figure A28.2. (3)

b) (i) Infant growth spurt.

(ii) The incline on the graph is steeper. (2)

c) His adolescent spurt would have occurred later than that of the female and his final adult height would have been greater than that of the female. (2)

d) (i) Her adolescent spurt would have occurred later and her final adult height would have been

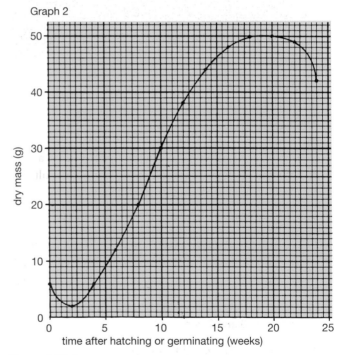

Graph 2

y-axis: dry mass (g)
x-axis: time after hatching or germinating (weeks)

Figure A28.1

smaller compared with an average woman of today.

(ii) The quality of life (e.g. availability of food and housing) was poorer 200 years ago so the average woman would not reach her full potential height. Slower and poorer growth would have held back puberty and the adolescent spurt until a year or two later than the young woman of today. (4)

4 a) (i) 60%. **(ii)** 28%. **(iii)** 55%. (1)

b) (i) 10%. **(ii)** 52%. **(iii)** 96%. (1)

c) (i) Brain. **(ii)** Reproductive organs.

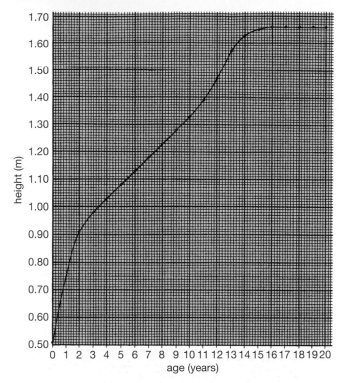

Figure A28.2

(iii) It allows plenty of time during a lengthy childhood for the person to learn useful survival skills from the parents before becoming an adult (and parent) him/herself. (3)

d) (i) Human males.

(ii) The reproductive organs remain dormant until about the age of 13 which is later than the female whose reproductive organs become active a year or two earlier. (2)

e) Between 0 and 2 years. (1)

f) (i) 16.5. **(ii)** 9. **(iii)** 1.8. (1)

g) Brain curve. The cranium protects the brain and has to develop at a similar rate. (1)

29 Genetic control

1 *E. coli* is able to obtain energy for cell division from glucose so there is no delay. *E. coli* is not able to gain energy from lactose. The lactose must first be digested by an enzyme. There is a delay because the gene that codes for the enzyme must first be switched on before the enzyme is made. (2)

2 **a) (i)** The regulator gene produces the <u>repressor</u> molecule.

(ii) The inducer molecule combines with the <u>repressor.</u>

(iii) When the operator is free, the structural gene is switched <u>on.</u> (3)

b) (i) The operon that codes for this enzyme only becomes free to do so when lactose, the inducer, is present to combine with the repressor.

(ii) It prevents energy and resources being wasted on the production of an enzyme whose substrate might be unavailable. (3)

c) No β-galactosidase will be produced since transcription and translation of the correct message will be impossible. (1)

d) (i) Regulator gene.

(ii) It no longer coded for the repressor so the operator and structural genes remained switched on. (2)

3 **a)** A mutated gene cannot code for the repressor so the operator and structural genes remained switched on. (2)

b) (i) Q. **(ii)** R. **(iii)** T. (3)

c) It shows up the problem sooner allowing earlier treatment and reducing the chance of brain damage. (2)

4 **a)** Excess phenylalanine is converted to tyrosine. (1)

b) The enzyme needed to promote this conversion is absent. (1)

c) It is converted to other metabolites. (1)

5 **a)** Zygote. (1)

b) Every cell from the 4-cell stage went on to develop into a normal live lamb, suggesting that all of each of these cells' genes were still switched on. (2)

c) They would take differentiated cells from a lamb and see if these could be made to act as zygotes and develop into more lambs. (3)

d) They are genetically identical and have all arisen from the same parental cell. (1)

6 See pages 260–263. Use should be made of information in Figures 29.4, 29.5 and 29.7. (10)

30 Hormonal influences on growth

1 **a)** Indole acetic acid. (1)

b) **(i)** 14.00 hours. **(ii)** After this time the rate of increase in length decreased rapidly. (2)

c) A very small quantity of IAA was still present and making some of the cells elongate. (1)

d) A tip from another coleoptile or an agar block containing IAA could have been placed on the decapitated coleoptile's cut surface. (2)

2 a) See Figure A30.1. (4)

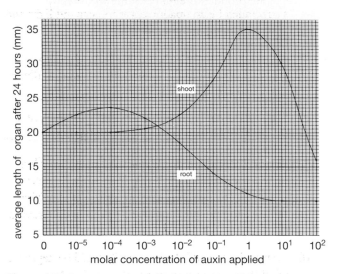

Figure A30.1

b) **(i)** 10^{-4} molar. **(ii)** 1 molar. (2)

c) **(i)** Inhibition. **(ii)** Stimulation. (2)

d) 10^2 molar. (1)

e) To increase the reliability of the results. (1)

f) **(i)** Auxin is destroyed by light.

   (ii) Temperature, volume of bathing solution. (3)

3 F, E, B, D, A, C. (1)

4 a) See Figure A30.2. (2)

b) **(i)** X. **(ii)** The average total length of the side buds and shoots per plant remained unchanged until day 8. (2)

c) It acts a control to check that plain lanolin has no effect on the plant. (1)

d) **(i)** Gibberellin.

   (ii) Drawing such a conclusion is justified because the average length of the side buds and shoots for group Z (with gibberellin) is even greater than control group W (untreated).

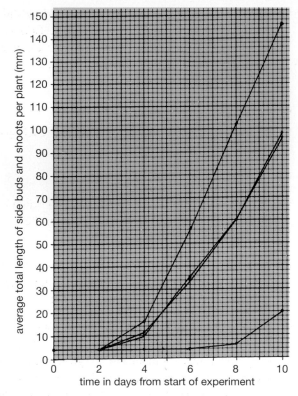

Figure A30.2

(iii) $\dfrac{\text{average change in length relative to control}}{\text{average increase in length of control}} \times \dfrac{100}{1}$

$$= \frac{42}{56} \times \frac{100}{1} \quad 75\% \ (3)$$

e) Perhaps the auxin supply in the lanolin was running out by day 10 and apical dominance was no longer being maintained. (1)

5 a) Y. (1)

b) Unlike leaf X, leaf Y has not undergone abscission and its lateral bud has remained dormant. (2)

6 a) As concentration of gibberellic acid increases (up to 1 mg), plant height continues to increase. (1)

b) **(i)** 1 mg. **(ii)** Several members of the tall variety could be grown, applying lanolin only to their shoot tips at 2 days. Their average height could be compared with that of the treated dwarf plants showing the maximum effect. (2)

c) **(i)** 0.0001 mg.

   (ii) It was too low a concentration to make the cells elongate. (2)

d) 0 mg. (1)

e) At 7 days the cells at the tip were still dividing and this did not affect height. At 22 days the cells were at the elongation stage of development and were able to elongate and increase the plant's height because gibberellic acid was present. (1)

7 a) Endosperms need GA to be able to digest starch. (1)

b) GA does not reverse an embryo's inability to digest starch. (1)

c) Embryo. (1)

d) It induces the production of α-amylase (a starch-digesting enzyme) by the aleurone layer in the endosperm. (1)

e) It prevents the seed grain from starting to germinate in adverse environmental conditions which would result in the death of the young plant. (1)

8 a) Two of the hormones (chemical messengers) produced by the pituitary gland affect human growth and development. Somatotrophin (growth hormone) accelerates amino acid transport into cells of soft tissues (e.g. muscle) and bones (e.g. femur and other long bones). This results in the rapid synthesis of tissue proteins required for growth of the body especially during the adolescent spurt.

Under-production of growth hormone during adolescence results in a form of dwarfism; over-production results in giantism.

Thyroid-stimulating hormone controls growth and activity of the thyroid gland which in turn produces a further hormone called thyroxine. This regulates the body's metabolism and therefore affects growth and development. (4)

b) Auxins are a group of plant hormones. Synthetic versions of auxins are produced commercially and used for a variety of purposes. Some are sprayed onto unfertilised flowers to induce fruit formation. This produces a crop of seedless fruit (e.g. grapes) which are all ripe at the same time.

Auxins are also used by fruit growers to delay the formation of the abscission layer in fruit stalks. This prevents the fruit (e.g. pears) falling before they are ripe.

Synthetic auxin is present in rooting powder. When applied to the cut ends of woody stems, it stimulates some of their cells to develop into roots. This makes the cuttings easy to grow.

Synthetic auxins are also used as selective weedkillers, especially on lawns. They are sprayed onto the plants and absorbed by the leaves. Broad-leaved weeds absorb so much of the hormone that they grow too quickly, exhaust their food reserves and die. The narrow-leaved grass plants are hardly affected. (6)

31 Effects of chemicals on growth

1 a) (i) and (ii) Each gas jar should have been rinsed out with concentrated nitric acid to remove traces of mineral salts. Each gas jar should have been surrounded by an opaque cover to keep out light and prevent growth of algae. (4)

b) The set-up in Figure 31.1 would be likely to give more reliable results because it uses several seedlings not just one plant. (1)

c) (i) A sign of unhealthy growth caused by lack of a chemical element needed for a metabolic process.

(ii) The younger, most recently formed ones. They have developed in the absence of the mineral element. The older ones are part of the original cutting taken from a healthy plant with access to the mineral. (3)

d) Eventually it would die because it would be deprived of all the mineral elements that it needs. (2)

2 a (i) See Figure A31.1.

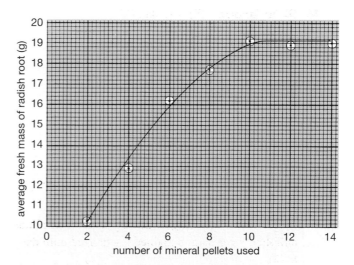

Figure A31.1

(ii) Increase in number of mineral pellets (up to 10) produces an increase in fresh mass of radish root. (4)

b) A set-up with no mineral pellets. (1)

c) So that the only mineral elements available to the plant are those supplied by the pellets. (1)

d) To ensure a continuous supply of minerals to the plants over the six-week period. (1)

e) The student pooled her results with those of four other students. (1)

f) **(i)** Dry mass is a more reliable indicator of growth than fresh mass because it is not affected by fluctuations in the organism's water content.

(ii) Each radish plant would be dried in a warm oven and weighed and then dried and reweighed until it was found to have reached constant dry mass. (3)

3 a) 1500 g. (1) **b)** Liver, eggs, spinach. (3)

4 a) X = 300; Y = 200. (2)

b) Some would be used for bone formation and therefore less would pass out in body wastes. (1)

5 a) D. (1)

b) In Asia their skin had received plenty of ultraviolet radiation forming vitamin D. (1)

c) The child would be bow-legged. (1)

d) Cod liver oil. (1)

e) It promotes the absorption of calcium and phosphate from the intestine and their uptake by bone. (2)

6 a) See Figure A31.2. (3)

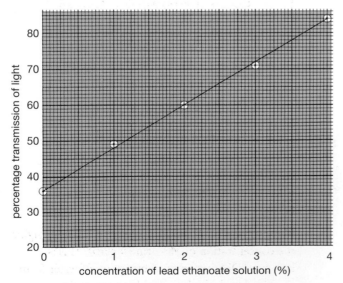

Figure A31.2

b) 66% approximately. (1) **c)** Apple juice. (1)

d) To maintain constant pH throughout and avoid the introduction of a second variable factor. (1)

e) **(i)** and **(ii)** Tube E would have been the lightest because the highest concentration of lead ethanoate would have inhibited the enzyme and very little catechol would have been converted to brown compounds. Tube A would have been the darkest since much of its catechol would have been converted by enzyme action to brown compounds in the absence of lead ethanote, the enzyme inhibitor. (2)

f) Inhibition of the enzyme is not complete even at the highest concentration of lead ethanoate used. (1)

g) Increasing concentration of lead ethanoate results in increasing inhibition of the enzyme catechol oxidase. (1)

7 a) **(i)** 38%.

(ii) 10%. (2)

b) **(i)** 62%.

(ii) 90%. (2)

c) It brings about a decrease in birth weight. (1)

8 See pages 292–294. (10)

32 Effect of light on growth

1 a) An etiolated bean seedling shows an increase in cell number but a decrease in dry mass. Whether this can be regarded as real growth is debatable. It depends upon which definition of growth is favoured. (2)

b) **(i)** Death. **(ii)** Eventually it would run out of food. (2)

c) **(i)** The existing shoots will probably turn yellowish and some leaves may die but the leaves and stems will not change in structure.

(ii) The new shoots will show etiolation with respect to both leaves and internodes. (2)

2 a) Leaf. (1) **b)** They should be given short nights (i.e. less than the critical 9 hours of darkness) up until a few weeks before Christmas and then be given the correct length of darkness repeatedly over several days to make them flower. (2)

3 a) A = short day (11 hours of darkness), B = long day (12 hours of light), C = short day (12 hours of darkness), D = day neutral. (4)

b) 12 hours of light and 12 hours of darkness. (1)

4 a) Length of photoperiod. (1) **b)** Long. (1)

c) Breeding would begin even although it was the wrong time of the year. (1)

d) Daylength is absolutely predictable since its pattern of change is repeated in exactly the same way year after year. Temperature is unpredictable since 'freak' conditions such as cool summers and mild winters occur. (1)

5 a) A direct relationship/correlation. (1)

b) Mass of testes would fail to increase. (1)

c) (i) April, May and June

(ii) The young are born during the summer when conditions are mild and food is plentiful. Their chance of surviving and growing into independent adults is much greater than it would have been had they been born during the winter. (3)

6 Photoperiodism is the name given to the response made by a living organism to a change in the period of daily illumination (photoperiod) to which it is exposed over several days or weeks.

Amongst plants, some only flower when they have been exposed repeatedly to photoperiods of a certain minimum number of hours of light. They are called long day (short night) plants e.g. spinach. They do not flower at photoperiods below the critical level. Other plants only flower when the photoperiod is less than a critical number of hours. They are called short day (long night) plants e.g. strawberry.

Flowering is thought to be initiated by a hormone produced in response to the appropriate photoperiod. The hormone is transported to cells in the apical meristems where it causes the genes that control leaf formation to be switched off and those that control the development of floral structures to be switched on. By responding to a photoperiod of a certain length, all the members of a species flower at around the same time. This allows cross-pollination to occur and results in an increase in variation amongst the offspring produced. It is of biological significance because natural selection acts on the members of the population and selects the fittest (those that are best adapted to the environment).

Many animals only breed at certain times of the year and their mating behaviour is triggered by the arrival of photoperiods of a certain length.

In long day breeders (e.g. rabbit), longer photoperiods in spring stimulate the hypothalamus in the animal's brain. A hormone is released which stimulates the pituitary gland. This releases further hormones which lead to an increase in size and activity of the sex organs, production of gametes and the onset of mating behaviour.

In a short day breeder (e.g. sheep), a similar sequence of events occurs but in response to the arrival of shorter photoperiods in autumn. In both cases the behaviour is of biological significance because it ensures that the offspring are born in the spring (following the rabbit's short gestation period or the sheep's long gestation period). The young animals will have several months of mild weather when food is abundant to grow and become strong before harsh winter conditions return.

The arrival of shorter photoperiods in autumn also triggers other responses in animals. Some hibernate e.g. hedgehog; others migrate e.g. swallow. In both cases the behaviour helps the animal to survive the winter. (10)

33 Physiological homeostasis

1 a) See Figure A33.1. (2)

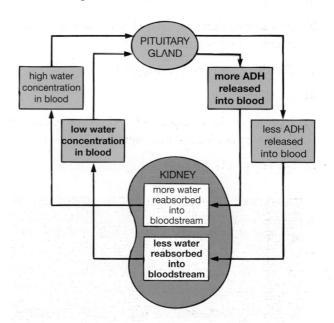

Figure A33.1

b) Antidiuretic hormone. It makes them more permeable to water. (2)

c) The higher the concentration of ADH in the bloodstream, the lower the volume of, and the higher the concentration of, the urine formed. This is because ADH promotes reabsorption of water from the kidney tubules into the bloodstream. (3)

d) If water is drunk, it will pass from the gut into the bloodstream and return the blood to its normal water concentration. (2)

2 a) (i) 07.00–08.00. (ii) Physiological homeostasis. (2)

b) **(i)** 08.00. **(ii)** They both increased.

(iii) It took a little time for the blood containing extra glucose to arrive at the pancreas and for the pancreas to respond and make more insulin. (3)

c) As the concentration of insulin increased. The concentration of fatty acids (the breakdown products of fat) decreased. This suggests that insulin suppressed the breakdown of fat to fatty acids. (1)

3 a) Osmoregulation involves the maintenance of the correct water concentration of the blood. This is effected homeostatically by the required volume of water being reabsorbed from glomerular filtrate. Since the urine formed contains most of the urea but much less water, the concentration of urea in urine is higher than that in glomerular filtrate. (1)

b) **(i)** 2. **(ii)** 70. (2)

c) **(i)** Glucose would have been present in the urine.

(ii) Insulin and a controlled diet. (2)

4 a) Blood plasma. (1)

b) **(i)** Adrenal gland. **(ii)** Kidney. (2)

c) Hormonal. (1)

d) Negative feedback control involving reduced secretion of aldosterone (and reduced reabsorption of salt). (1)

e) See Figure A33.2. (2)

f) **(i)** This promotes the absorption of extra salt needed to replace losses.

(ii) Anti-diuretic hormone (ADH). (2)

g) The person would continue to lose salt in sweat and so more and more aldosterone would be secreted to promote maximum reabsorption of salt from glomerular filtrate. However the body's salt content would still continue to decrease and eventually the homeostatic mechanism would break down. This would lead to death. (2)

5 a) Hypothalamus. (1)

b) It sends nerve impulses to them. (1)

c) **(i)** It would become constricted.

(ii) Less blood would flow to the skin surface so less heat would be lost by radiation. (2)

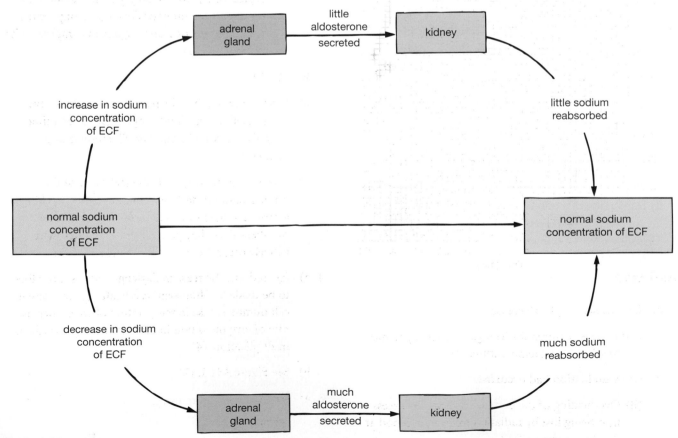

Figure A33.2

d) (i) It would increase its rate of sweat production.

(ii) When the liquid sweat coated the outside of the skin, excess body heat would be used to convert it to water vapour thereby cooling the body. (2)

6 a) (3) and **b)** See Figure A33.3. (4)

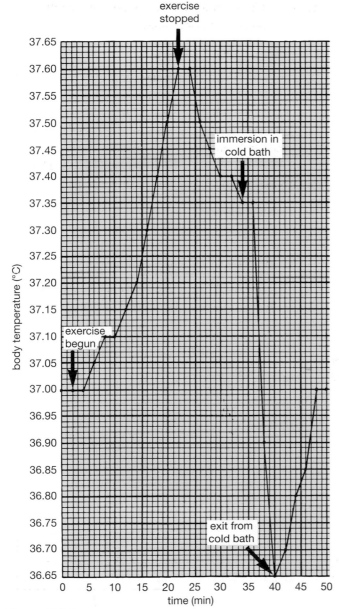

Figure A33.3

c) (i) Increase. **(ii)** Decrease.

(iii) Time is required for heat gain or loss by tissues to affect blood temperature. (3)

d) (i) Vasodilation had occurred.

(ii) Overheating of the body is corrected by excess heat being lost by radiation from extra blood at the skin surface. (2)

e) (i) D. **(ii)** Overcooling of the body is corrected by the heat generated by muscular contraction. (2)

7 See pages 309–310 and 311–315. Answer should take the form of a summary of the main points. Use should be made of information in Figure 33.5 and 33.15. (10)

34 Regulation of populations

1 a) $55/m^2$. (1) **b)** 35 pairs. (1)

c) To give a balanced overall picture which is unaffected by any unusual results. (1)

d) Once the <u>population</u> size of a species of <u>plant</u> or animal has reached the <u>carrying</u> capacity of the <u>environment</u>, it remains relatively <u>stable</u> despite short-term <u>oscillations</u> in number. (1)

2 Density-dependent = increased predation and shortage of food; density-independent = a thunderstorm and intense drought. (4)

3 a) (i) Predator = broken line, prey = solid line.

(ii) The solid line shows a trend of population fluctuations which are repeated on a smaller scale by the broken line after a time lag. This is typical of a predator-prey graph. The solid line represents the numerically larger group which would be true of a prey organism compared with its predator. (3)

b) 10. (1)

c) When prey numbers increase, the predators have access to abundant food supplies and after a time they also increase in number. The reverse also applies. (1)

d) If the unpalatable grass is the real cause of the rabbits' periodic decline in numbers then tracking is a more accurate way of describing the predators' numbers since they passively follow rather than actively cause the prey's decline. (1)

4 a) (i) and **(ii)** Increase in daylength allows more food to be made by photosynthesis leading to increase in cell numbers. Rise in temperature of water increases rate of enzyme action in cells promoting growth and multiplication. (4)

b) See Figure A34.1. (2)

5 a) (i) 35.

(ii) 33. (2)

Answers

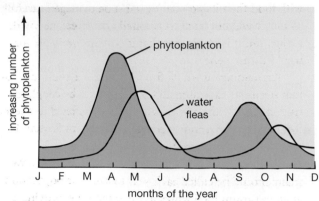

Figure A34.1

b) 100. (1)

c) (i) Lack of available prey to eat.

(ii) Fewer predators were eating them. (2)

6 a) (i) Limpet and chiton.

(ii) In the absence of starfish, barnacles increased in number and won out in the competition for the algae. (2)

b) It keeps the number of barnacles down. (1)

7 a) (i) Density-dependent.

(ii) It only affects the dense populations in the burrows but not the solitary individuals above ground. (2)

b) (i) See Figure A34.2.

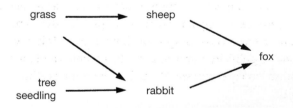

Figure A34.2

(ii) There will be more grass for the sheep to eat but the farm animals will suffer more attacks by foxes. (4)

8 See pages 322–324. Answer should take the form of a summary of the main points. (10)

35 Monitoring populations

1 a) 4 years. (1)

b) (i) Decrease in total number of plaice caught.

(ii) The stocks are being overfished. (2)

c) The '11 years' entry refers to one year only; the '12+ years' entry refers to the sum of several years. (1)

d) A greater number of younger fish were caught in years 4 and 5. (2)

e) If the catch continues to get younger, eventually there will be no adult fish left to produce future stocks. (2)

f) 70%. (1)

2 a) To make the results more reliable. (1)

b) To prevent the results from being biased. (1)

c) Take a large number of samples and calculate the average number of sheep ticks per square metre. Calculate the area of the field in square metres and multiply this by the average number of ticks per square metre. (2)

d) Move sheep to different grazing land until the ticks die out. Treat ticks with insecticide. (2)

3 a) Orange tree → scale insect → ladybird. (1)

b) (i) Y = predator.

(ii) X = prey.

(iii) Once Y arrived, the numbers of X dropped immediately. (3)

c) 12. (1) **d)** 1889. (1)

e) There were too few prey left to feed them all. (1)

f) As the prey numbers increased so, after a short time, did the predators. This caused the prey numbers to fall which in turn caused the predators to decrease and so on with the predators' curve always tracking that of the prey. (2)

g) (i) Most of the scale insects died so the crop of fruit was very good.

(ii) A new epidemic of scale insects soon followed.

(iii) Reintroduction of ladybirds. (3)

h) The same pattern as between 1900 and 1940. (1)

i) To make sure that their numbers remain so low that they pose no threat to the orange trees. (1)

4 a) As the concentration of pesticide increases, the percentage thickness of the shell decreases. (1)

b) (i) 15%. **(ii)** 25.

(iii) Zero success. All shells would be so thin that they would break during incubation. (4)

5 a) (i) The more animal material (especially fish) that the bird eats, the higher the concentration of pesticide residue in its muscle tissues.

(ii) Animals such as fish are situated further along the food chain than plants or invertebrates and therefore eat food which has already concentrated pesticide at several links in the chain. (3)

b) (i) The sea.

(ii) The sea dilutes the concentration of pesticide arriving in rivers before it enters the marine food chain. (2)

6 a) An inverse relationship exists. As the bacterial numbers increase they consume more and more of the dissolved oxygen which therefore decreases. (1)

b) (i) It is absent from the region of the river upstream from the sewage inflow.

(ii) It increases because essential factors for growth (such as food, oxygen, water etc.) are available.

(iii) It could have run out of food (or oxygen).

(iv) Density-dependent. (4)

c) (i) Lack of light for photosynthesis (owing to the dirty, cloudy state of the water).

(ii) Density-independent. The sudden reduction in light intensity in the murky water following the sewage inflow affects the population of algae regardless of whether the population is dense or sparse. (2)

d) In the nitrogen cycle the sequence of events is:

$$\text{ammonium compounds} \xrightarrow{\text{bacteria}} \text{nitrites} \xrightarrow{\text{bacteria}} \text{nitrates} \quad (1)$$

e) Nitrate ions are needed by plants to make protein. A rich supply of nitrate therefore promotes plant growth. (1)

f) Heavy rain. (1)

7 Scientists monitor populations of wild plant and animal species for a variety of reasons. If the species provides us with food then its population must be managed carefully so that it will not be overexploited and threatened with extinction. It is important that it remains available for use by future generations of mankind.

Populations of edible fish in the sea are monitored so that scientists can estimate how many can be caught by fishing fleets without stocks becoming depleted to a level from which they cannot recover. Red deer in Scotland tend to show uncontrolled population growth. By monitoring their numbers, scientists can estimate how many should be culled each year to keep the population at an optimum level while, at the same time, providing humans with a supply of venison.

Some species act as pests by posing a threat to mankind's health or economy. Scientists study epidemics caused by fungi which attack cereal crops. They are able to estimate the pest's rate of transmission and can warn farmers to spray their crops with fungicide before the pest arrives in their area.

Scientists also keep track of climatic conditions such as heavy rainfall which, in Africa, is closely followed by the emergence of swarms of locusts. Farmers can be warned in advance to prepare for an invasion by this plant-eating pest.

Some species of wild plants and animals act as indicators of environmental pollution. By monitoring populations of lichens, scientists can assess the level of air pollution affecting an area. An abundance of 'hairy' lichen species, for example, indicates that little or no poisonous sulphur dioxide gas is present in the air. An abundance of rat-tailed maggots in a river indicates a high level of pollution of the water by organic waste such as sewage.

Populations of endangered species are also monitored in an attempt to protect and conserve them. If it becomes widely known, for example, that there are only a few thousand tigers or rhinoceroses left in the world, publicity campaigns can be mounted and money raised to try to prevent them from becoming extinct. (10)

36 Succession in plant communities

1 a) Succession. (1)

b) (i) and **(ii)** X = habitat modification. Each new community can only colonise the habitat once it has been modified by the previous occupant. Y = extinction. Each community dies out when, having modified the habitat, it is succeeded by the next community. Z = self-perpetuation. The climax community is not replaced by another community but perpetuates itself indefinitely. (6)

2 **a)** 6. (1)

 b) **(i)** 8. **(ii)** G, H, I and J. **(iii)** 3.

   (iv) The soil was not sufficiently fertile. (4)

 c) The shrubs improved the soil fertility and provided the sheltered conditions needed by the tree seedlings which grew and shaded out the shrubs. (1)

3 d, b, e, a, c. (1)

4 **a)** See Figure A36.1. (1)

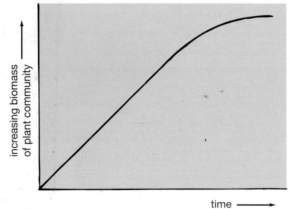

Figure A36.1

b) As succession occurs, the members of each successive community of plants are bigger, taller, bushier etc. and therefore the overall biomass increases accordingly. (1)

5 **a)** Primary succession, since this is colonisation of a barren area which has not been previously inhabited. (1)

 b) **(i)** If forms hummocks held together by the plant's roots.

   (ii) Marram grass.

   (iii) Its extensive roots stabilise more sand and its dead remains add humus to the sand.

   (iv) Sea holly, fescue grass and heather are able to succeed marram grass because the latter has formed a thin layer of soil. (6)

 c) D. (1)

6 See pages 341–342. Use should be made of information in Figure 36.3. (10)